ANIMAL HUSBANDRY AND EXTENSION

ANIMAL HUSBANDRY AND EXTENSION

Editor

Branko Albescu

scitus academics

Animal Husbandry and Extension

Edited by **Branko Albescu**

This Book is Printed on Acid-Free Paper. ♾

ISBN: 9781681179209

Library of Congress Control Number: 2017948949

SCITUS Academics LLC,

616, Corporate Way, Suite 2, 4766,

Valley Cottage, NY 10989

www.scitusacademics.com

Printed in United States of America.

Preface

Animal husbandry is a branch of agriculture concerned with the care and management of livestock. Animal husbandry deals with the feeding, breeding, housing and health care of livestock for getting maximum benefits. Since long time, animals like bees and silkworm, have been used by humans. Livestock has been used by humans for products like milk, eggs, meat, wool, silk, honey, etc. The increase in human population has increased the demand of these products. Hence, it is necessary to improve the management of livestock scientifically. Modern animal husbandry relies on production systems adapted to the type of land available. Subsistence farming is giving way to intensive animal farming in the more developed parts of the world, where for example beef cattle may be kept in high density feedlots, and thousands of chickens may be raised in broiler houses or batteries. More than 70 per cent of the world livestock population is in India and China. But it is interesting to note that the contribution to the world farm produce is only 25 per cent. Hence in addition to old practices of animal breeding and care, new techniques have to be used to improve quality and productivity. Animal husbandry extension will play an important role in changing the knowledge regarding animal husbandry management, skill in animal rearing practices and attitude towards newer technology of livestock owner which can lead to an increase in per animal milk production.

This volume 'Animal Husbandry and Extension' is an integrated text to provide educational concepts and self study guide for students, researchers, teachers, livestock extension specialists and administrators interested in the study of Animal science. The aim of this book is to identify and explain the causes and contributors to current problems in animal husbandry, especially those related to 'factory farming', and advance arguments that may contribute to its successful re-orientation.

Table of Contents

Chapter 1	Refining Housing, Husbandry and Care for Animals Used in Studies Involving Biotelemetry 1
Chapter 2	Double Muscling in Cattle: Genes, Husbandry, Carcasses and Meat 17
Chapter 3	Novel analytical technologies of Quality in the Sheep & Goat Dairy Sector 55
Chapter 4	Synbiotic as Feed Additives Relating to Animal Health and Performance 75
Chapter 5	Prevention and Control of Production Limiting Goat Diseases on Small Farms 97
Chapter 6	Comparison of Weight Gain, Milk Production, and Milk Composition of Iranian Mamasani Goat and its Cross with Saanen 105
Chapter 7	Hanwoo cattle: origin, domestication, breeding strategies and genomic selection 111
Chapter 8	Nutritional Influences on Skatole Formation and Skatole Metabolism in the Pig 125

Chapter 9 Assessment of Blood Concentration of Cardiac Troponin I (cTnI) in Healthy Calves and Stocker Calves Effected with Bovine Respiratory Disease Complex 151

Chapter 10 The genetic prehistory of domesticated cattle from their origin to the spread across Europe 167

Chapter 11 Livestock Helminths in a Changing Climate: Approaches and Restrictions to Meaningful Predictions 189

Chapter 12 Health and Welfare in Dutch Organic Laying Hens 207

Chapter 13 Mutilating Procedures, Management Practices, and Housing Conditions That May Affect the Welfare of Farm Animals: Implications for Welfare Research 229

Chapter 14 Animal Rights as a Mainstream Phenomenon 261

Chapter 15 Pet husbandry and infection control practices related to zoonotic disease risks in Ontario, Canada 281

Chapter 16 Exploration Feeding and Higher Space Allocation Improve Welfare of Growing-Finishing Pigs 307

Chapter 17 Cattle brucellosis in traditional livestock husbandry practice in Southern and Eastern Ethiopia, and its zoonotic implication 323

Chapter 18 **The role of the companion animal veterinary surgeon in behavioural husbandry 341**

Chapter 19 **The current state of welfare, housing and husbandry of the English pet rabbit population 343**

Chapter 20 **Animal Welfare in Relation to Standards in Organic Farming 371**

Chapter 21 **Factors Influencing Effective Use of Information and Communication Technology (ICT) Tools By Various Animal Husbandry (A.H.) Organizations in Andhra Pradesh 383**

Index **399**

CHAPTER 1

Refining Housing, Husbandry and Care for Animals Used in Studies Involving Biotelemetry

Penny Hawkins

Research Animals Department, Royal Society for the Prevention of Cruelty to Animals (RSPCA), Wilberforce Way, Southwater, West Sussex, RH13 9RS, UK

SIMPLE SUMMARY

Biotelemetry, the remote detection and measurement of an animal function or activity, is widely used in animal research. Biotelemetry devices transmit physiological or behavioural data and may be surgically implanted into animals, or externally attached. This can help to reduce animal numbers and improve welfare, e.g., if animals can be group housed and move freely instead of being tethered to a recording device. However, biotelemetry can also cause pain and distress to animals due to surgery, attachment, single housing and long term laboratory housing. This article explains how welfare and science can be improved by avoiding or minimising these harms.

ABSTRACT

Biotelemetry can contribute towards reducing animal numbers and suffering in disciplines including physiology, pharmacology and behavioural research. However, the technique can also cause harm to animals, making biotelemetry a 'refinement that needs refining'. Current welfare issues relating to the housing and husbandry of animals used in biotelemetry studies are single *vs.* group housing, provision of environmental enrichment, long term laboratory housing and use of telemetered data to help assess welfare. Animals may be singly housed because more than one device transmits on the same wavelength; due to concerns regarding damage to surgical sites; because they are wearing exteriorised jackets; or if monitoring systems can only record from individually housed animals. Much of this can be overcome by thoughtful experimental design and surgery refinements. Similarly, if biotelemetry studies preclude certain enrichment items, husbandry refinement protocols can be adapted to permit some environmental stimulation. Nevertheless, long-term laboratory housing raises welfare concerns and maximum durations should be defined. Telemetered data can be used to help assess welfare, helping to determine endpoints and refine future studies. The above measures will help to improve data quality as well as welfare, because experimental confounds due to physiological and psychological stress will be minimised.

Keywords: Three Rs; refinement; reduction; biotelemetry; environmental enrichment; animal husbandry; surgery; ethical review; animal welfare; welfare assessment

1. INTRODUCTION

The ethical decision-making process for research projects or procedures involving animals rests on a harm-benefit assessment, in which the likely harms (pain, suffering or distress experienced by animals) are 'weighed' against the potential benefits of the project. This assessment may be conducted by a number of different individuals or bodies, including the regulator, an ethics committee, the researcher or a combination of these. A good harm-benefit assessment involves setting out the potential harms to animals throughout their lives, including not only scientific procedures and their after effects, but also other aspects that may cause suffering such as early 'weaning', transport, husbandry restrictions (e.g., singly housing social animals) and some killing techniques. This approach also identifies opportunities to implement refinement, one of the Three Rs (Replacement of animal experiments with humane alternatives, Reduction of animal numbers to the minimum needed for statistical significance, and Refinement of animal husbandry and procedures to reduce suffering and improve welfare). Refinement can significantly reduce harms to animals, which will clearly impact on the harm-benefit assessment and have implications as to the justification for the project. However, refining one aspect of a procedure may have repercussions that affect welfare in other ways, so it is essential to conduct a thoughtful review of the animal's broader life experience when making decisions about appropriate refinements and how these will be implemented.

One example of a refinement that can give rise to wider welfare issues is biotelemetry, which can be defined as the remote detection and measurement of a human or animal function, activity or condition [1]. The first physiological signal telemetered from a conscious animal was recorded in 1963, when aortic blood flow was successfully transmitted from an exercising boxer dog at the San Diego Zoo hospital [2]. Since then, the use of biotelemetry has expanded hugely within a wide range of biological disciplines including physiology, pharmacology and behavioural studies in both the laboratory and field [3]. Its use has become so well-established that some regulatory bodies require data that can only be obtained using biotelemetry [4].

The use of biotelemetry to transmit physiological or behavioural data is often cited as a way of implementing two of the Three Rs; reduction and refinement. Biotelemetry can indeed reduce animal numbers, by enabling the collection of more, better quality data from each animal, e.g., because data can be gathered over longer periods, or multiple parameters can be recorded from the same individual [3,5]. Removing the need for tethers and restraint during recording, so that animals can behave more normally, is a refinement in that it can significantly reduce stress during data collection [5].

However, despite these benefits, the application of biotelemetry can also be a source of harms to animals. Surgical procedures to implant internal devices can cause pain and distress, as can the fitting of external devices, and the size, shape and weight of the device can be a welfare issue in smaller species [5]. Recording multiple parameters may add to the burden on the animal (e.g., if multiple sensors are implanted), and holding animals in the laboratory in the long term is also a cause for concern and debate as to acceptable maximum durations. Animals may be singly housed, either during recordings or permanently, which is a significant welfare issue for social species or strains.

Taking all of this into account, the use of biotelemetry within each project should in itself be subject to a harm-benefit assessment, in common with all other invasive procedures. It is essential to recognise this and regard biotelemetry as a 'refinement that may need refining' rather than a panacea. With this in mind, the RSPCA set up an expert Joint Working Group on Refinement (JWGR) in 2003 to produce good practice guidance reports on procedures involving biotelemetry [5] and housing and care for animals used in such projects [6]. This was within the context of the Society's work to promote effective ethical review of animal use and implementation of the Three Rs.

Ten years on, the use of telemetry has continued to increase, as have developments in device technology. This paper provides an update on some of the topics addressed in the JWGR on housing, husbandry and care for animals involved in biotelemetry studies. It does not address device attachment or implantation, but assumes that these will be fully refined in accordance with legal requirements to minimise suffering by employing effective pain management [7] and current good practice approaches to surgery such as the Laboratory Animal Science Association (LASA) asepsis guidelines [8].

Also over the last decade, the level of debate has increased within the scientific community about the scientific validity and translatability of many animal 'models'. Concerns that have come to the fore include the validity of 'extrapolating' between species, standards of experimental design and reporting, and confounding effects due to physiological and psychological responses to housing, husbandry and care that does not cater for animals' species-specific needs [9,10,11,12,13,14]. All of these issues have profound ethical implications with respect to the actual 'benefits' of animal research, how funding resources are spent, and whether the needs of patient groups are properly addressed. Reducing stress to animals by employing the approach to housing and care set out in this document will thus not only help to improve welfare, but will also enhance scientific validity.

2. HOUSING, HUSBANDRY AND CARE ISSUES ASSOCIATED WITH BIOTELEMETRY

A refined biotelemetry system should: allow animals to be housed in stable, compatible groups; not preclude environmental enrichment; be minimally invasive; and would not be interfered with by other animals in the group. In practice, however, internally and externally mounted devices currently both facilitate and present obstacles to these standards [5,6]. This is expanded upon in Section 2.1 and Section 2.2 below, which explain how to provide group housing and enrichment for animals on biotelemetry studies. Some instrumented animals are reused and housed long term, which is addressed in Section 2.3, while Section 2.4 provides examples of how available telemetered data can be used for an additional purpose: to help reduce suffering and improve welfare.

2.1. Group Housing

It is generally accepted that social species, including many commonly used laboratory mammals, should be housed in stable, compatible groups [7,15]. This varies with species, strain and sex, for example there is debate over the benefits of group housing hamsters and some strains of male mice. Singly housing social species or strains is likely to have significant negative consequences for animal welfare and will also affect data quality, if social isolation leads to physiological responses that affect results [16,17,18,19,20,21]. Despite this, a proportion of animals used in biotelemetry studies are still singly housed. The reasons that are given for this should always be challenged, as many of the perceived obstacles to group housing can be overcome, as explained below.

Reason 1: Devices all transmit on the same wavelength. Although considerable developments have been made in this area in recent years, many commercially available telemetry devices still transmit at the same frequency. It is obviously impossible to obtain meaningful data from several devices, on different animals, that are all transmitting at the same wavelength within the same pen or cage. If the use of single wavelength devices is unavoidable, solutions include:

1) The 'buddy' system, in which an instrumented animal is group housed with one or more uninstrumented individuals [20,21]. The fate of the buddy (or buddies) can be planned for in advance. Potential options are: (i) rehoming; (ii) use in another study, provided that the animals are suitable (e.g., of an appropriate age); or (iii) humane killing at the end of the study. Both options (i) and (ii) may involve changing established groups, in which case advice from animal technologists or the attending veterinarian can help to reduce stress to the animals. Option (iii) is an ethical issue, although it need not be an animal welfare issue if the killing technique is properly refined. The animals' tissues or organs should be used in other studies or for training purposes where appropriate, or facilities may sell surplus, euthanased animals for use as food for rehabilitated wildlife, raptors or 'exotic' companion animals. Some establishments have set up programmes in which surplus animals can be certified fit and rehomed to responsible carers [22,23] and there is a legal framework to support rehoming in many countries, provided that it is in the animal's best interests [7].

2) Devices that can be switched on and off with a magnetically actuated system *in situ*, *i.e.*, following implantation, and used one at a time in pair- or group-housed animals. This avoids the issue of deciding fates for 'ex-buddies', provided that recording data sequentially is compatible with the scientific protocol, or can be compensated for by the experimental design. Switching devices on and off can also extend

battery life, thus facilitating reduction in animal numbers because more data can be gathered from each animal (but see Section 2.3 below).

3) Instrumenting all animals, and then separating them for data recording sessions only. This allows data to be recorded at the same time from different individuals, but periods of social isolation (and being moved to a different enclosure, if applicable) will be highly stressful for some animals, with implications for welfare and data quality [20,21]. The suitability of this approach will depend upon factors such as the species, duration of isolation and whether animals can habituate to the protocol.

4) Use of data logging devices instead of transmitters, where data are recorded onto a microchip and downloaded once the device has been retrieved from the animal. Data can be retrieved following each session if external loggers are used, but there will clearly be a longer wait in the case of implanted loggers. Also, device failure may not be detected until the download is attempted, meaning that studies would have to be repeated.

At the time of writing, biotelemetry device manufacturers are working on technical innovations that will further facilitate pair and group housing, so it is good practice to keep up to date with developments in multi-wavelength devices and other innovations that can reduce the impact of telemetry on animals, e.g., smaller or lighter devices. A wider range of dual frequency and multi-channel devices has become available; initially for large animals such as dogs, but some are now marketed as small enough for use in rats. However, there are currently no commercially available, implantable, multi-wavelength devices that are suitable for mice.

Reason 2: Animals have undergone implantation surgery and are singly housed in case they damage one another's wounds. Surgical wound failure is painful and distressing, affecting both health and welfare and requiring either repair surgery or the euthanasia of the animal. However, the negative welfare impact of changing from social to permanent single housing should not be underestimated for social species. To look at this another way; wound damage or failure may occur, but the psychological distress due to changing to single housing will occur. In light of this, the ideal is to re-establish animals in their social groups as soon as possible following any kind of surgery, and this should also be the default position for biotelemetry [6].

Different facilities employ a variety of protocols with respect to regrouping after implantation surgery, ranging from immediate regrouping to waiting for a period of several days. In general, animals can be successfully regrouped after 24 hours [5] if steps are taken to ensure that:

- The pair or group was well-established and stable before surgery. If groups are stable before surgery, there is a greater chance of achieving harmonious regrouping afterwards. Drawing pairs or groups from littermates is an obvious option, which should be possible whether animals are bred either in-house or externally. Good liaison with the breeder will be necessary if animals are sourced from outside the establishment (liaison with any external breeding facilities to ensure consistency in housing and care conditions is also good practice, although if there are differences then the aim should be to 'level up' and implement the better husbandry protocols at both).

- If animals have undergone transport to the facility, they have had an adequate settling-in period to enable both recovery from any transport stress and acclimatisation to environmental changes and new caretakers. Absolute minimum settling-in periods before surgery have been suggested of a week for rodents, two weeks for dogs, and a month for non-human primates [6]. Note that animals should be allowed a minimum post-implantation recovery period of two weeks before conducting further procedures [20,21].

- Surgical procedures represent best practice with respect to asepsis, surgical approach, the competence of the surgeon and effective welfare assessment and pain management [5,8,24]. All of these will facilitate a more rapid return to normal behaviour, including social behaviour.

- Animals have recovered from anaesthesia so that they are capable of interacting appropriately with others.

- Wound closure has been fully refined, for example by the use of intradermal sutures [24]. This, together with effective pain management and careful asepsis, will stop animals paying excessive attention to their own wounds, reducing the risk of drawing the attention of others *via* visual or olfactory cues.

- Housing and care are also refined, including the provision of sufficient space and enrichment, which will shift animals' attention from their own and others' wounds and allow them to retreat from one another [6].

- There is adequate supervision from carers in the initial regrouping period, and for as long as is necessary to ensure that there is no persistent discomfort or pain. Behavioural assessments should also check that animals are interacting in an acceptable way, e.g., hierarchies are appropriate and there is no bullying [5,25].

There may be concerns about wound breakdown and bullying if group housing has not been usual practice, or if there is a proposal to regroup animals earlier following surgery. In these cases, a carefully-supervised pilot study, with advice from the attending veterinarian, could be undertaken. This should include and address all of the points in the above list.

One approach to reducing any negative effects of single housing that might arise is 'living apart together', in which animals have no (or minimal) physical contact with one another but can see, hear and/or smell other animals. Visual and olfactory contact is recommended for other species such as non-human primates, rabbits and dogs when separately housed [6,26,27,28]. However, the benefits of this limited contact can vary between species and strains. For example, female C57BL/6JOlaHsd mice benefit most from social housing after surgery, but if they are separated, limited contact through a grid partition is more stressful than housing them individually [29]. The literature on the effects of social *vs.* individual housing, including behavioural studies to evaluate whether particular species and strains benefit from limited contact, should be regularly reviewed to ensure that protocols reflect current knowledge about animal behaviour and welfare.

If single housing really is unavoidable for justifiable scientific reasons, animals should not undergo implantation surgery unless they are able to tolerate being housed on their own. A trial period should be set up to assess how well animals can adapt before moving on with the study.

Reason 3: Animals are wearing jackets or have exteriorised devices and are singly housed in case they interfere with these. Similar issues arise when group housing animals with devices that are partly or entirely external as when following surgery. Some species can be group housed while fitted with external devices without problems, but with other species there may be concerns that animals could damage one another's jackets or devices, or that stress or injuries could occur due to excessive attention to these. The basic approaches to overcoming this are the same as those for group housing post-surgery, *i.e.*, maintaining stable groups, refining housing and care, and ensuring appropriate supervision.

Other measures that can be undertaken to facilitate group housing for animals with jackets, harnesses or exteriorised devices include ensuring that animals are fully habituated to the device and/or attachment, are trained to accept it and are also accustomed to other animals wearing it. Animals may habituate to relatively large exteriorised devices if they are given the opportunity. For example, mice with headpieces measuring 15 by 20 mm have been successfully housed in groups of 12, with no injuries to the animals or significant damage to any of the devices. This was facilitated by forming and maintaining stable groups and by adding non-working devices to the cage for the mice to explore and manipulate. Although some of the mice tried to bite others' devices for a period of 15 to 20 minutes when first re-grouped after the headpieces were fitted, this was not sustained and the animals quickly lost interest [30].

For studies where jackets are required, trials can be conducted using old jackets and dummy devices to help with habituation. If the only problem lies with animals undoing fastenings, these can simply be made more robust. Some animals (e.g., some groups of dogs, particularly very young ones) may persist in damaging one another's jackets and/or devices, so that it really is impossible to group house them. In such cases, any welfare impact can be reduced by minimising the duration of single housing and implementing 'living apart together' for those species that will benefit from this (see above). It should also be possible to allow socialisation time for singly housed animals in a shared area, with adequate supervision and activities to distract from one another's jackets. Given the wide availability and increasing use of jacketed electrocardiogram (ECG) telemetry systems in large animals, achieving group housing for these animals is a very important goal for animal welfare.

Reason 4: Cage-type monitoring systems that can only record from singly housed animals are being used. Automated, cage-based monitoring systems are sometimes used for behavioural monitoring, for example in phenotyping screens or in neuroscience studies. These may operate using video, by recording food and water consumption, by recording and analysing an animal's movements within a cage placed on a sensor platform, or by a combination of these. Some of these monitoring systems cannot record from or track multiple animals, which necessitates single housing. It is somewhat paradoxical for a behavioural monitoring and characterisation system to require animal housing that is inherently stressful and liable to lead to confounding results, so the use of such systems should be critically questioned on both animal welfare and scientific grounds.

There are some cage-based monitoring systems already available that use transponder technology to recognise individuals, or that include software capable of distinguishing between animals. Where possible, these should be used in preference to those that only permit single housing. As with biotelemetry device technology, it is important to monitor ongoing developments and regularly check whether there is an alternative product or option that permits group housing for social animals.

2.2. Providing Environmental Enrichment

Animals with exteriorised devices are sometimes subject to restrictions on environmental enrichment for fear that they will become entangled or entrapped in some enrichment items (e.g., some nesting materials or refuges). Concerns that animals might hurt or injure themselves on some items can also result in reduced provision of enrichment for animals on biotelemetry studies.

Simply removing the enrichment item is not an appropriate solution, but in some cases there may be justification for adjusting enrichment protocols on veterinary and animal welfare grounds. For example, entanglement in nesting materials can be a real problem for rodents with head caps or cannulae, so some nesting materials would not be appropriate for them. However, it is often possible to cater for the animals' needs in other ways, such as providing non-tangling nesting material and refuges with wide entrances that are designed so that animals cannot climb on top of them and catch exteriorised devices on the cage roof (Figure 1).

(a) (b)

Figure 1. This nest box is designed for rats with external devices. (a) The box has a wide entrance and is bolted to the cage lid so that the animal cannot move it. The corners are extended to meet the side of the cage at the front left and back right of the box to prevent animals from squeezing between the box and cage walls. (b) The adapted box *in situ*.

Besides concerns about potential injury, some animals may be denied enrichment (including interaction sessions with carers) in case physiological responses to the stimulation will affect data quality [31]. This is unlikely to be the case; it is more likely that isolation and boredom will cause more stress and serious welfare problems [6,20,21], with negative effects on the science.

2.3. Long Term Housing in the Laboratory and Reuse

Instrumented animals are often reused in biotelemetry studies, for example in order to avoid conducting surgical procedures using naïve animals. As a consequence of this, larger species such as dogs and primates may be housed in the laboratory for some time (there are legal constraints on reuse in many countries, including Member States of the European Union. These include requirements relating to the severity of the previous procedures and the health and fitness of the animal [7]). Battery life is an inevitable constraint on the duration of an animals' stay in the laboratory, but the primary consideration should be animal welfare. Some facilities implement fixed time limits after which animals will no longer be reused, which may run to a couple of years or, exceptionally, six years or more. Others do not implement a fixed limit but regularly review the individual animals and how they respond to long term housing in the laboratory [31]. Regardless of whether time limits are in place, it is good practice to monitor animals for signs that they are becoming 'institutionalised' and losing the ability to cope with their environment. An appropriately tailored welfare assessment scheme, including indicators that animals are experiencing chronic adverse effects, is essential in this regard. Refining housing and care will also help to ensure that animals will be comfortable in their environment in the long term [6].

Enrichment is a husbandry refinement that can also supply additional indicators of an animal's welfare state; for example, animals who are becoming depressed or uncomfortable may stop making properly constructed nests, or climbing onto platforms, or they may generally become less interested in their enrichment items and one another. Welfare assessment protocols should be adequately tailored to animals on long-term studies and should ensure that subtle signs of suffering can be detected before more significant welfare problems occur [25].

There is also an ethical dilemma associated with long-term housing and reuse, in which the requirement to minimise the use of naïve animals has to be balanced against setting limits on the number of times animals are re-used, and on how long they should live in research facilities. The attending veterinarian, others with animal welfare expertise and well-informed ethics or animal care and use committees can play a useful role in considering and weighing these issues, e.g., by helping to decide on humane endpoints such as behaviours that indicate chronic distress, on absolute limits on the number of times animals are reused, or on the length of time that individuals stay at the facility. Input from the attending veterinarian is an essential part of the decision-making process, and is a legal requirement in many countries [7].

2.4. Using Telemetered Physiological or Behavioural Data to Help Assess Wellbeing

Many variables that are monitored and recorded in biotelemetry studies are also useful in assessing animal welfare. For example, heart rate, blood pressure and body temperature can all act as indicators that animals may be experiencing pain, suffering or distress, in conjunction with other criteria such as behaviour, body weight, *etc.* Physiological data can be obtained from animals with devices in place for scientific purposes (so there are no additional harms to experimental subjects) and used to evaluate the impact of disturbances (e.g., building work), or different housing and care protocols [5]. Some examples of ways in which practitioners are using biotelemetry to assess welfare include (on the basis of an online survey conducted by the author):

- using telemetered data to help determine humane endpoints;
- reusing implanted animals in welfare studies, e.g., to provide physiological correlates with behaviour;
- using telemetered data to assess stress due to restraint and transport;
- comparing responses to different housing, enrichments and interactions with humans;
- monitoring recovery from surgery using telemetered data; and
- using physiological data to help compare the effectiveness of different analgesic agents.

Opportunities to use telemetered data to reduce the suffering and improve the welfare of the animals on the study—or other animals—should be identified and taken up wherever possible. For example, discussion between the researcher and the ethics or animal care and use committee at the project planning stage may identify the potential to use the data in this way, or this may arise at mid-term or retrospective review of a project [31].

3. RECOMMENDATIONS

Whatever the species of animal, purpose of the study or nature of the biotelemetry system, the aim should be to provide a good standard of housing and care for animals on biotelemetry studies, as for any other project. Although biotelemetry can present some challenges to refining housing and care, many of these can be overcome by flexible thinking, keeping up to date with developments and being prepared to resource new equipment and approaches, as set out in the recommendations below.

- Recognise that although biotelemetry can facilitate reduction and refinement, the technique has the potential to cause suffering—it is a 'refinement that needs refining'. There is a need to apply a harm-benefit assessment to procedures involving biotelemetry and to ensure that these are fully refined.
- Keep up to date with technical developments such as wider availability of multi-frequency devices and smaller implants. If there are unmet technological needs that could help to refine the use of biotelemetry (e.g., multi-frequency, implantable devices for mice), communicate with device manufacturers to make them aware of the demand.
- Ensure that surgical procedures are conducted according to current good practice with respect to the competence of the surgeon, surgical approach, welfare assessment, pain management, wound closure and asepsis.
- Provide group housing for social animals used in biotelemetry studies as the default; ensure that animals are only singly housed if there is sound scientific or veterinary justification.
- If periods of single housing are unavoidable during data collection, keep these to a minimum and monitor animals carefully during reintroductions.
- Provide environmental enrichment as the default, thinking creatively about alternative approaches if some items would not be appropriate.
- Critically consider the pros and cons of automated observation systems that necessitate single housing, taking the welfare and scientific implications into account.
- Set a limit for maximum duration of laboratory housing for long term studies, with clear criteria for behavioural assessments and humane endpoints. Seek advice on this from the attending veterinarian, experienced animal technologists, and/or local ethics, animal care and use committee or animal welfare body as appropriate.

ACKNOWLEDGMENTS

Many thanks to Dusty Karazan and Cindy Argue for a very helpful discussion; also to members of Compmed, the Laboratory Animal Refinement and Enrichment Forum (LAREF) and the LinkedIn Laboratory

Animal Science Group for sharing information about their current practices with me. Thanks also to Maggy Jennings, Sophie Masneuf and Tony Webb for their helpful comments during the drafting of this paper.

REFERENCES

1. Dictionary and Thesaurus – Merriam-Webster Online. Available online: http://www.merriam-webster.com/ (accessed on 13 December 2013).
2. Sarazan, R.D.; Schweitz, K.T.R. Standing on the shoulders of giants: Dean Franklin and his remarkable contributions to physiological measurements in animals. *Adv. Physiol. Educ.* 2009, *33*, 144–156.
3. Grenwis, J.E. Recent advances in telemetry promote further progress in reduction and refinement. NC3Rs Invited Articles, 2010. Available online: http://www.nc3rs.org.uk/news.asp?id=16 (accessed on 13 December 2013).
4. Anonymous. ICHS7A Safety Pharmacology Studies for Human Pharmaceuticals. 2000. Available online: http://www.ich.org/fileadmin/Public_Web_Site/ICH_Products/ Guidelines/Safety/S7A/Step4/S7A_Guideline.pdf (accessed on 13 December 2013).
5. Morton, D.B.; Hawkins, P.; Bevan, R.; Heath, K.; Kirkwood, J.; Pearce, P.; Scott, L.; Whelan, G.; Webb, A. Refinements in telemetry procedures: Seventh report of the BVA(AWF)/FRAME/ RSPCA/UFAW Joint Working Group on Refinement, Part A. *Lab. Anim.* 2003, *37*, 261–299.
6. Hawkins, P.; Morton, D.B.; Bevan, R.; Heath, K.; Kirkwood, J.; Pearce, P.; Scott, L.; Whelan, G.; Webb, A. Husbandry refinements for rats, mice, dogs and non-human primates used in telemetry procedures: Seventh report of the BVA(AWF)/FRAME/RSPCA/UFAW Joint Working Group on Refinement, Part B. *Lab. Anim.* 2004, *38*, 1–10.
7. European Commission. Directive 2010/63/EU of the European Parliament and of the Council of 22 September 2010 on the Protection of Animals used for Scientific Purposes. *Off. J. Eur. Union* 2010, *L 276*, 33–79.
8. Guiding Principles for Preparing for Undertaking Aseptic Surgery. A Report by the LASA Education, Training and Ethics Section. Jennings, M., Berdoy, M., Eds.; Laboratory Animal Science Association (LASA): Tamworth, UK, 2010; Available online: http://www. lasa.co.uk/publications.html (accessed on 13 December 2013).
9. van der Staay, F.J.; Arndt, S.S.; Nordquist, R.E. Evaluation of animal models of neurobehavioral disorders. *Behav. Brain. Funct.* 2009, *5*.
10. Kilkenny, C.; Parsons, N.; Kadyszewski, E.; Festing, M.F.W.; Cuthill, I.C.; Fry, D.; Hutton, J.; Altman, D.G. Survey of the quality of experimental design, statistical analysis and reporting of research using animals. *PLoS ONE* 2009, *4*.
11. van der Worp, H.B.; Howells, D.W.; Sena, E.S.; Porritt, M.J.; Rewell, S.; O'Collins, V.; Macleod, M.R. Can animal models of disease reliably inform human studies? *PLoS Med.* 2010, *7*.
12. Editorial Staff of Annals of the New York Academy of Sciences. *Animal Models: Their Value in Predicting Drug Efficacy and Toxicity*; Wiley-Blackwell: Hoboken, NJ, USA, 2011; Volume 1245.
13. CAMARADES. *Why is CAMARADES Needed?* 2011. Available online: http://www.camarades. info/ index_files/Why%20CAMARADES.htm (accessed on 29 January 2014).
14. Hawkins, P. Facts and demonstrations: Exploring the effects of enrichment on data quality. *The Enrichment Record*2014, 12–21. Available online: http://enrichmentrecord.com/current-issue/ (accessed on 18 February 2014).
15. National Research Council. *Guide for the Care and Use of Laboratory Animals*, 8th ed.; National Academies Press: Washington, DC, USA, 2011.

16. Pérez, C.; Canal, J.R.; Domínguez, E.; Campillo, J.E.; Guillén, M.; Torres, M.D. Individual housing influences certain biochemical parameters in the rat. *Lab. Anim.* 1997, *31*, 357–361.
17. Späni, D.; Arras, M.; König, B.; Rülicke, T. Higher heart rate of laboratory mice housed individually *vs* in pairs. *Lab. Anim.* 2003, *37*, 54–62.
18. Verwer, C.M.; van der Ven, L.T.M.; van den Bos, R.; Hendriksen, C.F.M. Effects of housing condition on experimental outcome in a reproduction toxicity study. *Reg. Tox. Pharmacol.* 2007, *48*, 184–193.
19. Karelina, K.; Norman, G.J.; Zhang, N.; Morris, J.S.; Peng, H.; DeVries, A.C. Social isolation alters neuroinflammatory response to stroke. *PNAS* 2009, *106*, 5895–5900.
20. Meijer, M.K.; Kramer, K.; Remie, R.; Spruijt, B.M.; van Zutphen, L.F.M.; Baumans, V. Effect of routine procedures on physiological parameters in mice kept under different husbandry procedures. *Anim. Welf.* 2006, *15*, 31–38.
21. Meijer, M.K.; Sommer, R.; Spruijt, B.M.; van Zutphen, L.F.M. Influence of environmental enrichment and handling on the acute stress response in individually housed mice. *Lab. Anim.* 2007, *41*, 161–173.
22. Laboratory Animal Science Association (LASA). *Laboratory Animal Science Association Guidance on the Rehoming of Laboratory Dogs. A Report Based on a LASA Working Party and LASA Meeting on Rehoming Laboratory Animals*; Jennings, M., Howard, B., Eds.; LASA: Tamworth, UK, 2004; Available online: http://www.lasa.co.uk/publications.html (accessed on 13 December 2013).
23. Hawkins, P.; Hubrecht, R.; Buckwell, A.; Cubitt, S.; Howard, B.; Jackson, A. *Refining Rabbit Care: A Resource for Those Working With Rabbits in Research*; RSPCA: Southwater, UK, 2008. Available online: http://www.rspca.org.uk/researchrabbits (accessed on 13 December 2013).
24. Wolfensohn, S.; Lloyd, M. Introduction to surgery and surgical techniques. In *Handbook of Laboratory Animal Management and Welfare*, 4th ed.; Wolfensohn, S., Lloyd, M., Eds.; Wiley-Blackwell: Chichester, UK, 2013.
25. Hawkins, P.; Morton, D.B.; Burman, O.; Dennison, N.; Honess, P.; Jennings, M.; Lane, S.; Middleton, V.; Roughan, J.V.; Wells, S.; Westwood, K. A guide to defining and implementing protocols for the welfare assessment of laboratory animals: Eleventh report of the BVA(AWF)/FRAME/RSPCA/UFAW Joint Working Group on Refinement. *Lab. Anim.* 2011, *45*, 1–13.
26. Seaman, S.C. *Laboratory Rabbit Housing: An Investigation of the Social and Physical Environment*, A Summary of the Report to the UFAW Pharmaceutical Housing and Husbandry Steering Committee (PHHSC), Based on a Ph.D. Thesis, 2002. Available online: http://www.ufaw.org.uk/pdf/phhsc-schol1-summary.pdf (accessed on 13 December 2013).
27. Buchanan-Smith, H. Marmosets and tamarins. In *The UFAW Handbook on the Care and Management of Laboratory and Other Research Animals*, 8th ed.; Hubrecht, R., Kirkwood, J., Eds.; Wiley-Blackwell: Chichester, UK, 2010; pp. 543–563.
28. MacArthur Clark, J.; Pomeroy, C.J. The laboratory dog. In *The UFAW Handbook on the Care and Management of Laboratory and Other Research Animals*, 8th ed.; Hubrecht, R., Kirkwood, J., Eds.; Wiley-Blackwell: Chichester, UK, 2010; pp. 432–452.
29. Van Loo, P.L.P.; Kuin, N.; Sommer, R.; Avsaroglu, H.; Pham, T.; Baumans, V. Impact of 'living apart together' on postoperative recovery of mice compared with social and individual housing. *Lab. Anim.* 2007, *41*, 441–455.
30. Masneuf, S.; University of Zurich, Zurich, Switzerland. Personal observations. 2011.
31. LASA/RSPCA. *Guiding Principles on Good Practice for Ethical Review Processes*, 2nd ed. 2010. Available online: http://www.lasa.co.uk/publications.html (accessed on 13 December 2013).

CHAPTER 2

Double Muscling in Cattle: Genes, Husbandry, Carcasses and Meat

Leo O. Fiems

Animal Sciences Unit, The Institute for Agricultural and Fisheries Research (ILVO), Scheldeweg 68, B-9090 Melle, Belgium

SIMPLE SUMMARY

Selection for an increased meatiness in beef cattle has resulted in double-muscled (DM) animals, owing to the inactivation of the myostatin gene. These animals are characterized by an excellent conformation and an extremely high carcass yield, coinciding with a reduced organ mass. As a consequence, voluntary feed intake is reduced, but feed efficiency is considerably improved, although maintenance requirements are not clearly reduced. DM animals are more susceptible to respiratory disease, stress and dystocia, requiring extra attention for accommodation and welfare. Carcasses of DM animals are very lean, and intramuscular fat content is low. The fatty acid profile is different when compared with non-DM animals, containing less saturated fatty acids. Collagen content of the meat is lower, so that meat from double-muscled animals is mostly more tender. However, meat tenderness, color and juiciness are not always improved. A different metabolism as a consequence of faster glycolytic myofibers can be partly responsible for this phenomenon. DM animals are interesting for the producer and butcher, and beneficial for the consumer, if an appropriate nutrition and accommodation, and adequate slaughter conditions are taken into account.

ABSTRACT

Molecular biology has enabled the identification of the mechanisms whereby inactive myostatin increases skeletal muscle growth in double-muscled (DM) animals. Myostatin is a secreted growth differentiation factor belonging to the transforming growth factor-β superfamily. Mutations make the myostatin gene inactive, resulting in muscle hypertrophy. The relationship between the different characteristics of DM cattle are defined with possible consequences for livestock husbandry. The extremely high carcass yield of DM animals coincides with a reduction in the size of most vital organs. As a consequence, DM animals may be more susceptible to respiratory disease, urolithiasis, lameness, nutritional stress, heat stress and dystocia, resulting in a lower robustness. Their feed intake capacity is reduced, necessitating a diet with a greater nutrient density. The modified myofiber type is responsible for a lower capillary density, and it induces a more glycolytic metabolism. There are associated changes for the living animal and post-mortem metabolism alterations, requiring appropriate slaughter conditions to maintain a high meat quality. Intramuscular fat content is low, and it is characterized by more unsaturated fatty acids, providing healthier meat for the consumer. It may not always be easy to find a balance between the different disciplines underlying the livestock husbandry of DM animals to realize a good performance and health and meat quality.

Keywords: double muscling; myostatin; carcass quality; organ size; locomotion; reproduction; stress susceptibility; meat quality

1. INTRODUCTION

There is a persisting trend to improve carcass quality in specialized beef breeds [1,2,3]. A higher meat yield and more lean meat are desirable for the meat industry [4]. A smaller fifth quarter is more interesting for both the producer and the butcher, resulting in more edible meat cuts. It is also beneficial for the consumer, as leaner meat may be favorable with regard to coronary and heart diseases [5].

Double-muscled (DM) animals produce very lean meat. They can be classified as an extreme type of late maturing animals [6]. They are often defined as animals with muscle hypertrophy. However, there is also some muscle hyperplasia besides muscle hypertrophy. Several beef breeds are characterized by the double-muscling phenomenon: A lot of research was done in the 1970s with Charolais DM animals [7]. Other typical breeds for the presence of DM animals are Piedmontese [8] and Belgian Blue breeds [9]. Animals with the double-muscled character are also reported for Asturiana de los Valles [10] and Marchigiana [11] breeds.

According to Darwin [12] in his book "The Origin of Species", "it cannot be supposed that all the breeds were suddenly produced as perfect and as useful as we now see them; indeed, in many cases, we know that this has not been their history. The key is man's power of accumulative selection; nature gives successive variations: man adds them up in certain directions useful to him. In this sense he may be said to have made for himself useful breeds". The exploitation of the double-muscling phenomenon is clearly the effort of man to improve the meatiness of beef cattle. The Belgian Blue breed is a typical example of the exploitation of the myostatin gene. A review of the characteristics of DM animals is extant [13], but this paper aims to relate different characteristics of double muscling to feed intake, health, reproduction, histology and metabolism, and carcass and meat quality, and defines the consequences for livestock husbandry. In referring to DM animals it is inferred that they are homozygote for myostatin; if heterozygote animals are involved, it is explicitly mentioned.

2. ORIGIN OF DM ANIMALS: THE UNRAVELING OF THE MYOSTATIN GENE AND ITS EXPRESSION

Double muscling in cattle has been reported for more than a century [14]. However, due to the difficulties at parturition and the use of caesarean operations to assist calf birth, double muscling was only fully exploited after the Second World War, mainly because of the availability of anesthesia, antibiotics and new surgical methods.

Charlier *et al.* [15] have mapped a locus, referred to as mh (muscular hypertrophy), on bovine Chromosome 2, confirming the validity in the Belgian Blue population of the monogenic model involving an autosomal mh locus. The MSTN (myostatin) gene has been identified as the cause of the double-muscled phenotype in cattle [16]. An 11 base-pare deletion of the protein is responsible for muscular hypertrophy in Belgian Blue DM (BBDM) cattle. This naturally occurring mutation results in an inactivation of the MSTN gene and provokes a massive increase in the skeletal muscle mass. McPherron and Lee [17] reported the similarity in phenotypes of DM cattle and mice, carrying a targeted deletion of the MSTN gene (MSTN null mice, MSTN knock-out mice), suggesting that MSTN performs the same biological function in these two species. This similarity in biological processes between cattle and mice has been confirmed [18]. This statement means that MSTN may be a useful target for genetic manipulation in other farm animals. MSTN is a member of the transforming growth factor-β superfamily that encompasses a large group of secreted growth and differentiation factors playing important roles in regulating development and tissue homeostasis [17]. Gerrard *et al.* [19] showed that the presence of a co-twinned fetus with a lower genetic potential for muscle development reduced the capacity of heavily muscled fetuses to develop muscle mass, and they concluded that blood-borne factors regulate muscle hypertrophy in fetal cattle. MSTN expression is detected in the myotome early in myogenesis through to adult skeletal muscle, where it may control fiber number and size during embryonic, fetal, and postnatal myogenesis [9]. The existence of circulating forms of MSTN was reported [20,21]. MSTN is secreted under a latent form due to its association with inhibitory proteins [22,23]. Members of the bone morphogenetic protein-1 (BMP)/tolloid family of metalloproteinases can activate latent MSTN [24]. MSTN signals through a transforming growth factor-β/activin/nodal-like pathway and binds to activin IIB receptors, and, to a lesser extent, to activin IIA receptors, and partners with a Type I receptor, either activin receptor-like kinase 4 (ALK4 or ActRIB) or ALK5, which in turn initiates the intracellular signaling cascade by phosphorylation of the receptor-regulated proteins SMAD2 and SMAD3 [20,25]. MSTN potently antagonizes BMP7 but not BMP2 by competing for BMP7 binding to the ActRIIB type II receptor [25]. MSTN also auto-regulates its expression by feedback loop through a SMAD7 dependent mechanism [26]. However, muscle hypertrophy observed in constitutive MSTN knock-out mice is primarily due to myofiber hypertrophy, while that observed in MSTN-null cattle results primarily from hyperplasia [27]. The investigation of the protein alterations in the *semitendinosus* muscle in response to an 11-bp deletion in the MSTN gene showed that thirteen proteins are significantly altered between BBDM and non-DM animal muscles. Eight proteins are related to the contractile apparatus, two proteins are involved in metabolic pathways, and three proteins are significantly altered including sarcosin, sarcoplasmic reticulum 53 kDa glycoprotein and 20 kDa heat shock protein [27].

McPherron and Lee [17] also sequenced the MSTN gene in the Piedmontese breed. The similar map positions of the MSTN gene and the muscular hypertrophy locus and the identification of relatively severe mutations in the MSTN gene of two different DM cattle breeds suggest that these mutations are responsible for the DM phenotype. Mutations in Belgian Blue and Piedmontese DM cattle are different [9]. The entire MSTN coding sequence in ten European cattle breeds was determined [28]. The observation that in at least eight of the ten studied breeds double muscling involves five independent MSTN mutations indicates that the number of genes susceptible to affect muscular development in a comparable way is likely to be limited in cattle.

The role of MSTN in the regulation of cell growth and cell death occurs in concert with insulin-like growth factor (IGF)-I [29]. It has been shown that the maximum expression of muscle IGF-II is delayed in DM animals during fetal development [30]. A differential regulation of the IGF-II gene transcription between DM and normal fetuses was also reported [31]. Insulin-like growth factor binding proteins (IGFBP-3 and IGFBP-5) affect processes downstream the receptor-mediated SMAD phosphorylation, enabling MSTN to suppress proliferation of porcine embryonic myogenic cells [32]. The role of IGF and IGFBP was also confirmed in mice [33]. MSTN null mice have higher circulating IGF-I levels than wild-type mice, whereas IGFBP-1 and IGFBP-2 levels were lower. Furthermore, it is assumed that the relative IGF-binding capacity was potentially lower in MSTN null mice. Muscle regulatory factors (MRF4, Myf5, MyoD and myogenin) play a key role in muscle development and maturation [34,35]. A distinct pivotal function in muscle development is attributed to each of these muscle regulatory factors. The expression of MyoD in fetal bovine muscles is up-regulated by the loss of functional MSTN [36,37,38]. Higher levels of MyoD expression throughout primary and secondary fiber formation were found in BBDM fetuses compared to non-DM fetuses [37,38]. Higher expressions of MRF4, Myf5 and MyoD were found in DM than in non-DM Japanese Shorthorn cattle, but the effect was different between muscles [39]. Furthermore, the effect of MSTN on MRF4 expression may be more specific for fast type muscle fibers. Myf5 and MyoD expressions were increased in faster and slower muscles, respectively, indicating a different role of each muscle regulatory factor. On the other hand, calcineurin signaling plays an important role in the hypertrophy response through decreases in MSTN expression [40].

3. CONSEQUENCES OF THE INACTIVATION OF MYOSTATIN ON CONFORMATION AND CARCASS QUALITY

The repercussions of double muscling in cattle husbandry will be elaborated in this paper. Figure 1 shows the impact of double muscling on some important animal characteristics. Several living animal characteristics and post-mortem issues are affected by MSTN. Interactions can occur, so that the mechanisms of the MSTN inactivation act in a coordinated manner.

3.1. Impact of Double Muscling on Conformation and Carcass Quality

An improved animal conformation means increased meat to other tissue ratio. DM animals are characterized by a better conformation (Figure 1(A)) and a hypermuscularity. These properties result in more compact animals. In fetuses of similar crown-rump lengths, double muscling increases body weight and the weight of individual muscles [41]. At carcass level, blockiness is significantly higher for DM animals [42]. Muscle hypertrophy is not similar for all muscles [43]. The variation ranges from 8 to 51% in the forequarters and from 9 to 34% in the hindquarters [44]. This hypermuscularity also occurs at the expense of bones. The limb bones are undersized in DM animals. Decrease in bone weight between DM and non-DM animals varies from 2.9 to 9.6%. Muscle hypertrophy and reduced bone weight result in an improved carcass quality (Figure 1(B)). Dressing percentage is increased in comparison with non-DM animals, concomitant with a higher lean meat content in the carcass and a lower content of fat and bone [45,46]. This means that carcass grading is different between DM and non-DM animals. Carcasses from DM animals are mostly classified as superior and excellent [46], according to the European beef carcass classification scale [47].

DM animals not only have a higher carcass yield, they also have a higher cutability or more expensive cuts in the meat yield. This can be demonstrated by comparing Belgian Blue bulls with a DM or a normal genotype, weighing 600 kg at slaughter. Dressing averages 70 and 64%, respectively, resulting in a cold carcass weight of 420 and 384 kg, respectively. Owing to the higher lean meat content in the carcass (76 *vs.* 65%), total meat amounts to 319 and 250 kg, respectively, being 28% more for DM animals. From this total meat yield, 171 and 123 kg are higher-priced roasts and steaks, while 148 and 127 kg are lower-priced cuts for stewing or boiling, or for hamburgers, being 39 and 17% more for DM animals within each class of retail cuts, respectively. More information about the effect of double muscling on carcass characteristics is reported by Arthur [13].

3.2. Impact of Double Muscling on Organ Size

Several internal organs have a reduced size in DM animals compared to non-DM animals (Figure 1(C)). Charolais DM bulls, slaughtered at 15 or 20 months, were characterized by lower weights (% reductions) for spleen (±30%), liver (±20%), heart (±20%), blood (14 and 20%, respectively) and lungs (10 and 24%, respectively), compared to non-DM bulls [48]. Similar results were found for BBDM bulls [44], with reductions varying from 10% for the adrenal glands to 15% for the heart, 17% for the liver and the digestive tract, 19% for the lungs, 37% for the spleen and 51% for the thymus. Vermorel *et al.* [49] expressed organ size as g per kg empty body weight and found a reduction of 18% for the digestive tract, 16% for the liver and the kidneys and 14% for the heart. The weights of several organs (e.g., liver, kidney, heart and digestive tract) were also significantly reduced in mice, by 12–20%, when a MSTN mutation was introduced [50]. The lower liver weight in MSTN null mice was explained by the negative feedback of hepatic-produced IGF-I on circulating GH, which in turn regulates liver size [33]. The reduced digestive tract in DM animals is clearly demonstrated by a higher ratio of empty body weight to body weight. This ratio amounted to 90 and 93%, respectively in BBDM cows [51] and bulls [52]. Ratios of 72% [53], 79–86% [54] and 84% [55] have been reported for non-DM animals.

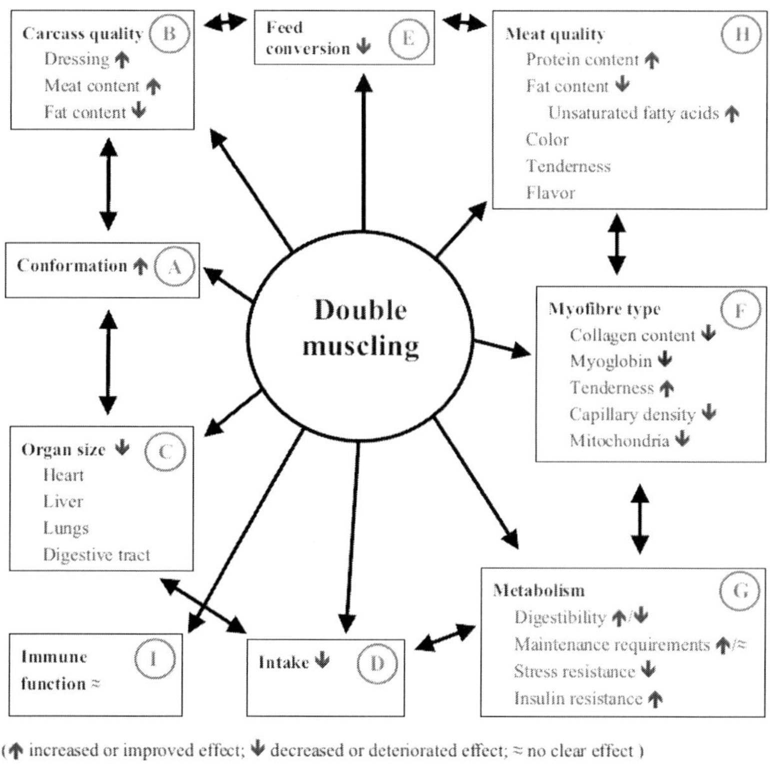

Figure 1. Impact of double muscling on some important animal characteristics.

Besides internal organs, skin weight is also reduced in DM animals, ranging from ±20% [44,48] to 29% [49]. The reduced weight of internal organs, skin, head and legs results in a large reduction of the fifth quarter and yields a high dressing percentage, as mentioned above. In a comparison involving 10 beef cattle breeds, Asturiana de los Valles yielded the best carcass quality due to the DM condition [56].

The reduction in lung weight not only affects the weight of the fifth quarter and carcass weight, but it may also induce a negative effect on respiratory functions in DM animals. Bovine lungs are characterized by about 25% of lung volume per unit of body weight as compared to the mammalian mean [57]. Comparing horses with steers, horses have nearly two times larger lung volumes, a 100% larger cardiac output, and a 40% higher hemoglobin concentration [58]. The small airway resistance is significantly higher in DM calves than non-DM, which may explain the sensitivity of this cattle type to severe bronchopneumonia [59]. Furthermore, DM cattle appear to be predisposed to develop alveolar hypoxia and hypoxemia in comparison with conventional cattle [60,61]. The quantity of β-adrenoceptors in the tracheal smooth muscle membranes is lower in DM calves ($P < 0.05$) than in non-DM calves, while the density of muscarinic receptors is about 40% higher ($P < 0.01$). Inflammation of the respiratory tract in DM calves may be complicated by an imbalance between the cholinergic bronchoconstrictor and the β-adrenergic bronchodilator components of the autonomic nervous system [62], which may help to explain their sensitivity to respiratory diseases. The production of viscous purulent exudate in response to respiratory diseases often exceeds the removal by expectoration in cattle, so that the airways are obstructed and gaseous exchange is impaired [63]. Consequently, DM animals seem handicapped when there is a risk of respiratory diseases. Therefore, extra attention should be paid to the housing of DM animals. DM calves housed in outdoors hutches are mostly associated with a decreased risk of pneumonia compared to those in indoor confinement [64].

Tyler and Ramsey [65] found that immunoglobulin G absorption during the first 18 h was slower in hypoxic than in normoxic newborn calves. However, absorption rate increased dramatically after 18 h, and immunoglobulin G levels did not diminish until 42–48 h. The absorptive capacity was unaffected by hypoxia. Extension of the absorptive period without affecting the absorptive capacity was accomplished by the diminished rate of absorption during hypoxia. Therefore, it is assumed that immunoglobulin absorption in DM calves is not negatively affected due to their higher sensitivity to hypoxia.

The aforementioned smaller heart [44,48,49] acts in concert with the smaller lungs. Cardiac performance, capability, and reserve are reduced in

DM animals compared to non-DM ones [66]. The lower cardiac performance, together with the predisposition to develop alveolar hypoxia and hypoxemia, is critical for the oxygen transport in DM cattle. Furthermore, endurance of DM cattle is lower, leading more quickly to exhaustion after severe exercise [67], which may even terminate in sudden death [68]. This phenomenon is similar to results found in peroxisome proliferator-activated receptor γ coactivator 1α (PGC-1α) knock-out mice. These mice exhibit a shift from oxidative Type I and IIA toward fast glycolytic fibers. They have a reduced endurance capacity and exhibit fiber damage and elevated markers of inflammation following treadmill running. These animals also have a reduction in mitochondrial gene expression [69]. Mice with a deletion of the PGC-1α gene may be similar to DM animals, because of an increased number of fast-twitch muscle fibers and a reduced number of mitochondria (see further). Subjecting these mice to the task of running to exhaustion, using a treadmill, resulted in a significantly lower fatigue resistance index than in wild-type mice. Furthermore, the hearts of the PGC-1α null mice were unable to mount an appropriate chronotropic response to exercise and other physiologic stimuli that activate β-adrenergic input to the heart [70]. They also traveled a significantly shorter distance. This behavior is in accordance with the reduced locomotor activity of DM animals (see further). This exhaustion is explained by a lower blood hematocrit count [67] and hemoglobin level [71], although Gustin *et al.* [72] doubted whether the greater sensitivity of DM cattle to exercise could be explained by hematological factors. Nevertheless, the robustness of DM cattle seems to be reduced, so that extra precautions are necessary with regard to animal welfare.

A consequence of the smaller digestive tract is reduced feed intake capacity (Figure 1(D)), as reported by several authors [6,73,74]. It is possible that a reduced feed intake capacity is also partly responsible for the reduced liver weight in DM animals, besides the growth hormone axis [33]. A reduced liver weight in DM animals is in line with the significantly decreased liver mass observed in steers after a nutrient deprivation [75]. When concentrate intake was reduced to about 55% of the *ad lib* intake, liver weight was decreased by 25% after a 90 kg live-weight loss. However, the relationship between reduced intake level and liver weight is not clear. An increased nutrient density of the diet is necessary for DM animals to counterbalance the reduced feed intake capacity. Furthermore, a large dietary protein concentration is necessary for the high lean meat deposition (Figure 1(B)); albeit that dietary protein is more efficiently converted into body protein. DM animals are known to be slow "turning over", depositing large amounts of muscle protein with a greater capacity for protein synthesis than non-DM animals [76]. It is questionable if the reduced digestive tract exerts an effect on digestibility (Figure 1(G)). Clinquart *et al.* [73] found a lower dry matter digestibility (−4.6%; $P < 0.01$) in BBDM bulls compared to Holstein bulls and a tendency for a lower digestibility (−2.7%) in BBDM than in non-DM bulls. This was not confirmed in another study [74], where a tendency for a higher dry matter digestibility (1.1%; $P = 0.10$) was found for BBDM bulls compared to Belgian Blue non-DM bulls, resulting in a higher dietary energy content. However, feed intake was significantly lower for the DM genotype. When a correction was made for a different feed intake, the energy value was similar for both genotypes. A similar dry matter digestibility for DM and non-DM cattle was also reported by Vermorel *et al.* [49]. From these results it can be concluded that feed energy evaluation is not different between DM and non-DM cattle. Specific energy and protein requirements have been deduced for BBDM animals [77] based on a similar digestibility.

The reduced feed intake capacity also means that DM animals are incapable of using low quality diets efficiently. Dry matter intake is reduced and the energy content per kg dry matter is also reduced, so that there is a double negative effect on energy supply. For these reasons high-energy diets are often fed to DM animals.

Feeding BBDM cows below their requirements resulted in a mobilization of adipose tissue, but the protein balance was also negative [78]. Mobilization of body protein does not result in an efficient use of protein and can be explained by the energy demand for urea synthesis in the liver [79]. Similar results were found in bulls when they were quantitatively restricted. A similar feed restriction was more dramatic for live-weight gain and energy efficiency in DM bulls than in non-DM animals. Feed conversion was lower for Charolais DM animals than for Charolais non-DM and Friesian bulls when fed to appetite, while the opposite result was observed when fed at 75% of *ad lib* intake [80]. When a good quality diet is freely available feed conversion is lower for DM animals (Figure 1(E)), because of their reduced fat deposition. A significant interaction between diet type (concentrate or roughage based diets) and beef type (DM *vs.* non DM animals) with regard to voluntary intake was reported [81]. An interaction between nutrition and MSTN status for growth rate and myofiber characteristics has been confirmed in heterozygote lambs with a MSTN mutation [82]. Genotype-nutrition interactions cannot be isolated from body composition [83]. When *ad lib* energy intake is greater than the requirements for protein retention capacity and associated lipid deposition, a restricted energy intake reduces lipid deposition. When energy intake is just sufficient for protein retention, a restricted energy intake reduces lipid deposition and protein retention. DM animals are an example of animals with an extreme protein deposition. These findings emphasize the importance of an appropriate diet to realize a high performance in DM finishing animals.

Besides specific protein and energy requirements for DM animals [77], it cannot be excluded that some mineral requirements may also be increased. In a study sacrificing seven Belgian Blue newborn calves, seventeen bulls and twenty-four cows with diverging parities, all with DM genotype, the phosphorus content in the empty body was determined. Increasing empty body weight had no significant effect on phosphorus concentration [84], while other findings [85,86,87] were unanimous about a decreasing phosphorus accretion. It can be concluded that DM animals probably have a higher phosphorus requirement than non-DM animals, especially at a heavier weight. Diets with low levels of calcium and high levels of phosphorus are known to be calculogenic [88]. Males are at greater risks of obstruction due to urinary calculi than females, because of smaller urethras [89] so that stones pass less easily. Apart from the reduced weight of the kidneys in DM animals [44], Ouhayoun and Arnal [90] found that the medulla and the cortex were reduced as well. Furthermore, the number of urinary tubes and the diameter of the Bowman's capsules were also reduced. This may reduce plasma ultrafiltration, so that DM bulls are more predisposed to urolithiasis.

There is also need for an increased dietary selenium level. Guyot *et al.* [91] reported that 0.5 ppm dietary selenium may be advised for BBDM cattle to optimize health and performance, while 0.1 ppm is recommended [92]. The DM phenomenon may result in an increased susceptibility to white muscle disease or muscular dystrophy [93], explaining the higher need for selenium.

Plasma creatine phosphokinase is a good indicator during the active stage of myodegeneration when the myofibers are disintegrating [94]. Plasma creatine phosphokinase in DM animals parallels that of non-DM animals for the first 18 hours of fasting, but continues to increase steadily throughout the fast, so that the difference between genotypes is significant by 30 hours [68]. A higher creatine phosphokinase activity in DM animals has also been reported [42,68], although the difference was not always significant [95]. These findings may confirm the predisposition of DM cattle for muscle degeneration.

3.3. Consequences of Double Muscling for Locomotion

According to Wolff's law, bones will adapt to their loads. Increased loads on a particular bone will result in a remodeling, so that the bone will become stronger to resist the extra load. The aforementioned smaller bone content in DM animals may have consequences for their legs. As more body weight should be carried by the reduced skeleton [44,48], bones are more highly loaded. Hendricks *et al.* [96] reported a smaller metacarpal length in 13–15 month old Angus DM bulls than in non-DM animals. Cortices tended to be thinner. Maximum shearing stress and modulus of elasticity were 15 and 18% lower, respectively, in DM than in non-DM bulls, but due to a large variation and a low number of animals, differences were not significant. The length of *humerus* and *femur* in DM Charolais bulls were significantly reduced by 4 to 5%, while femur circumference was reduced by 7 to 10% [48]. Shahin *et al.* [97] dissected DM and non-DM cows at slaughter. They found that DM cows have a lower proportion of bone in the carcass, in the entire forelimb as well as in the entire hind limb. *Humerus, carpus* and *ox coxa* percentages of total bone weight were significantly lower in DM cows. Moreover, work in mice showed a decrease in force generation of muscles from MSTN null mutants [98,99,100]. These MSTN null mutant mice were also characterized by increased degenerative changes in the intervertebral disc [101]. Mendias *et al.* [98] investigated the role of MSTN on the properties of tendons. Although the mass of the *tibialis anterior* and *soleus* muscles of MSTN null mice were 72 and 82% greater, respectively, than those of wild-type mice, tendons were 40 and 44% smaller, respectively, resulting in a 64 and 69% decrease in the tendon/muscle mass. The cross-sectional area of the *tibialis anterior* tendons was 50% smaller. MSTN null mice also showed a 14-fold-greater stiffness. The stiffness of tendons is a critical factor in determining the damage to muscle fibers during lengthening contractions. Consequently, the loss of MSTN had a major impact on the mechanical properties of the tendons.

Bone development is a fine-tuned process. The cells of maturing cartilage enlarge at the expense of the surrounding matrix, and the matrix itself becomes mineralized by the deposition of calcium phosphate as hydroxyapatite crystals. However, the main components of the matrix are Type I collagen fibers. The hardness of bone originates from the calcified matrix, while the strength comes from the embedded collagen fibers. Dying chondrocytes leave large empty cavities, which are invaded by osteoclasts and blood vessels, while osteoblasts deposit bone matrix. Chondrogenic differentiation and cell death involves a signaling pathway mediated by BMP-2, BMP-4 and BMP-7 [102]. Blood vessels are necessary for the transport of the nutrients for the survival of the bone. The ossification of the epiphysis may be retarded in the absence of cartilage canals [103]. Merino *et al.* [104] provided evidence for a stage-dependent specificity of the chondrogenic function of activins in the course of limb development. Activin βA and βB subunits and activin type IIA and IIB receptors exhibit a regulated pattern of expression in the different steps of joint formation, indicative of the involvement in digit arthrogenesis. We already mentioned that MSTN signals through a transforming growth factor-β/activin/nodal-like pathway. MSTN potently antagonizes BMP7, which is involved in chondrogenic differentiation and cell death. BMP receptor signaling is required for early development and creation of multiple tissues, but also for ongoing maintenance of articular cartilage after birth [105].

Rebbapragada *et al.* [27] speculated that BMP-dependent patterning of bone and cartilage within the limbs might be refined by MSTN. It may become clear that deleted MSTN has an adverse effect on cartilage and bone formation. The cartilage content in the 10th rib cut is reduced in DM animals [106].

The hooves of cattle are important for locomotion. The corium is part of the hoof containing blood vessels and nerves. At the heel, a pad of fatty tissue forms the digital cushion, playing an important role in blood circulation. It draws blood out of the foot into circulation as the heel makes contact with the ground when the animal moves. Therefore, exercise is important for new hoof horn formation, as it maintains an optimal blood circulation through the foot.

DM animals are deviant from non-DM animals, as their tissues contain less collagen. Genes encoding Type I and III collagen are down-regulated in DM muscles [107]. Consequently, bone strength may be reduced. Furthermore, capillary density is lower in DM animals. This may be critical for the nutrient support in bones and hooves. If the increased phosphorus requirement of heavier DM-animals is not fulfilled, it may be detrimental for bone formation. Therefore, precautions with regard to nutrition and confinement of DM cattle are necessary. It is remarkable that the energy cost of standing is 25% higher (P = 0.01) in BBDM than in non-DM bulls [49]. As a consequence, time spent standing is 13% lower in DM bulls. This may be detrimental for the blood circulation in the hooves and the formation of new horn. This is confirmed by the undersized limb bones in DM-animals [44] and the smaller circumference of the canon bone [108] and the *femur* [48]. Inhibition of collagen synthesis in connective tissues may also attenuate mechanical and structural adaptation of tendons and ligaments.

Double muscling may be a predisposing factor to arthrogryposis [109]. Belgian Blue and Charolais breeds have a high incidence of arthrogryposis. It is a hereditary disorder caused by mutation. Genetic testing for this disorder, and six other congenital defects (congenital muscular dystonia I and II, crooked tail syndrome, dwarfism, hamartoma, and prolonged gestation) is frequently conducted in BBDM cattle to eliminate these defects, so that important economic losses can be reduced. In breeding bulls it is advised to test the presence of these mutations.

In pens with concrete slatted floors compared to other bedding types, both the percentage of animals with hoof lesions and the severity of the lesions are increased [110]. In beef cattle, movement may be reduced on slippery floors. Movement difficulties can cause irregular hoof wearing and lameness [111]. It looks like DM cattle are more susceptible to leg injuries. In a beef production experiment with Piedmontese DM bulls housed in fully slatted-floor pens, half of the animals showed signs of lameness in the course of the finishing period [112].

As mentioned already, DM animals often receive high-energy diets with a low amount of fibrous components. These diets may provoke subacute ruminal acidosis with lameness as a consequence. Therefore, fibrousness of the diet is important for finishing DM animals fed high-energy diets. A minimum structural value of 0.6/kg dietary dry matter for BBDM beef bulls is desirable [113].

3.4. Impact of Double Muscling on Reproduction Performance

The previously mentioned reduction of the skeleton in DM animals, and especially the reduced development of the hip bone [48] may have repercussions on reproduction performance. Short *et al.* [114] reported a linearly decreased pelvic area ($P < 0.05$) when the MSTN allele was included in Piedmontese cattle ($P < 0.05$). Pelvic opening of DM dams was 10 and 6% lower ($P < 0.05$) than in non-DM Charolais [115] or crossbred cows [116], respectively, so that the incidence of dystocia and perinatal mortality was higher in DM cows [116]. The heavier fetal weight of BBDM cattle compared to non-DM cattle [117], rising up to 50 kg in BBDM calves at birth [118] may increase dystocia. The ratio of calf birth weight to dam weight averaged 9.0% for primiparous BBDM cows [119] *vs.* 8.3% ($P < 0.001$) for multiparous BBDM cows [120], compared to 7.5 and 6.5% for primiparous and multiparous Holstein cows, respectively [121]. The discrepancy between the underdeveloped maternal reproductive tract and the increased calf birth weight has made caesarean section an elective operation [122]. Calves born after an early surgical removal have a good vitality, blood pH and lactatemia at birth [123] close to those of eutocial calves, without negative effect on colostral immunoglobulin absorption compared to delayed surgical removal (delayed parturition after severe calf pulling). A drawback of the application of caesarean section is the significant reduction in subsequent pregnancy rate [124] resulting in an enlarged calving interval. A field study showed that the calving interval was significantly longer for BBDM cows, averaging 435 days compared to 393 days for dairy cows [125]. Moreover, each of the first three calving intervals investigated was longer for DM cows. This is longer than the average reported in the literature: three-hundred and seventy days in Angus cows [126], 365–368 days in Angus-Hereford and Charolais-Maine Anjou crossbred cows [127] and 377 days for Simmental cows, 387 days for Salers cows, 381 days Limousin cows and 393 days for Hereford cows [128]. Reduced fertility after caesarean operations may occur as a consequence of an increased incidence of uterine adhesions that affect the ovary or uterine tube and hinder involution [129]. In addition, the scar tissue formation within the uterine wall may increase the risk of abortion during subsequent pregnancies, limiting expansion of the uterus and/or hindering nutrition of the fetus. DM cows are also less able to carry fetus to term: implantation is difficult, leading to an increased embryonic mortality rate [130].

A longer calving interval cannot solely be attributed to caesarean section. The semen quality is significantly poorer for BBDM than for Holstein bulls [131]. Nevertheless, egg quality was better in BBDM cows than in Holstein cows [132]. However, although not significantly different, the number of transferable embryos per embryo recovery session was reduced by 31%, from 6.1 in Holstein cows to 4.2 in BBDM cows, while the number of unfertilized oocytes per embryo recovery session was increased by 39%, from 2.8 to 3.9. These results are in accordance with those obtained with mice, showing a reduced litter number in MSTN immunized mice compared to control mice, while embryo development was normal in both groups [133]. This was explained by a significantly lower number of developing follicles in ovaries of MSTN immunized mice. According to *in vitro* results of Ciarmela *et al.* [134] MSTN reduces the total number of pregnant human myometrial cells in a time- and dose-dependent manner, suggesting some functions in myometrial smooth muscle cells. Their findings also demonstrate that MSTN is expressed at higher levels only in some phases of the reproductive cycle. The pattern of MSTN expression during the estrous cycle may be regulated by steroid hormones.

From experiments with golden hamsters, a possible role of MSTN in placental development and function cannot be excluded [135]. Active MSTN may immunomodulate the uterine microenvironment. MSTN gene expression level declined as pregnancy progressed, to reach its nadir during late pregnancy. The amount of active MSTN peptide also declined, suggesting that MSTN is more active in early pregnancy, the phase with the implantation of the embryo. Therefore, the release of MSTN into the uterine environment is beneficial to the developing conceptus. It is possible that the inactivation of MSTN in DM cattle may compromise fetal development.

Activins influence several physiological processes in adults, among others reproduction in the male and the female [136,137]. Muttukrishna *et al.* [138] described the pituitary ovarian axis, where activin A has an endocrine and activin B has an autocrine/paracrine role on regulating pituitary FSH secretion. In the case of early pregnancy loss they referred to significantly lower levels of activin A when the miscarriage was complete. Maternally derived activin subunits, predominantly βA, are specifically up-regulated in the uterine luminal epithelium during pre-implantation embryo development, and subsequently in the endometrium, during the critical development events of gastrulation and organogenesis [139]. It is suggested that activins secreted by the oviduct and uterus could facilitate the development of the embryo prior to implantation. Because MSTN and activins are structurally related, it should not be amazing that MSTN may affect bovine fertility and reproduction.

Heat stress can negatively affect aspects of reproduction [140]. DM cattle are more susceptible to heat stress than non-DM cattle. Rectal temperature of cattle increases with rising ambient temperature, but the difference between DM and non-DM animals becomes larger with increasing ambient temperature [141]. The lower capacity of DM animals for heat dissipation is due to their blockiness and the smaller body exchange surface per body mass unit. Body heat is generated in proportion to muscle mass or volume, while body heat is dissipated in proportion to body surface area. Furthermore, DM cattle are characterized by a lower capillary density [142]. Blood flow to the skin is positively correlated to the sweating rate [143], and sweating is an important thermoregulatory mechanism to dissipate excess body heat. As the ease with which metabolic heat is moved to the skin for dissipation depends on the amount of heat traveling in the blood from the body core to the skin, the reduced capillary density is a disadvantage for DM animals. Reynolds *et al.* [144] reported the effect of heat stress in pregnant cows. On day 100 of gestation non-DM Hereford cows were confined in a heated barn at 16 °C or in individual stalls with a diurnal temperature regimen of 28 °C for 12 h and 36 °C for 12 h and a dewpoint of 21 °C. Uterine and umbilical blood flows were estimated on day 169. They were reduced by 34 and 23%, respectively, in heat stressed cows, and fetal weights amounted to 82% of those of non-stressed cows. In any case, capillarity is reduced in DM cattle. Additional heat stress may further reduce blood flow in these animals, which may increase the detrimental effect on fetal development.

Besides over the body surface, heat loss can also occur via the lungs. About 15% of the endogenous heat is lost directly from the body core via the respiratory tract in cattle [145]. Consequently, the reduced lung capacity, as mentioned before, is another disadvantage for DM animals. Furthermore, cattle that had severe respiratory disease early in life will have a decreased ability to regulate their heat load. Because DM cattle are more sensitive to severe bronchopneumonia [59], relatively more DM animals will have to contend with heat stress.

3.5. Influence of Double Muscling on Histological Aspects of Skeletal Muscle

Growth and development is a complex process involving the cell cycle regulation with progression through the G1, S, G2 and M phases [146]. Muscle fibers are issued from myoblasts which proliferate, then fuse to form myotubes, and finally differentiate into muscle fibers. At least two generations of cells are involved in bovine myogenesis. These cells are different by the expression of the different myosin heavy chain isoforms and by their size. Cells from the primary and secondary generations proliferate until 180 days after conception. The total fiber number is fixed from this stage on. Therefore, the fetal stage is crucial for skeletal muscle development. The first generation of myoblasts is completely differentiated from the end of the second trimester. The second generation achieves its differentiation at the end of fetal life. The first generation of myoblasts will give rise to Type I fibers after birth, while the second generation of myoblasts will be converted mainly into fast fibers, but also slow fibers depending on the muscle type [147]. The cell cycle progression is affected by biologically active MSTN, with an arrest in the G1 and G2 phases. Consequently, cells are not allowed to make the transition to the S phase where DNA synthesis occurs, and the M phase where cell division occurs, respectively. At a cellular level, MSTN affects cell cycling by controlling the entry of satellite cells into the S phase [148,149,150]. Intact MSTN is thus a regulator in the development of fast glycolytic myofibers [151]. MSTN expression increases during myoblast differentiation [38].

In DM animals with inactive MSTN, muscle fiber number [19,41,96, 142,152,153] and the number of fast glycolytic fibers were significantly increased [142,153,154,155,156]; the same was found in MSTN knockout mice [157], emphasizing an extensive hyperplasia. This can be explained by the lack of inhibition of myoblast proliferation and differentiation by MSTN. MSTN expression was detected in bovine embryos from day 15 of pregnancy and increased from day 31 onwards [158] and is regulated throughout pregnancy [159]. At 100 days after conception, BBDM fetuses already present muscle hypertrophy, which originates from a higher myoblast proliferation [160]. Accordingly, MSTN expression was found in the latest differentiating cells from the second generation [159]. Furthermore, the *semitendinosus* muscle contained a smaller proportion of primary fibers and a higher proportion of secondary fibers, resulting principally in fast fibers in adult muscle. This greater proliferation is accompanied by a delay in the differentiation phase [147,161]. Picard *et al.* [162] found that differentiation of muscle fibers was slower in DM fetuses, particularly during the first two-thirds of gestation. DM animals showed a slower rate of differentiation at 90 days after conception in first generation cells, while differentiation was mainly retarded in second generation cells at 130 days after conception. Differentiation of first generation cells was no longer different between DM and non-DM animals. Cells of the *longissimus* muscle of DM fetuses displayed a greater proliferating capacity and faster proliferating and alignment phases, but this advance is apparently lost after alignment, and the fusion of the DM myoblasts is delayed [163]. As a consequence, muscles of adult DM animals exhibit a higher proportion of fast glycolytic fibers, simultaneously with a higher total number of fibers compared to muscles of non-DM cattle breeds [151] (Figure 1(F)). The overall increase in muscle mass, which is due to an increase in the number of muscle fibers (hyperplasia) and to a lesser extent to fiber enlargement (hypertrophy), differs between muscles. Muscle fibers tended to be larger in the *rectus abdominis* muscle, while fibers in the *semitendinsus* muscle, especially Type I fibers, of DM animals were smaller [155]. The myofiber cross-sectional area within the fiber type was smaller in DM than in non-DM bulls. Due to the higher occurrence of fast glycolytic fibers, mean fiber size is larger in DM animals. The modified histological aspects may have some repercussions. Muscles with fast glycolytic fibers are characterized by an increased fiber sectional area and glycogen content, and a reduction of the content of myoglobin, fat, collagen, the number of mitochondria, and the capillary density [164]. Myoglobin is one of the first identified gene targets of calcineurin, and is expressed in highly-oxidative Type IIA fibers [40]. Calcineurin activation is sufficient to induce the slow fiber gene regulatory program *in vivo*, but additional signals may be required for skeletal muscle hypertrophy [165]. Transcriptional co-activator peroxisome proliferator-activated receptor γ coactivator 1α (PGC-1α) is a principal factor regulating muscle fiber type determination in skeletal muscle. PGC-1α is expressed preferentially in muscle enriched in Type I fibers [166].

3.6. Influence of Double Muscling on Metabolism

The aforementioned shift towards faster glycolytic myofibers in DM animals (Figure 1(F)) is responsible for a shift from an oxidative to a more glycolytic metabolism [164]. This coincided with significantly less aerobic and more anaerobic fiber ratios in DM compared to non-DM Belgian Blue bulls [155]. The anaerobic or glycolytic metabolism is in agreement with the reduced number of mitochondria in fast glycolytic myofibers. Considerable evidence shows that PGC-1α is sufficient to promote mitochondrial biogenesis and regulate mitochondrial respiratory capacity [70]. The increased number of fast-twitch muscle fibers in DM animals and the relationship between this myofiber type and PGC-1α [166] may lead to fewer mitochondria. It is noted that the lower number of mitochondria is not due a lower expression of the mitochondria themselves, but to the glycolytic muscle type containing less mitochondria [167]. Energy is generated in the mitochondria from the breakdown of glucose or fatty acids and the conversion to ATP in the Krebs cycle. Oxygen is of prime importance for muscle energy metabolism in the mitochondria. Factors involved in meeting the oxygen requirements are blood flow, arterial oxygen concentrations, tissue extraction rates, myoglobin content and mitochondrial volume [168]. After activation in the cytoplasm fatty acids are oxidized in the mitochondria through a cycle of different reactions mediated by enzymes. However, as the number of mitochondria is reduced in DM animals, there is less opportunity for fatty acid oxidation. This is in accordance with the low fat content in DM animals, so that the amount of fatty acids that should be oxidized after breakdown is very limited. Furthermore, due to the reduction of the capillary density, the oxygen supply is also reduced and the elimination rate of lactate may be impaired. The hemoglobin in the red blood cells and myoglobin inside muscle cells act in tandem for oxygen transport. However, myoglobin content is lower in DM than in non-DM finishing cattle [46,155]. The hemoglobin content in blood was not different between DM and non-DM dairy animals weighing between 40 and 680 kg [72], while the hemoglobin content tended to be lower in DM calves compared to Friesian neonatal calves, and the difference was even significant in calves between 2 and 28 days of age [61]. As a consequence, lactate can accumulate, especially after exercise [169]. Even in rest, blood lactate concentrations are significantly higher in DM animals, although this was not always confirmed [46].

Fast twitch glycolytic muscles have a much lower protein turnover than slow-twitch oxidative muscles with more Type I fibers [170]. Based on the indirect calorimetry technique to measure metabolic rate in MSTN null mice, total and resting oxygen consumption were 14% and 8% higher, respectively, compared with wild-type mice, but their body weight was also higher [171]. However, MSTN null mice had lower ($P < 0.01$) rates of total and resting oxygen consumption if results were expressed as a function of body weight. Results of Guo *et al.* [172] confirmed that MSTN null mice have a lower metabolic rate when normalized to total body weight. Tu *et al.* [173] used double knockout mice and also found a lower metabolic rate compared with controls, if oxygen consumption was expressed as a function of lean mass weight. This means that maintenance requirements and feed intake capacity may both be modified in the same direction (Figure 1(G)).

Visceral organs are considerably smaller in DM than in non-DM animals, as previously reported (Figure 1(C)). Ferrell [174] reported that the mass of these organs was highly correlated with energy expenditure. Even though visceral organs constitute less than 10% of body mass, they account for 40 to 50% of total energy expenditure. Therefore, it can be assumed that DM animals have a lower energy expenditure than non-DM animals. However, the daily energy expenditure of BBDM bulls was similar to that of non-DM bulls [49]. These results are not in agreement with previous findings [175], based on indirect calorimetry measurement in DM and phenotypically normal bulls. The metabolic rate was significantly higher in DM bulls and found to be 415.0 *vs.* 365.9 kJ/kg body weight daily. Moreover, this is not in accordance with the slower turnover rate of fast glycolytic myofibers, the lower feed intake capacity, and the reduced liver size of DM animals.

MSTN null mice have an improved glucose metabolism and an increased insulin sensitivity than non-transgenic control mice [173]. A higher insulin sensitivity, which was also reported for BBDM bulls [176], may be responsible for a reduced adipogenesis. BBDM calves have lower basal glucose concentrations and insulin peak concentrations, and higher glucose elimination rates than Holstein-Friesian calves [177]. MSTN up-regulates several genes involved in the regulation of the glucose metabolism [178]. The improved insulin sensitivity of DM animals is due to the AMP activated protein kinase for the promotion of glycolysis, as shown in C2C12 mouse skeletal muscle cells and L6 rat skeletal muscle cells. MSTN specifically inhibits glucose uptake into BeWo cells, a placental cell line, suggesting that MSTN may control glucose metabolism within the placenta [179]. This inhibition may be withdrawn in case of inactive MSTN. Significantly lower basal glucose and insulin levels after an overnight fast were reported for BBDM calves (4.3 ± 0.67 mmol/L and 5.1 ± 3.56 mU/mL, respectively), compared to 5.9 ± 0.86 mmol/L and 8.7 ± 5.81 mU/ml for Holstein calves [180]. After infusion of 150 mg glucose per kg body weight, a comparable increase in glucose was seen for BBDM (1.8 ± 0.63 mmol/L) and HF calves (1.6 ± 0.56 mmol/L). The increase of insulin was significantly higher in HF (27.6 ± 17.43 mU/mL) than in BBDM (18.9 ± 18.9 mU/mL) calves. As a result, the insulinogenic index, as a measure of early insulin secretion, was significantly higher in Holstein than in BBDM calves. The higher insulinogenic index of Holstein calves may be an indication of a lower insulin sensitivity, as HF calves have to produce more insulin to maintain normal glucose levels. Lower serum insulin levels were also reported for DM Piedmontese bulls in comparison with non-DM Piedmontese and Friesian bulls [181]. Istasse *et al.* [182] found a tendency for a higher plasma insulin level in BBDM bulls than in Holstein bulls at a younger age, but the opposite was found in older animals. Muscle fiber Type I coincides with a higher level of sensitivity to insulin since the increase in mitochondrial protein expression is demonstrated in response to insulin. In accordance, skeletal muscle glucose uptake is regulated through insulin-dependent regulation of mitochondrial gene expression. Insulin stimulates the biogenesis of mitochondria in muscle cells [183]. As mentioned earlier, PGC-1α drives the formation of slow-twitch muscle fibers in skeletal muscle [164]. PGC-1α null mice were less susceptible to diet-induced insulin resistance [70]. This phenomenon corresponds with the reduced insulin resistance in DM cattle.

3.7. Influence of Double Muscling on Stress Susceptibility

DM cattle have more problems with peaks of physical stress than non-DM cattle (Figure 1(G)). As mentioned previously, forced exercise leads to exhaustion more rapidly [67,68]. These findings were confirmed by Ploquin *et al.* [184], where an endurance test resulted in differential effects on redox homeostasis and mitochondrial function. The muscle fiber type and the lower capillary density lead to an accumulation of lactic acid. Another consequence of the reduced capillary density is the difficulty for DM animals to eliminate their body heat. Heat is transported from inside to the body surface through blood vessels [145]. Furthermore, body surface per unit of body weight is also lower in DM animals, so that they are more susceptible to heat stress [141]. The reduced body area combined with a lower capillary density will decrease evaporative cooling.

As DM animals are characterized by a reduced body fat content, we may expect a more severe effect of cold stress. Homeothermic animals cope with cold environmental temperatures by their ability to make the necessary physiological and metabolic adaptations. They may increase heat production, that is fuelled by an enhanced feed intake and/or mobilize their body reserves. Protein turnover may also contribute to cold-induced thermogenesis. As mentioned above in case of DM animals, feed intake capacity is reduced. The ability for tissue mobilization is clearly reduced [78]. Furthermore, their increased content of fast glycolytic myofibers, suggests a lower protein turnover [170]. Lammoglia *et al.* [185] compared the effect of a cold environment in newborn Piedmontese crossbred calves that contained at least one copy of the Piedmontese MSTN allele with non-DM Hereford crossbred calves. At 5 h of age, calves were placed in a 0 °C room for 140 minutes, and rectal temperatures and blood samples were obtained at 10- and 20-minute intervals. Breed × time interactions were observed for rectal temperature and serum concentrations of cortisol and glucose in newborn calves. Differences in rectal temperatures and blood concentrations of glucose and cortisol suggest possible differences in thermogenic mechanisms between calves with muscular hypertrophy and normal-muscled calves. Mice with a deletion of the PGC-1α gene exhibited a marked drop in core body temperature compared to the wild-type controls when subjected to cold exposure (4 °C) for 5 h [70]. The deletion of the PGC-1α gene, resulting in more fast-twitch muscle fibers, may show a lot of similarities with DM animals. However, body temperatures of MSTN null mice and wild-type mice either under normal conditions or in response to a 60-minute cold tolerance test were similar [171].

The effect of nutritional stress was studied in DM and non-DM Aberdeen Angus bulls [186]. Animals were fed *ad libitum* or received 125 or 85% of their maintenance requirement in two equal meals per day for 18 days. Blood samples were collected at 4-h intervals during the next 2 days. Glucose concentrations tended ($P < 0.10$) to be different between genotypes at 85% intake of their maintenance requirement. Non-esterified fatty acid fluctuations in DM animals responded less rapidly after the meals, suggesting a lower ability to mobilize fatty acids in response to short term alterations in energy availability. The weak response of non-esterified fatty acids after energy restriction was confirmed with BBDM cows. Energy restrictions up to 30% of the requirements during a 140-day indoor period did not significantly affect blood concentration of non-esterified fatty acids at the end of the restriction period [78].

MSTN inactivation attenuates the changes with acute daily psychological stress [187]. Three days of daily restraint of 1 h resulted in a lower decrease in body mass in MSTN null mice than in wild-type mice, while there were no decreases in muscle mass of *tibialis anterior* and *soleus* in contradiction with wild-type mice. MSTN null mice showed a non-significant increase in serum levels of corticosterone and glucose with daily restraint, while serum levels were significantly increased in wild-type mice.

Apart from their susceptibility to lameness and respiratory disorders, there is no clear indication that DM animals have a reduced immune function. Reductions in total body fat are correlated with impaired immunity, and experimental reductions in body fat can impair antibody formation [188]. Therefore, the reduced body fat content in DM animals may suggest such an impaired immune function. Furthermore, MSTN expression was detected in spleens, suggesting that it may play an important role in modulating immune system function during conditions of stress although the precise immune function of MSTN is unknown [189]. Consequently, it can be assumed that the inactive MSTN in DM animals may exert a negative effect on the immune function (Figure 1(I)). However, there is no evidence of any deficiency in the immune response of DM calves based on the lymphocyte function when compared with non-DM calves [190].

3.8. Effect of Double Muscling on Meat Quality

Several authors have made allusion to the repercussion of a modified muscle fiber type on meat quality [168,191,192]; (Figure 1(F)). There are several modes of action to affect meat quality. We already mentioned the reduced myoglobin content and its impact on oxygen transport. Myoglobin also affects meat color. It is well known that DM animals yield paler meat compared to non-DM animals [46,147]. The glycolytic fibers can result in a faster decline of pH *post-mortem* [155]. Strath *et al.* [193] compared the glucose pool size in phenotypically categorized bulls as either DM animals or normal to moderate in muscling. Glucose pool size was significantly less in DM cattle, which was expressed by a higher ultimate pH [42]. A reduced glucose pool and a faster glycolysis may accelerate glucose depletion in DM animals, especially in stress situations, so that they may be slightly predisposed to produce DFD meat. An increased frequency of DFD can occur as a consequence of heat stress [194], and the increased susceptibility of DM animals for heat stress may be an additional disadvantage for DM animals. Some authors mentioned an adverse meat quality in DM animals due to a pale and two-toned color in deeper parts of the hindquarter [195,196]. A very fast grey discoloration shortly after cutting is typical for this phenomenon. At 5 h *post-mortem*, the temperature of the inner *biceps femoris* was still greater than 35 °C, while the pH was decreased to less than 5.5. Initial L* values (lightness) amounted to 36.9 and 50.1 for the outer and the inner *biceps femoris*, respectively, and were significantly different, illustrating the problem of two-toning. L* values increased during a 10 day display, but the increase was most significant for the inner *biceps femoris*, while a* values (redness) decreased. Hot boning exerted a positive effect on pH decline. At 5 h *post-mortem* the pH averaged 5.93 for the hot boned inner *biceps femoris*: compared to 5.42 for the cold boned inner *biceps femoris*. There was no period of heat toughening as a consequence of the hot boning, while a duration of heat toughening of 3.9 h was estimated for the cold boning [195]. Traditional chilling of large beef muscles is detrimental for color stability, as the rapid pH decline with higher meat temperatures in the deep muscle causes more protein denaturation, and a lower color stability and water holding capacity [197]. *Post-mortem* temperature evolution showed a tendency for higher values ($P < 0.10$) in muscles of DM animals [198]. The temperature was probably higher in DM animals because of the large muscle mass. Carcass weights of 600 kg are not exceptional for BBDM cows. The combination of a temperature greater than 35 °C and a pH of less than 6 (heat shortening), causing early exhaustion of proteolytic activity, may lead to increased toughness [199].

The higher muscle temperature in carcasses of DM animals means that the metabolism may continue at a faster rate than in non-DM animals, resulting in a lower muscle pH. Batjoens *et al.* [155] observed a more rapid pH fall *post-mortem* in DM animals. During *post-mortem* pH decline values were higher in DM carcasses compared to non-DM carcasses in experiments of Pringle *et al.* [198]. These higher pH values are in accordance with the small but significantly higher ultimate pH for DM animals [42]. Steen *et al.* [200] reported pH values of 5.54 at 3 h *post-mortem* for DM animals, while other DM animals had an ultimate pH of 5.86, so that this meat could be regarded as dry, firm and dark. The increased levels of denatured creatine phosphokinase, 34 kDa and 36 kDa components, found in SDS-PAGE patterns of the myofibrillar fraction, may be related to the higher rate of pH decline [42].

The general conclusion from most studies is that DM cattle have more tender meat than non-DM animals [13,201]. Fast glycolytic fibers imply a lower collagen content. Collagen is associated with the background meat toughness, which may partly explain the lower toughness of meat from DM animals. A lower collagen content for DM animals was confirmed by several experiments [42,202,203,204,205]. Soluble collagen content, expressed as a percentage of total collagen, was mostly not affected [204,205], although a higher collagen solubility [202] or a less mature collagen [203] was also reported for DM animals. Nevertheless, meat of DM animals was not always more tender than that of non-DM animals [42,156,206]. Within BBDM bulls, Warner-Bratzler shear force value can also differ ($P < 0.1$) due to a different livestock management [207]. The shear force value was reduced by 26% when bulls were finished after a feed restriction period of 6 months yielding a daily gain of 0.5 kg, compared to bulls fed permanently on a concentrate diet, although their body weight was similar, but they were 4.5 months older at slaughter. In any case, results from different experiments [42,200,207,208] show a large variation in shear force value of meat from DM animals. This emphasizes the need for appropriate slaughter conditions. Indeed, overall tenderness is also dependent on slaughter conditions (the rigor period) and the extent of proteolysis. Pre-slaughter stress of fasting for 48 h resulted in a significantly higher pH in DM cattle, provoking some animals with dark cutting [68]. Different calpain and calpastatin activities between BBDM and non-DM animals were reported [42]. μ-Calpain was lower in DM bulls, indicating a lower proteolytic activity, which can explain the lower meat tenderness. However, calpastatin was also lower in DM bulls, but the μ-calpain/calpastatin ratio was also lower in DM animals, suggesting a reduced efficiency of μ-calpain. Only μ-calpain activity in *longissimus* and *biceps femoris* muscles was lower in Piedmontese DM bulls compared to non-DM Angus steers [198]. Besides, cathepsin B and L concentrations in the *longissimus* muscle on 1 and 8 days *post-mortem* were also lower in BBDM bulls [42]. These results were not confirmed by Caballero *et al.* [10] who found that the activity of cathepsins B, B + L, D and H in the cytosolic compartment of the *longissimus* muscle was higher in DM bulls of the Asturiana de los Valles breed at 3 and 7 days *post-mortem* than in non-DM animals. At 14 days *post-mortem* enzyme activity tended to be lower in DM animals with significant differences for cathepsin D, meaning that the tenderisation rate was faster in DM animals. Steen *et al.* [200] found that calpain and calpastatin activity in the *longissimus* muscle varied considerably within BBDM bulls between years, although animals were finished on similar diets and slaughtered in similar conditions. Average sarcomere length amounted to 1.76 μm, ranging from 1.26 to 2.26 μm. It was similar to the length reported by Uytterhaeghen *et al.* [42]. The latter authors confirmed a larger variation of sarcomere length within DM animals. Animals were fasted for 20–24 h prior to slaughter in both experiments. Taking into account that fasting for 48 h may have a negative effect on meat quality, measured in the *triceps longus* muscle [68], it is not clear if fasting for 24 h can also exert a negative effect on meat tenderness. Sañudo *et al.* [205] reported a significant interaction between breed and slaughter weight. DM animals had the lowest shear force values at a live weight of 300–350 kg, while values were higher at 530–560 kg.

Management of the finishing animals was mostly not mentioned. However, pre-slaughter growth rate of beef cattle can affect meat tenderness. A rapid growth rate results in an elevation of both muscle protein synthesis and degradation in the growing animal and an increased proteolysis *post-mortem* and an improved ultimate tenderness [199]. Furthermore, the divergence between the authors can be partly explained by the different muscles studied.

Twenty-nine and 49% of the variation in juiciness in the *longissimus* and the *semitendinosus* muscles, respectively, can be explained by the DJ-1 protein in Salers [209]. DJ-1 protein is associated with cancer and Parkinson's disease [210]. The cytoplasmic protein DJ-1 has a protective function and changed significantly in abundance during the early *post-mortem* storage period. It acts against oxidative stress and cell death. The relationship between DJ-1 protein and meat quality traits is scarcely reported, but *post-mortem* changes of DJ-1 protein are correlated to Warner-Bratzler shear force, drip loss, and meat color [211]. Muscles with a lighter meat had increased abundance of DJ-1 protein [212], but its effect on tenderness was not clearly demonstrated [213]. Chelh *et al.* [214] showed an up-regulation of DJ-1 protein in MSTN-null mice. Consequently, juiciness may be negatively correlated with this protein. However, juiciness was not significantly different between DM and non-DM British South Devon cattle, but there was a trend for a lower juiciness in DM animals [206].

Meat from DM animals is characterized by a low intramuscular fat content [46,206,215,216], resulting in an intramuscular fat content of less than 1%. Besides the lower fat content, fatty acid composition is also different from intramuscular fat in non-DM animals. The concentration of polyunsaturated fatty acids (PUFA) is increased, while the concentration of saturated fatty acids (SFA) is decreased, resulting in a higher PUFA/SFA ratio [206,215,216]. This is in accordance with the fact that cell membranes are relatively more important as a fat source in DM animals. They are mainly composed of phospholipids, and phospholipids are particularly rich in PUFA's. Furthermore, it has been shown that the metabolism of n-6 and n-3 fatty acid is affected by the MSTN gene and its effect is most evident in n-3 PUFAs of the intramuscular fat compared to intermuscular and subcutaneous fat [217]. Although the absolute content of total n-3 fatty acids is lower in meat of DM animals, they show greater deposition rates of n-3 elongation and desaturation products. This PUFA/SFA ratio together with a decreased ratio of n-6/n-3 PUFA's [215] means that DM animals yield healthy meat.

4. CONCLUSIONS

DM animals may provide some advantages for farmers, meat industry and consumers, such as an efficient conversion of feed into high valuable carcasses, a tender meat with a low fat content, indicating a meat that is healthier than normal. However, DM animals are also characterized to some extent by a lower robustness. They are more susceptible to dystocia, respiratory disease, lameness, muscle degeneration, and heat stress. Therefore, some extra precautions are needed to preserve animal welfare: appropriate nutrition to reduce nutritional stress, proper confinement to reduce respiratory disease and locomotion problems, shade during summer grazing to reduce heat stress, or elective use of caesarean section to reduce calf loss as a consequence of dystocia. Furthermore, adequate slaughter conditions are required to prevent cold shortening and heat toughening. The maturation rate of meat of DM animals is faster. Although meat of DM animals has a lower collagen content it is not always more tender than that of non-DM animals. The lower intramuscular fat content of DM animals may reduce meat flavor. It may not always be easy to find a balance between the different disciplines underlying the livestock husbandry of DM animals to obtain good performance and health, and lean meat with acceptable tenderness, flavor and juiciness.

Knowledge of the DM phenomenon has considerably increased during the last decade owing to molecular biology. It is clear that mechanisms due to the inactivation of the MSTN gene work in a coordinated manner. This means that highly qualified actors are necessary to realize good animal performance and high valuable carcasses and meat pieces. Further research may help to fine-tune the subsequent processes from birth of DM animals to the supply of healthy meat for the consumer.

REFERENCES

1. Berg, R.T.; Walters, L.E. The meat animal: Changes and challenges. *J. Anim. Sci.* 1983, *57*, 133–146.
2. Kempster, A.J. Carcass and meat quality research to meet market needs. *Anim. Prod.* 1989, *48*, 483–496.
3. Vanek, J.K.; Watts, M.J.; Brester, G.W. Carcass quality and genetic selection in the beef industry. *J. Agric. Res. Econ.*2008, *33*, 349–363.
4. Kempster, A.J.; Cuthbertson, A.; Harrington, G. The relationship between conformation and the yield and distribution of lean meat in the carcasses of British pigs, cattle and sheep: A review. *Meat Sci.* 1982, *6*, 37–53.
5. Luciano, F.B. The impacts of lean red meat consumption on human health: A review. *J. Food* 2009, *7*, 143–151.
6. Geay, Y.; Robelin, J.; Vermorel, M.; Beranger, C. Muscle hypertrophy and energy utilization in cattle: The double muscled as an extreme or a deviant animal. In *Muscle Hypertrophy of*

Genetic Origin and Its Use to Improve Beef Production; King, J.W.B., Menissier, F., Eds.; Martinus Nijhoff Publishers: The Hague, The Netherlands, 1982; pp. 74–87.

7. Vissac, B. L'hypertrophie musculaire d'origine génétique ou caractère culard. *Ann. Génét. Sél. Anim.* 1972, *4*, 87–97.
8. Guo, K.J.; Liu, F.H.; Lu, L.; Biagini, D.; Lazzaroni, C.; Zoccarato, I. Productive performance of double-muscled Piemontese culled cows in finishing period. *Agric. Sci. China* 2010, *9*, 721–728.
9. Kambadur, R.; Sharma, M.; Smith, T.P.L.; Bass, J.J. Mutations in myostatin (*GDF8*) in double-muscled Belgian Blue and Piedmontese cattle. *Genome Res.* 1997, *7*, 910–915.
10. Caballero, B.; Sierra, V.; Oliván, M.; Vega-Naredo, I.; Tomás-Zapico, C.; Alvarez-García, O.; Tolivia, D.; Hardeland, R.; Rodríguez-Colunga, M.J.; Coto-Montes, A. Activity of cathepsins during beef aging related to mutations in the myostatin gene. *J. Sci. Food Agric.* 2007, *87*, 192–199.
11. Marchitelli, C.; Savarese, M.C.; Crisà, A.; Nardone, A.; Marsan, P.A.; Valentini, A. Double muscling in Marchigiana beef breed is caused by a stop codon in the third exon of myostatin gene. *Mamm. Genome* 2003, *14*, 392–395.
12. Darwin, C. *The Origin of Species by Means of Natural Selection*; John Murray: London, UK, 1859.
13. Arthur, P.F. Double muscling in cattle: A review. *Austr. J. Agric. Res.* 1995, *46*, 1493–1515.
14. Kaiser. Uber die sogenannten doppellendigen Rinder. *Landw. Jhrb.* 1888, *17*, 387–403.
15. Charlier, C.; Coppieters, W.; Farnir, F.; Grobet, L.; Leroy, P.L.; Michaux, C.; Mni, M.; Schwers, A.; Vanmanshoven, P.; Hanset, R.; Georges, M. The *mh* gene causing double-muscling in cattle maps to bovine Chromosome 2. *Mamm. Genome* 1995, *6*, 788–792.
16. Grobet, L.; Royo Martin, L.J.; Poncelet, D.; Pirottin, D.; Brouwers, B.; Riquet, J.; Schoeberlein, A.; Dunner, S.; Menissier, F.; Massabanda, J.; Fries, R.; Hanset, R.; Georges, M. A deletion in the myostatin gene causes double muscled phenotype in cattle. *Nat. Genet.* 1997, *17*, 71–74.
17. McPherron, A.C.; Lee, S.J. Double muscling in cattle due to mutations in the myostatin gene. *Proc. Natl. Acad. Sci. USA* 1997, *94*, 12457–12461.
18. Chelh, I.; Picard, B.; Hocquette, J.-F.; Cassar-Malek, I. Myostatin inactivation induces a similar muscle molecular signature in double-muscled cattle as in mice. *Animal* 2010, *5*, 278–286.
19. Gerrard, D.E.; Grant, A.L.; Anderson, D.B.; Lemenager, R.P.; Judge, M.D. *In-vivo* analysis of serum-borne growth factors in developing co-twinned fetuses. *J. Anim. Sci.* 1995, *73*, 1689–1693.
20. Lee, S.J.; McPherron, A.C. Regulation of myostatin activity and muscle growth. *Proc. Natl. Acad. Sci. USA* 2001, *98*, 9306–9311.
21. Hill, J.J.; Quiu, Y.; Hewick, R.N.; Wolfman, N.M. Regulation of myostatin *in vivo* by GASP-1: A novel protein with protease inhibitor and follistatin domains. *Mol. Endocrinol.* 2003, *17*, 1144–1154.
22. Zimmers, T.A.; Davies, M.V.; Koniaris, L.G.; Haynes, P.; Esquela, A.F.; Tomkinson, K.N.; McPherron, A.C.; Wolfman, N.M.; Lee, S.J. Induction of cachexia in mice by systemically administered myostatin. *Science* 2002, *296*, 1486–1488.
23. Joulia-Ekaza, D.; Cabello, G. Myostatin regulation of muscle development: Molecular basis, natural mutations, physiopathological aspects. *Exp. Cell Res.* 2006, *312*, 2401–2414.
24. Wolfman, N.M.; McPherron, A.C.; Pappano, W.N.; Davies, M.V.; Song, K.; Tomkinson, K.N.; Wright, J.F.; Zhao, L.; Sebald, S.M.; Greenspan, D.S.; Lee, S.J. Activation of latent myostatin by the BMP-1/tolloid family of metalloproteinases. *Proc. Natl. Acad. Sci. USA* 2003, *100*, 15842–15846.
25. Rebbapragada, A.; Benchabane, H.; Wrana, J.L.; Celeste, A.J.; Attisano, L. Myostatin signals through a transforming growth factor β-like signaling pathway to block adipogenesis. *Mol. Cell. Biol.* 2003, *23*, 7230–7242.
26. Forbes, D.; Jackman, M.; Bishop, A.; Thomas, M.; Kambadur, R.; Sharma, M. Myostatin auto-regulates its expression by feedback loop through Smad7 dependent mechanism. *J. Cell. Physiol.* 2006, *206*, 264–272.

27. Bouley, J.; Meunier, B.; Chambon, C.; De Smet, S.; Hocquette, J.F.; Picard, B. Proteomic analysis of bovine skeletal muscle hypertrophy. *Proteomics* 2005, *5*, 490–500.
28. Grobet, L.; Poncelet, D.; Royo, L.J.; Brouwers, B.; Pirottin, D.; Michaux, C.; Menissier, F.; Zanotti, M.; Dunner, S.; Georges, M. Molecular definition of an allelic series of mutations disrupting the myostatin function and causing double-muscling in cattle. *Mamm. Genome* 1998, *9*, 210–213.
29. Yang, W.; Zhang, Y.; Li, Y.; Wu, Z.; Zhu, D. Myostatin induces cyclin D1 degradation to cause cell cycle arrest through a phosphatidylinositol 3-kinase/AKT/GSK-3β pathway and is antagonized by insulin-like growth factor 1. *J. Biol. Chem.* 2007, *282*, 3799–3808.
30. Gerrard, D.E.; Grant, A.L. Insulin-like growth factor-II expression in developing skeletal muscle of double muscled and normal cattle. *Dom. Anim. Endocrinol.* 1994, *11*, 339–347.
31. Listrat, A.; Belair, L.; Picard, B.; Boulle, N.; Geay, Y.; Djiane, J.; Jammes, H. Insulin-like growth factor II (IGF-II) mRNA expression during skeletal muscle development of double-muscled and normal bovine fetuses. *Reprod. Nutr. Dev.* 1999, *39*, 113–124.
32. Kamanga-Sollo, E.; Pampusch, M.S.; White, M.E.; Hathaway, M.R.; Dayton, W.R. Insulin-like growth factor binding protein (IGFBP)-3 and IGFBP-5 mediate TGF-h- and myostatin-induced suppression of proliferation in porcine embryonic myogenic cell cultures. *Exp. Cell Res.* 2005, *311*, 167–176.
33. Williams, N.G.; Interlichia, J.P.; Jackson, M.F.; Hwang, D.; Cohen, P.; Rodgers, B.D. Endocrine actions of myostatin: Systemic regulation of the IGF and IGF binding protein axis. *Endocrinology* 2011, *152*, 172–180.
34. Ludolph, D.C.; Komeczny, S.F. Transcription factor families: Muscling in on the myogenic program. *FASEB J.* 1995, *9*, 1595–1604.
35. Arnold, H.H.; Braun, T. Genetics of muscle determination and development. *Curr. Top. Dev. Biol.* 2000, *48*, 129–164.
36. Bass, J.; Oldham, J.; Sharma, M.; Kambadur, R. Growth factors controlling muscle development. *Dom. Anim. Endocrinol.* 1999, *17*, 191–197.
37. Oldham, J.; Martyn, J.A.K.; Sharma, M.; Jeanplong, F.; Kambadur, R.; Bass, J.J. Molecular expression of myostatin and MyoD is greater in double-muscled than normal-muscled cattle fetuses. *Am. J. Physiol.* 2001, *280*, R1488–R1493.
38. Langley, B.; Thomas, M.; Bishop, A.; Sharma, M.; Gilmour, S.; Kambadur, R. Myostatin inhibits myoblast differentiation by down-regulating MyoD expression. *J. Biol. Chem.* 2002, *277*, 49831–49840.
39. Muroya, S.; Watanabe, K.; Hayashi, S.; Miyake, M.; Konashi, S.; Sato, Y.; Takahashi, M.; Kawahata, S.; Yoshikawa, Y.; Aso, H.; Chikuni, K.; Yamaguchi, T. Muscle type-specific effect of myostatin deficiency on myogenic regulatory factor expression in adult double-muscled Japanese Shorthorn cattle. *Anim. Sci. J.* 2009, *80*, 678–685.
40. Michel, R.N.; Dunn, S.E.; Chin, E.R. Calcineurin and skeletal muscle growth. *Proc. Nutr. Soc.* 2004, *63*, 341–349.
41. Ashmore, C.R.; Parker, W.; Stokes, H.; Doerr, L. Comparative aspects of muscle fiber types in fetuses of the normal and "double-muscled" cattle. *Growth* 1974, *38*, 501–506.
42. Uytterhaegen, L.; Claeys, E.; Demeyer, D.; Lippens, M.; Fiems, L.O.; Boucqué, C.V.; Van de Voorde, G.; Bastiaens, A. Effects of double-muscling on carcass quality, beef tenderness and myofibrillar protein degradation in Belgian Blue White bulls. *Meat Sci.* 1994, *38*, 255–267.
43. Dumont, B.L. Carcass composition and muscle structure in hypertrophied animals. In *Muscle Hypertrophy of Genetic Origin and Its Use to Improve Beef Production*; King, J.W.B., Menissier, F., Eds.; Martinus Nijhoff Publishers: The Hague, The Netherlands, 1982; pp. 111–133.
44. Ansay, M.; Hanset, R. Anatomical, physiological and biochemical differences between conventional and double-muscled cattle in the Belgian Blue and White Breed. *Livest. Prod. Sci.* 1979, *6*, 5–13.
45. Arthur, P.F.; Makarechian, M.; Price, M.A.; Berg, R.T. Heterosis, maternal and direct effects in double-muscled and normal cattle: II. Carcass traits of young bulls. *J. Anim. Sci.* 1989, *67*, 911–919.

46. Fiems, L.O.; Van Hoof, J.; Uytterhaegen, L.; Boucqué, C.V.; Demeyer, D.I. Comparative quality of meat from double-muscled and normal beef cattle. In *Expression of tissue proteinases and regulation of protein degradation as related to meat quality*; Ouali, A., Demeyer, D.I., Smulders, F.J.M., Eds.; Ecceamst: Utrecht, The Netherlands, 1995; pp. 381–393.
47. Determining the Community Scale for the Classification of Carcasses of Adult Bovine Animals. Council Regulation (EEC) No 1026/91 of 22 April 1991, Amending Regulation Council Regulation (EEC) No 1208/81. *Official J. Eur. Comm.*1991, *L106*, 2–3.
48. Vissac, B. Étude du caractère culard. II. Incidence du caractère culard sur la morphologie générale des bovins. *Ann. Zootech.* 1968, *17*, 77–101.
49. Vermorel, M.; Ortigues, I.; Vernet, J.; Geay, Y.; Jailler, R.; Jailler, R.; Baumont, R.; Hocquette, J.F.; Deswysen, A.G. Energy metabolism in normal and double-muscled Belgian Blue calves in relation with body composition and organ size. In *Energy Metabolism of Farm Animals*; Aguilera, J.F., Ed.; Consejo Superior de Investigaciones: Granada, Spain, 1994; pp. 209–211.
50. Bünger, L.; Ott, G.; Varga, L.; Schlote, W.; Rehfeldt, C.; Renne, U.; Williams, J.L.; Hill, W.G. Marker-assisted introgression of the compact mutant myostatin allele MstnCmpt-dl1Abc into a mouse line with extreme growth effects on body composition and muscularity. *Genet. Res.* 2004, *84*, 161–173.
51. Fiems, L.O.; Van Caelenbergh, W.; Vanacker, J.M.; De Campeneere, S.; Seynaeve, M. Prediction of empty body composition of double-muscled beef cows. *Livest. Prod. Sci.* 2005, *92*, 249–259.
52. De Campeneere, S.; Fiems, L.O.; De Paepe, M.; Vanacker, J.M.; Boucqué, C.V. Compositional data on Belgian Blue double-muscled bulls. *Anim. Res.* 2001, *50*, 43–55.
53. Yan, T.; Patterson, D.C.; Mayne, C.S.; Agnew, R.E.; Porter, M.G. Prediction of empty body weight and composition from live weight and other live animal measurements in lactating dairy cows. *J. Agric. Sci. Camb.* 2009, *147*, 241–252.
54. Andrew, S.M.; Waldo, D.R.; Erdman, R.A. Direct analysis of body composition of dairy cows at three physiological stages. *J. Dairy Sci.* 1994, *77*, 3022–3033.
55. Wright, I.A.; Russel A.J.F. Partition of fat, body composition and body condition score in mature cows. *Anim. Prod.*1984, *38*, 23–32.
56. Piedrafita, J.; Quintanilla, R.; Sañudo, C.; Olleta, J.L.; Campo, M.M.; Panea, B.; Renand, G.; Turin, F.; Jabet, S.; Osoro, K.; Oliván, M.C.; Noval, G.; García, P.; García, M.D.; Oliver, M.A.; Gispert, M.; Serra, X.; Espejo, M.; García, S.; López, M.; Izquierdo, M. Carcass quality of 10 beef cattle breeds of the Southwest of Europe in their typical production systems. *Livest. Prod. Sci.* 2003, *82*, 1–13.
57. Weekly, L.B.; Veit, H.P. Potential morphologic and physiologic factors that may predispose the bovine lung to respiratory diseases. *Compend. Contin. Educ. Pract. Vet.* 1995, *17*, 974–982.
58. Jones, J.H.; Longworth, K.E.; Lindholm, A.; Conley, K.E.; Karas, R.H.; Kayar, S.R.; Taylor, C.R. Oxygen transport during exercise in large mammals. I. Adaptive variation in oxygen demand. *J. Appl. Physiol.* 1989, *67*, 862–870.
59. Gustin, P.; Lomba, F.; Bakima, J.; Lekeux, P.; Van de Woestijne, K.P. Partitioning of pulmonary resistance in calves. *J. Appl. Physiol.* 1987, *62*, 1826–1831.
60. Gustin, P.; Dhem, A.R.; Lomba, F.; Lekeux, P. Cardio-pulmonary function values in double-muscled cattle during muscular exercise. *Vet. Res. Comm.* 1988, *12*, 407–416.
61. Cambier, C.; Clerbaux, T.; Detry, B.; Beerens, D.; Frans, A.; Gustin, P. Blood oxygen binding in double-muscled calves and dairy calves with conventional muscle conformation. *Am. J. Vet. Res.* 2000, *61*, 299–304.
62. Roets, E.; Burvenich, C.; Roberts, M. Muscarinic receptor subtypes, β-adrenoceptors and cAMP in the tracheal smooth muscle of conventional and double-muscled calves. *Vet. Res. Comm.* 1992, *16*, 465–476.
63. Cusack, P.M.V.; McMeniman, N.; Lean, I.J. The medicine and epidemiology of bovine respiratory disease in feedlots. *Aust. Vet. J.* 2003, *81*, 480–487
64. Fiems, L.; De Campeneere, S.; Cottyn, B.; Boucqué, C. Effect of housing on the performance of double-muscled rearing calves. *Ann. Zootech.* 1998, *47*, 207–214.

65. Tyler, H.; Ramsey, H. Hypoxia in neonatal calves: Effect on intestinal transport of immunoglobulins. *J. Dairy Sci.*1991, *74*, 1952–1956.
66. Amory, H.; McEntee, K.; Linden, A.S.; Desmecht, D.J.M.; Beduin, J.M.L.; D'Orio, V.; Lekeux, P.M. Comparison of the cardiac pumping capability and cardiac pumping reserve in double-muscled and conventional calves. *Can. J. Physiol. Pharmacol.* 1993, *71*, 946–951.
67. Monin, G.; Boccard, R. Caractéristiques physiologiques respiratoires des bovins culards. *Ann. Génét. Sél. Anim.* 1974, *6*, 187–193.
68. Holmes, J.H.G.; Ashmore, C.R.; Robinson, D.W. Effects of stress on cattle with hereditary muscular hypertrophy. *J. Anim. Sci.* 1973, *36*, 684–694.
69. Handschin, C.; Chin, S.; Li, P.; Liu, F.; Maratos-Flier, E.; LeBrasseur, N.K.; Yan, Z.; Spiegelman, B.M. Skeletal muscle fiber-type switching, exercise intolerance, and myopathy in PGC-1 α muscle-specific knock-out animals. *J. Biol. Chem.*2007, *282*, 30014–30021.
70. Leone, T.C.; Lehman, J.J.; Finck, B.N.; Schaeffer, P.J.; Wende, A.R.; Boudina, S.; Courtois, M.; Wozniak, D.F.; Sambandam, N.; Bernal-Mizrachi, C.; Chen, Z.; Holloszy, J.O.; Medeiros, D.M.; Schmidt, R.E.; Saffitz, J.E.; Abel, E.D.; Semenkovich, C.F.; Kelly, D.P. PGC-1α-deficiency causes multi-system energy metabolic derangements: Muscle dysfunction, abnormal weight control and hepatic steatosis. *PLoS Biol.* 2005, *3*.
71. Clerbaux, T.; Gustin, P.; Detry, B.; Cao, M.L.; Frans, A. Comparative study of the oxyhaemoglobin dissociation curve of four mammals: Man, dog, horse and cattle. *Comp. Biochem. Physiol.* 1993, *106A*, 687–694.
72. Gustin, P.; Clerbaux, T.; Willems, E.; Lekeux, P.; Lomba, F.; Frans, A. Oxygen transport properties of blood in two different bovine breeds. *Comp. Biochem. Physiol.* 1988, *89A*, 553–558.
73. Clinquart, A.; Van Eenaeme, C.; Mayombo, A.P.; Gauthier, S.; Istasse, L. Plasma hormones and metabolites in cattle in relation to breed (Belgian Blue *vs.* Holstein) and conformation (double-muscled *vs.* dual-purpose type). *Vet. Res. Comm.* 1995, *19*, 185–194.
74. Fiems, L.O.; Cottyn, B.G.; Boucqué, C.V.; Bogaerts, D.F.; Van Eenaeme, C.; Vanacker, J.M. Effect of beef type, body weight and dietary protein content on voluntary feed intake, digestibility, blood and urine metabolites and nitrogen retention. *J. Anim. Physiol. Anim. Nutr.* 1997, *77*, 1–9.
75. Sainz, R.D.; Bentley, B.E. Visceral organ mass and cellularity in growth-restricted and refed beef steers. *J. Anim. Sci.*1997, *75*, 1229–1236.
76. Bjercke, R.J.; Goll, D.E.; Robson, R.M.; Dutson, T.R. Relative roles of polysomes and cytoplasmic enzymes in regulating bovine skeletal muscle protein synthesis. *J. Anim. Sci.* 1984, *59*, 684–696.
77. De Campeneere, S.; Fiems, L.O.; Boucqué, C.V. Energy and protein requirements of Belgian Blue double-muscled bulls. *Anim. Feed Sci. Technol.* 2001, *90*, 153–167.
78. Fiems, L.O.; Vanacker, J.M.; De Boever, J.L.; Van Caelenbergh, W.; Aerts, J.M.; De Brabander, D.L. Effect of energy restriction and re-alimentation in Belgian Blue double-muscled beef cows on digestibility and metabolites. *J. Anim. Physiol. Anim. Nutr.* 2007, *91*, 54–61.
79. McBride, B.W.; Kelly, J.M. Energy cost of absorption and metabolism in the ruminant gastrointestinal tract and liver: A review. *J. Anim. Sci.* 1990, *68*, 2997–3010.
80. Vermorel, M.; Bouvier, J.C.; Geay, Y. The effect of genotype (normal and double-muscled Charolais and Friesian) on energy utilization by growing cattle at 2 and 16 months of age. In *Energy Metabolism of Farm Animals*; Vermorel, M., Ed.; G.de Bussac: Clermont-Ferrand, France, 1976; pp. 217–220.
81. Fiems, L.O.; Moermans, R.J.; Boucqué, C.V.; Cottyn, B.G.; De Campeneere, S. Voluntary feed intake in dual-purpose and double-muscled Belgian White-Blue bulls. In *Regulation of Feed Intake*; van der Heide, D., Huisman, E.A., Kanis, E., Osse, J.W.M., Verstegen, M.W.A., Eds.; CAB International: Wallingford, UK, 1999; pp. 21–25.
82. Haynes, F.E.M.; Greenwood, P.L.; McDonagh, M.B.; Oddy, V.H. Myostatin allelic status interacts with level of nutrition to affect growth, composition, and myofiber characteristics of lambs. *J. Anim. Sci.* 2012, *90*, 456–465.
83. Geay, Y.; Robelin, J. Variation of meat production capacity in cattle due to genotype and level of feeding: Genotype-nutrition interaction. *Livest. Prod. Sci.* 1979, *6*, 263–276.

84. Fiems, L.O.; De Campeneere, S.; Vanacker, J.M.; De Boever, J.L.; De Brabander, D.L. Are Belgian Blue double-muscled animals deviant for their phosphorus requirements? In Proceedings of the30th Meeting Dutch Speaking Nutrition Researchers, Merelbeke, Belgium, 8 April 2005; pp. 49–50.
85. AFRC. A reappraisal of the calcium and phosphorus requirements of sheep and cattle. *Nutr. Abst. Rev. B* 1991, *61*, 573–612.
86. NRC, Nutrient Requirements of Dairy Cattle, 7th ed.National Academy Press: Washington, DC, USA, 2001.
87. Meschy, F. Recommandations d'apport en phosphore absorbé chez les ruminants. *Renc. Rech. Ruminants* 2002, *9*, 279–285.
88. Robbins, J.D.; Oltjen, R.R.; Cabell, C.A.; Dolnick, E.H. Influence of varying levels of dietary minerals on the development of urolithiasis, hair growth, and weight gains in rats. *J. Nutr.* 1965, *85*, 355–361.
89. Ewoldt, J.M.; Jones, M.L.; Miesner, M.D. Surgery of obstructive urolithiasis in ruminants. *Vet. Clin. Food Anim.* 2008, *24*, 455–465.
90. Ouhayoun, J.; Arnal, T. Étude du caractère culard. IV. Anatomie microscopique comparée du rein de males Charolais normaux et culards. *Ann. Génét. Sél. Anim.* 1969, *1*, 101–108.
91. Guyot, H.; Spring, P.; Andrieu, S.; Rollin, F. Comparative responses to sodium selenite and organic selenium supplements in Belgian Blue cows and calves. *Livest. Sci.* 2007, *111*, 259–263.
92. NRC, *Nutrient Requirements of Beef Cattle*, 7th ed.; National Academy Press: Washington, DC, USA, 2000.
93. Allen, W.M. New developments in muscle pathology: Nutritional myopathies including "muscular dystrophy" or "white muscle disease". *Vet. Sci. Comm.* 1977, *1*, 243–250.
94. Anderson, P.H.; Berrett, S.; Patterson, D.S.P. The significance of elevated plasma creatinephosphokinase activity in muscle disease of cattle. *J. Comp. Path.* 1976, *86*, 531–538.
95. Ashmore, C.R.; Robinson, D.W. Hereditary muscular hypertrophy in the bovine. 1. Histological and biochemical characterization. *Proc. Soc. Exp. Biol. Med.* 1969, *132*, 548–554.
96. Hendricks, H.B.; Aberle, E.D.; Jones, D.J.; Martin, T.G. Muscle fiber type, rigor development and bone strength in double muscled cattle. *J. Anim. Sci.* 1973, *37*, 1305–1311.
97. Shahin, K.A.; Berg, R.T.; Price, M.A. Muscle and bone distribution in mature normal and double muscled cows. *Livest. Prod. Sci.* 1991, *28*, 291–303.
98. Mendias, C.L.; Marcin, J.E.; Calerdon, D.R.; Faulkner, J.A. Contractile properties of EDL and soleus muscles of myostatin deficient mice. *J. Appl. Physiol.* 2006, *101*, 898–905.
99. Amthor, H.; Macharia, R.; Navarrete, R.; Schuelke, M.; Brown, S.C.; Otto, A.; Voit, T.; Muntoni, F.; Vrbova, G.; Partridge, T.; Zammit, P.; Bunger, L.; Patel, K. Lack of myostatin results in excessive muscle growth but impaired force generation. *Proc. Natl. Acad. Sci. USA* 2007, *104*, 1835–1840.
100. Matsakas, A.; Macharia, R.; Otto, A.; Elashry, M.I.; Mouisel, E.; Romanello, V.; Sartori, R.; Amthor, H.; Sandri, M.; Narkar, V.; Patel, K. Exercise training attenuates the hypermuscular phenotype and restores skeletal muscle function in the myostatin null mouse. *Exp. Physiol.* 2012, *97*, 125–140.
101. Hamrick, M.W.; Pennington, C.; Byron, C.D. Bone architecture and disc degeneration in the lumbar spine of mice lacking GDF-8 (myostatin). *J. Orthop. Res.* 2003, *21*, 1025–1032.
102. Zou, H.; Wieser, R.; Massagué, J.; Niswander, L. Distinct roles of type I bone morphogenetic protein receptors in the formation and differentiation of cartilage. *Genes Dev.* 1997, *11*, 2191–2203.
103. Blumer, M.J.F.; Longato, S.; Fritsch, H. Structure, formation and role of cartilage canals in the developing bone. *Ann. Anat.* 2008, *190*, 305–315.
104. Merino, R.; Macias, D.; Gañan, Y.; Rodriguez-Leon, J.; Economides, A.N.; Rodriguez-Esteban, C.; Izpisua-Belmonte, J.C.; Hurle, J.M. Control of digit formation by activin signalling. *Development* 1999, *126*, 2161–2170.

105. Rountree, R.B.; Schoor, M.; Chen, H.; Marks, M.E.; Harley, V.; Mishina, Y.; Kingsley, D.M. BMP receptor signaling is required for postnatal maintenance of articular cartilage. *PLoS Biol.* 2004, *2*, 1815–1827.
106. Mason, I.L. Symptoms of muscular hypertrophy in heterozygous steers. *Anim. Prod.* 1963, *5*, 57–65.
107. Cassar-Malek, I.; Passelaigue, F.; Bernard, C.; Léger, J.; Hocquette, J.F. Target genes of myostatin loss-of-function in muscles of late bovine fetuses. *BMC Genomics* 2007, *8*.
108. Nott, C.F.G.; Rollins, W.C.; Tanaka, M. Effects of the m gene for muscular hypertrophy on conformation at one year of age in beef cattle. *Ann. Génét. Sél. Anim.* 1980, *12*, 143–155.
109. Anderson, D.E.; Desrochers, A.; St. Jean, G. Management of Tendon Disorders in Cattle. *Vet. Clin. Food Anim.* 2008, *24*, 551–566.
110. Wechsler, B. Floor quality and space allowance in intensive beef production. *Anim. Welfare* 2011, *20*, 497–503.
111. Design Recommendations of Beef Cattle Housing; Report of the CIGR Section II; Working Group No. 14; Cattle Housing, 2nd ed. CIGR, East Lansing, MI, USA, 2004.
112. Schiavon, S.; Tagliapietra, F.; Dal Maso, M.; Bailoni, L.; Bittante, G. Effects of low-protein diets and rumen-protected conjugated linoleic acid on production and carcass traits of growing double-muscled Piemontese bulls. *J. Anim. Sci.*2010, *88*, 3372–3383.
113. De Campeneere, S.; Fiems, L.O.; De Brabander, D.L. Determination of a standard for physical structure requirement for Belgian Blue bulls. *J. Anim. Feed Sci.* 2004, *13* Suppl. 1, 623–626.
114. Short, R.E.; MacNeil, M.D.; Grosz, M.D.; Gerrard, D.E.; Grings, E.E. Pleiotropic effects in Hereford, Limousin, and Piedmontese F2 crossbred calves of genes controlling muscularity including the Piedmontese myostatin allele. *J. Anim. Sci.* 2002, *80*, 1–11.
115. Vissac, B.; Ménissier, F.; Perreau, B. Étude du caractère culard. VII. Croissance et musculature des femelles, déséquilibre morphologique au vêlage. *Ann. Génét. Sél. Anim.* 1973, *5*, 23–38.
116. Arthur, P.F.; Makarechian, M.; Price, M.A. Incidence of dystocia and perinatal calf mortality resulting from reciprocal crossing of double-muscled and normal cattle. *Can. Vet. J.* 1988, *29*, 163–167.
117. Mao, W.H.; Albrecht, E.; Teuscher, F.; Yang, Q.; Zhao, R.Q.; Wegner, J. Growth- and breed-related changes of fetal development in cattle. *Asian Aust. J. Anim. Sci.* 2008, *21*, 640–647.
118. Fiems, L.O.; De Campeneere, S.; Van Caelenbergh, W.; Boucqué, C.V. Relationship between dam and calf characteristics with regard to dystocia in Belgian Blue double-muscled cows. *Anim. Sci.* 2001, *72*, 389–394.
119. Fiems, L.O.; De Brabander, D.L. Optimum growth rate of Belgian Blue double-muscled replacement heifers. *South Afr. J. Anim. Sci.* 2009, *39* Suppl. 1, 6–10.
120. Fiems, L.O.; De Brabander, D.L. Relation between calf birth weight and dam weight in Belgian Blue double-muscled cattle. In Proceedings of the 60th Annual Meeting of the European Association for Animal Production, Barcelona, Spain, 24–27 August 2009; p. 96.
121. Johanson, J.M.; Berger, P.J. Birth weight as a predictor of calving ease and perinatal mortality in Holstein cattle. *J. Dairy Sci.* 2003, *86*, 3745–3755.
122. Kolkman, I.; De Vliegher, S.; Hoflack, G.; Van Aert, M.; Laureyns, J.; Lips, D.; de Kruif, A.; Opsomer, G. Protocol of the caesarean section as performed in daily bovine practice in Belgium. *Reprod. Dom. Anim.* 2007, *42*, 583–589.
123. Vermorel, M.; Vernet, J.; Saido, S.; Dardillat, C.; Demigne, C.; Davicco, M.J. Energy metabolism and thermoregulation in the newborn calf; Effect of calving conditions. *Can. J. Anim. Sci.* 1989, *69*, 113–122.
124. Patterson, D.J.; Bellows, R.A.; Burfening, P.J. Effects of caesarean section, retained placenta and vaginal or uterine prolapse on subsequent fertility in beef cattle. *J. Anim. Sci.* 1981, *53*, 916–921.
125. Hanzen, C.; Laurent, Y.; Ward, W.R. Comparison of reproductive performance in Belgian dairy and beef cattle. *Theriogenology* 1994, *41*, 1099–1114.
126. Frazier, E.L.; Sprott, L.R.; Sanders, J.O.; Dahm, P.F.; Crouch, J.R.; Turner, J.W. Sire marbling score expected progeny difference and weaning weight maternal expected progeny

difference associations with age at first calving and calving interval in Angus beef cattle. *J. Anim. Sci.* 1999, *77*, 1322–1328.
127. McCartney, D.; Basarab, J.A.; Okine, E.K.; Baron, V.S.; Depalme, A.J. Alternative fall and winter feeding systems for spring calving beef cows. *Can. J. Anim. Sci.* 2004, *84*, 511–522.
128. Pilarczyk, R.; Wójcik, J. Comparison of body weight and reproduction performance in cows of various beef breeds managed under equal conditions in West Pomerania. *Arch. Tierz.* 2008, *51*, 318–328.
129. Vermunt, J.J. The caesarean operation in cattle: A review. *Iran. J. Vet. Surg.* 2008, *3* Suppl. 1, 82–100.
130. Chupin, D. Analysis of reproduction problems in double muscle females. In *Muscle Hypertrophy of Genetic Origin and Its Use to Improve Beef Production*; King, J.W.B., Menissier, F., Eds.; Martinus Nijhoff Publishers: The Hague, The Netherlands, 1982; pp. 575–584.
131. Hoflack, G.; Opsomer, G.; Van Soom, A.; Maes, D.; de Kruif, A.; Duchateau, L. Comparison of sperm quality of Belgian Blue and Holstein Friesian bulls. *Theriogenology* 2006, *66*, 1834–1846.
132. Leroy, J.L.M.R.; Opsomer, G.; De Vliegher, S.; Vanholder, T.; Goossens, L.; Geldhof, A.; Bols, P.E.J.; de Kruif, A.; Van Soom, A. Comparison of embryo quality in high-yielding dairy cows, in dairy heifers and in beef cows. *Theriogenology* 2005, *64*, 2022–2036.
133. Liang, Y.C.; Yeh, J.Y.; Ou, B.R. Effect of maternal myostatin antibody on offspring growth performance and body composition in mice. *J. Exp. Biol.* 2007, *210*, 477–483.
134. Ciarmela, P.; Wiater, E.; Smith, S.M.; Vale, W. Myostatin in rat uterus and myometrial cells. *Endocrinology* 2009, *150*, 906–914.
135. Wong, C.L.; Huang, Y.Y.; Ho, W.K.; Poon, H.K.; Cheung, P.L.; O, W.S.; Chow, P.H. Growth-differentiation factor-8 (GDF-8) in the uterus: Its identification and functional significance in the golden hamster. *Reprod. Biol. Endocrinol.*2009, *7*.
136. Ethier, J.F.; Findlay, J.K. Roles of activin and its signal transduction mechanisms in reproductive tissues. *Reproduction*2001, *121*, 667–675.
137. Young, J.M.; Henderson, S.; Souza, C.; Ludlow, H.; Groome, N.; McNeilly, A.S. Activin B is produced early in antral follicular development and suppresses thecal androgen production. *Reproduction* 2012, *143*, 637–650
138. Muttukrishna, S.; Tannetta, D.; Groome, N.; Sargent, I. Activin and follistatin in female reproduction. *Mol. Cell. Endocrinol.* 2004, *225*, 45–56.
139. Jones, R.L.; Kaitu'u-Lino, T.J.; Nie, G.; Sanchez-Partida, L.G.; Findlay, J.K.; Salamonsen, L.A. Complex expression patterns support potential roles for maternally derived activins in the establishment of pregnancy in mouse. *Reproduction* 2006, *132*, 799–810.
140. Wolfenson, D.; Roth, Z.; Meidan, R. Impaired reproduction in heat-stressed cattle: Basic and applied aspects. *Anim. Reprod. Sci.* 2000, *60*, 535–547.
141. Halipré, A. Étude du caractere culard. X. Sensibilité des bovins culards au stress thermique. *Ann. Génét. Sél. Anim.*1973, *5*, 441–449.
142. Stavaux, D.; Art, T.; Entee, K.M.; Reznick, M.; Lekeux, P. Muscle fibre type and size, and muscle capillary density in young double-muscled Blue Belgian cattle. *J. Vet. Med. A* 1994, *41*, 229–236.
143. Blazquez, N.B.; Long, S.E.; Mayhew, T.M.; Perry, G.C.; Prescott, N.J.; Wathes, C.M. Rate of discharge and morphology of sweat glands in the perineal, lumbodorsal and scrotal skin of cattle. *Res. Vet. Sci.* 1994, *57*, 277–284.
144. Reynolds, L.P.; Ferrell, C.L.; Nienaber, J.A.; Ford, S.P. Effects of chronic environmental heat stress on blood flow and nutrient uptake of the gravid bovine uterus and foetus. *J. Agric. Sci.* 1985, *104*, 289–297.
145. Finch, V.A. Body temperature in beef cattle: Its control and relevance to production in the tropics. *J. Anim. Sci.* 1986, *62*, 531–542.
146. Schafer, K.A. The cell cycle: A review. *Vet. Pathol.* 1998, *35*, 461–478.
147. Gagnière, H.; Ménissier, F.; Geay, Y.; Picard, B. Influence of genotype on contractile protein differentiation in different bovine muscles during foetal life. *Ann. Zootech.* 2000, *49*, 405–423.

148. Thomas, M.; Langley, B.; Berry, C.; Sharma, M.; Kirk, S.; Bass, J.; Kambadur, R. Myostatin, a negative regulator of muscle growth, functions by inhibiting myoblast proliferation. *J. Biol. Chem.* 2000, *275*, 40235–40243.
149. Rios, R.; Carneiro, I.; Arce, V.M.; Devesa, J. Myostatin is an inhibitor of myogenic differentiation. *Am. J. Physiol. Cell Physiol.* 2002, *282*, C993–C999.
150. Amthor, H.; Nicholas, G.; McKinnell, I.; Kemp, C.F.; Sharma, M.; Kambadur, R.; Patel, K. Follistatin complexes myostatin and antagonises myostatin-mediated inhibition of myogenesis. *Dev. Biol.* 2004, *270*, 19–30.
151. Hayashi, S.; Miyake, M.; Watanabe, K.; Aso, H.; Hayashi, S.; Ohwada, S.; Yamaguchi, T. Myostatin preferentially down-regulates the expression of fast 2x myosin heavy chain in cattle. *Proc. Jpn. Acad. B* 2008, *84*, 354–362.
152. Albrecht, E.; Teuscher, F.; Ender, K.; Wegner, J. Growth- and breed-related changes of muscle bundle structure in cattle. *J. Anim. Sci.* 2006, *84*, 2959–2964.
153. Wegner, J.; Albrecht, E.; Fiedler, I.; Teuscher, F.; Papstein, H.J.; Ender, K. Growth- and breed-related changes of muscle fiber characteristics in cattle. *J. Anim. Sci.* 2000, *78*, 1485–1496.
154. West, R.L. Red to white fiber ratios as an index of double muscling in beef cattle. *J. Anim. Sci.* 1974, *38*, 1165–1175.
155. Batjoens, P.; Fiems, L.O.; Van Hoof, J.; Van Vooren, T.; Vereecke, D. Myofibre composition and metabolic aspects in different strains of Belgian white-blue bulls and their relation to meat colour. In Proceedings of the37th International Congress of Meat Science and Technology, Kulmbach, Germany, 1–6 September 1991; pp. 324–327.
156. Picard, B.; Hocquette, J.F.; Bornes, F.; Brazi, S.; Vermorel, M.; Geay, Y. Muscle metabolism in normal and double muscled calves: Fiber characteristics and glucose transport rate. In *Energy Metabolism of Farm Animals*; Aguilera, J.F., Ed.; Consejo Superior de Investigaciones: Granada, Spain, 1994; pp. 97–100.
157. Girgenrath, S.; Song, K.; Whittemore, L.A. Loss of myostatin expression alters fiber-type distribution and expression of myosin heavy chain isoforms in slow- and fast-type skeletal muscle. *Muscle Nerve* 2005, *31*, 34–40.
158. Kambadur, R.; Bishop, A.; Salerno, M.S.; McCroskery, S.; Sharma, M. Role of myostatin in muscle growth. In *Muscle Development of Livestock Animals*; te Pas, M.F.W., Everts, M.E., Haagsman, H.P., Eds.; CAB International: Wallingford, UK, 2004; pp. 297–316.
159. Deveaux, V.; Picard, B.; Bouley, J.; Cassar-Malek, I. Location of myostatin expression during bovine myogenesis *in vivo* and *in vitro*. *Reprod. Nut. Dev.* 2003, *43*, 527–542.
160. Deveaux, V.; Cassar-Malek, I.; Picard, B. Comparison of contractile characteristics of muscle from Holstein and double-muscled Belgian Blue foetuses. *Comp. Biochem. Physiol. A Physiol.* 2001, *131*, 21–29.
161. Gagnière, H.; Picard, B.; Jurie, C.; Geay, Y. Comparative study of metabolic differentiation of foetal muscle in normal and double-muscled cattle. *Meat Sci.* 1997, *45*, 145–152.
162. Picard, B.; Gagnière, H.; Robelin, J.; Geay, Y. Comparison of the foetal development of muscle in normal and double-muscled cattle. *J. Muscle Res. Cell Motil.* 1995, *16*, 629–639.
163. Picard, B.; Depreux, F.; Geay, Y. Muscle differentiation of normal and double-muscled bovine foetal myoblasts in primary culture. *Basic Appl. Myol.* 1998, *8*, 197–203.
164. Picard, B.; Lefaucheur, L.; Berri, C.; Duclos, M.J. Muscle fibre ontogenesis in farm animal species. *Reprod. Nutr. Dev.*2002, *42*, 415–431.
165. Naya, F.J.; Mercer, B.; Shelton, J.; Richardson, J.A.; Williams, R.S.; Olson, E.N. Stimulation of slow skeletal muscle fiber gene expression by calcineurin *in vivo*. *J. Biol. Chem.* 2000, *275*, 4545–4548.
166. Lin, J.; Wu, H.; Tarr, P.T.; Zhang, C.Y.; Wu, Z.; Boss, O.; Michael, L.F.; Puigserver, P.; Isotani, E.; Olson, E.N.; Lowell, B.B.; Bassel-Duby, R.; Spiegelman, B.M. Transcriptional co-activator PGC-1α drives the formation of slow-twitch muscle fibres. *Nature* 2002, *418*, 797–801.
167. Puddick, J.; Martinus, R.D. Comparative proteomics of skeletal muscle mitochondria from myostatin-null mice. *Cell Biol. Int. Rep.* 2011, *18*, 35–41.
168. Hocquette, J.F.; Ortigues-Marty, I.; Pethick, D.; Herpin, P.; Fernandez, X. Nutritional and hormonal regulation of energy metabolism in skeletal muscles of meat-producing animals. *Livest. Prod. Sci.* 1998, *56*, 115–143.

169. Holmes, J.H.G.; Robinson, D.W.; Ashmore, C.R. Blood lactic acid and behaviour in cattle with hereditary muscular hypertrophy. *J. Anim. Sci.* 1972, *35*, 1011–1013.
170. Garlick, P.J.; Maltin, C.A.; Baillie, A.G.; Delday, M.I.; Grubb, D.A. Fiber-type composition of nine rat muscles. II. Relationship to protein turnover. *Am. J. Physiol. Endocrinol. Metab.* 1989, *257*, E828–E832.
171. McPherron, A.C.; Lee, S.J. Suppression of body fat accumulation in myostatin-deficient mice. *J. Clin. Invest.* 2002, *109*, 595–601.
172. Guo, T.; Jou, W.; Chanturiya, T.; Portas, J.; Gavrilova, O.; McPherron, A.C. Myostatin Inhibition in muscle, but not adipose tissue, decreases fat mass and improves insulin sensitivity. *PLoS ONE* 2009, *4*.
173. Tu, P.; Bhasin, S.; Hruz, P.W.; Herbst, K.L.; Castellani, L.W.; Hua, N.; Hamilton, J.A.; Guo, W. Genetic disruption of myostatin reduces the development of proatherogenic dyslipidemia and atherogenic lesions in *Ldlr* null mice. *Diabetes* 2009, *58*, 1739–1748.
174. Ferrell, C.L. Contribution of visceral organs to animal energy expenditures. *J. Anim. Sci.* 1988, *66*, 23–34.
175. Strath, R.A.; Basarab, J.A.; Thompson, J.R.; Berg, R.T. Thyroid hormone concentrations: Some kinetic parameters of triiodothyronine and metabolic rate in "double-muscled" cattle. *Can. J. Anim. Sci.* 1982, *62*, 381–395.
176. Hocquette, J.F.; Bas, P.; Bauchart, D.; Vermorel, M.; Geay, Y. Fat partitioning and biochemical characteristics of fatty tissues in relation to plasma metabolites and hormones in normal and double-muscled young growing bulls. *Comp. Biochem. Physiol. A* 1999, *122*, 127–138.
177. Bossaert, P.; Leroy, J.L.M.R.; De Campeneere, S.; De Vliegher, S.; Opsomer, G. Differences in the glucose-induced insulin response and the peripheral insulin responsiveness between neonatal calves of the Belgian Blue, Holstein-Friesian, and East Flemish breeds. *J. Dairy Sci.* 2009, *92*, 4404–4411.
178. Chen, Y.; Ye, J.; Cao, L.; Zhang, Y.; Xia, W.; Zhu, D. Myostatin regulates glucose metabolism via the AMP-activated protein kinase pathway in skeletal muscle cells. *Int. J. Biochem. Cell Biol.* 2010, *42*, 2072–2081.
179. Antony, N.; Bass, J.J.; McMahon, C.D.; Mitchell, M.D. Myostatin regulates glucose uptake in BeWo cells. *Am. J. Physiol. Endocrinol. Metab.* 2007, *293*, E1296–E1302.
180. Van Eetvelde, M.; Kamal, M.M.; Fiems, L.; Opsomer, G. Pancreatic b-cell function differs between newborn Belgian Blue and Holstein calves. In Proceedings16th Annual Conferenceof theEuropean Society for Domestic Animal Reproduction, Dublin, Ireland, 29 August–1 September 2012.
181. Toscano Paganoa, G.T.; Lazzaronia, C.; Pagano, P.G.; Galeazzi, D.; Biagini, D. A study on muscular hypertrophy in cattle: Serum concentrations of IGF-1, cortisol, insulin and testosterone. *Livest. Prod. Sci.* 2001, *70*, 235–239.
182. Istasse, L.; Van Eenaeme, C.; Evrard, P.; Gabriel, A.; Baldwin, P.; Maghuin-Rogister, G.; Bienfait, J.M. Animal performance, plasma hormones and metabolites in Holstein and Belgian Blue growing-fattening bulls. *J. Anim. Sci.*1990, *68*, 2666–2673.
183. Pajak, B.; Pawlikowska, P.; Cassar-Malek, I.; Picard, B.; Hocquette, J.F.; Orzechowski, A. Abundance of some skeletal muscle mitochondrial proteins is associated with increased blood serum insulin in bovine fetuses. *Res. Vet. Sci.* 2010, *89*, 445–450.
184. Ploquin, C.; Chabi, B.; Fouret, G.; Vernus, B.; Feillet-Coudray, C.; Coudray, C.; Bonnieu, A.; Ramonatxo, C. Lack of myostatin alters intermyofibrillar mitochondria activity, unbalances redox status and impairs tolerance to chronic repetitive contractions in muscle. *Am. J. Physiol. Endocrinol. Metab.* 2012, *302*, 1000–1008.
185. Lammoglia, M.A.; Bellows, R.A.; Grings, E.E.; Bergman, J.W. Effects of prepartum supplementary fat and muscle hypertrophy genotype on cold tolerance in newborn calves. *J. Anim. Sci.* 1999, *77*, 2227–2233.
186. Holmes, J.H.G.; Robinson, D.W. Hereditary muscular hypertrophy in the bovine: Metabolic response to nutritional stress. *J. Anim. Sci.* 1970, *31*, 776–780.
187. Allen, D.L.; McCall, G.E.; Loh, A.S.; Madden, M.C.; Mehan, R.S. Acute daily psychological stress causes increased atrophic gene expression and myostatin-dependent muscle atrophy. *Am. J. Physiol. Regul. Integr. Comp. Physiol.* 2010, *299*, R889–R898.

188. Demas, G.E. The energetics of immunity: A neuroendocrine link between energy balance and immune function. *Horm. Behav.* 2004, *45*, 173–180.
189. Helterline, D.L.I.; Garikipati, D.; Stenkamp, D.L.; Rodgers, B.D. Embryonic and tissue-specific regulation of myostatin-1 and -2 gene expression in zebrafish. *Gen. Comp. Endocrinol.* 2007, *151*, 90–97.
190. Michaux, C.; Brochier, B.; Roupain, J.; Pastoret, P.P.; Hanset, R.; Barta, O. Comparaison de l'immunocompétence des veaux culards et mixtes de race Blanc-Bleu-belge au moyen du test de transformation lymphoblastique. *Ann. Med. Vet.* 1984, *128*, 559–564.
191. Klont, R.E.; Brocks, L.; Eikelenboom, G. Muscle fibre type and meat quality. *Meat Sci.* 1998, *49* Suppl. 1, 219–229.
192. Guillemin, N.; Cassar-Malek, I.; Hocquette, J.F.; Jurie, C.; Micol, D.; Listrat, A.; Levéziel, H.; Renand, G.; Picard, B. La maîtrise de la tendreté de la viande bovine: Identification de marqueurs biologiques. *INRA Prod. Anim.* 2009, *22*, 331–344.
193. Strath, R.A.; Thompson, J.R.; Christopherson, R.J. Effect of fasting on plasma glucose kinetics in double-muscled cattle. *Can. J. Anim. Sci.* 1982, *62*, 955–958.
194. Mitlöhner, F.M.; Galyean, M.L.; McGlone, J.J. Shade effects on performance, carcass traits, physiology, and behavior of heat-stressed feedlot heifers. *J. Anim. Sci.* 2002, *80*, 2043–2050.
195. De Smet, S.; Lescouhier, S.; Claeys, E. Two-toned colour in the biceps femoris muscle in relation to post mortem pH and temperature fall in Belgian Blue beef. In Proceedings of the53th International Congress of Meat Science and Technology, Cape Town, South Africa, 10–15 August 2008.
196. De Boever, M.; Lescouhier, S.; De Smet, S. Influence of post mortem glycolysis and cooling on colour stability in different muscles of Belgian Blue beef. In Proceedings of the54th International Congress of Meat Science and Technology, Copenhagen, Denmark, 16–21 August 2009; pp. 772–776.
197. Sammel, L.M.; Hunt, M.C.; Kropf, D.H.; Hachmeister, K.A.; Kastner, C.L.; Johnson, D.E. Influence of chemical characteristics of beef inside and outside semimembranosus on color traits. *J. Food Sci.* 2002, *67*, 1323–1330.
198. Pringle, T.D.; West, R.L.; Williams, S.E.; Johnson, D.D. *The Role of the Calpain/Calpastatin System in Muscle Hypertrophy Associated with Double-Muscling in Beef; Annual Report*; Department of Animal and Dairy Science, University of Georgia: Athens, GA, USA, 1995; pp. 61–65.
199. Thompson, J. Managing meat tenderness. *Meat Sci.* 2002, *62*, 295–308.
200. Steen, D.; Claeys, E.; Uytterhaegen, L.; De Smet, S.; Demeyer, D. Early post-mortem conditions and the calpain/calpastatin system in relation to tenderness of double-muscled beef. *Meat Sci.* 1997, *45*, 307–319.
201. Wheeler, T.L.; Shackelford, S.D.; Casas, E.; Cundiff, L.V.; Koohmaraie, M. The effects of Piedmontese inheritance and myostatin genotype on the palatability of longissimus thoracis, gluteus medius, semimembranosus, and biceps femoris. *J. Anim. Sci.* 2001, *79*, 3069–3074.
202. Boccard, R.; Dumont, B.L.; Schmitt, O. Caractéristiques différentielles du tissu conjonctif des bovins normaux et "culards". *Gen. Sel. Evol.* 1969, *1*, 178.
203. Bailey, A.J.; Enser, M.B.; Dransfield, E.; Restall, D.J.; Avery, N.C. Muscle and adipose tissue from normal and double muscled cattle: Collagen types, muscle fiber diameter, fat cell size and fatty acid composition and organoleptic properties. In *Muscle Hypertrophy of Genetic Origin and its Use to Improve Beef Production*; King, J.W.B., Menissier, F., Eds.; Martinus Nijhoff Publishers: The Hague, The Netherlands, 1982; pp. 178–203.
204. Ngapo, T.M.; Berge, P.; Culioli, J.; Dransfield, E.; De Smet, S.; Claeys, E. Perimysial collagen crosslinking and meat tenderness in Belgian Blue double-muscled cattle. *Meat Sci.* 2002, *61*, 91–102.
205. Sañudo, C.; Macie, E.S.; Olleta, J.L.; Villarroel, M.; Panea, B.; Albertí, P. The effects of slaughter weight, breed type and ageing time on beef meat quality using two different texture devices. *Meat Sci.* 2004, *66*, 925–932.

206. Wiener, P.; Woolliams, J.A.; Frank-Lawale, A.; Ryan, M.; Richardson, R.I.; Nute, G.R.; Wood, J.D.; Homer, D.; Williams, J.L. The effects of a mutation in the myostatin gene on meat and carcass quality. *Meat Sci.* 2009, *83*, 127–134.
207. Hornick, J.L.; Van Eenaeme, C.; Clinquart, A.; Diez, M.; Istasse, L. Different periods of feed restriction before compensatory growth in Belgian Blue Bulls: I. Animal performance, nitrogen balance, meat characteristics, and fat composition. *J. Anim. Sci.* 1998, *76*, 249–259.
208. Fiems, L.O.; De Campeneere, S.; Van Caelenbergh, W.; De Boever, J.L.; Vanacker, J.M. Carcass and meat quality in double-muscled Belgian Blue bulls and cows. *Meat Sci.* 2002, *63*, 345–352.
209. Picard, B.; Jurie, C.; Garcia-Launay, F.; Meteau, K.; Agabriel, J.; Micol, D. Marqueurs des qualités sensorielles de la viande bovine en race Salers. *Renc. Rech. Ruminants* 2011, *18*, 181–184.
210. Clements, C.M.; McNally, R.S.; Conti, B.J.; Mak, T.W.; Ting, J.P.Y. DJ-1, a cancer- and Parkinson's disease-associated protein, stabilizes the antioxidant transcriptional master regulator Nrf2. *Proc. Natl. Acad. Sci. USA* 2006, *103*, 15091–15096.
211. Jia, X.; Ekman, M.; Grove, H.; Færgestad, E.M.; Aass, L.; Hildrum, K.I.; Hollung, K. Proteome changes in bovine longissimus thoracis muscle during the early postmortem storage period. *J. Proteome Res.* 2007, *6*, 2720–2731.
212. Picard, B.; Berri, C.; Lefaucheur, L.; Molette, C.; Sayd, T.; Terlouw, C. Skeletal muscle proteomics in livestock production. *Briefings Funct. Genomics* 2010, *9*, 259–278.
213. Jia, X.; Veiseth-Kent, E.; Grove, H.; Kuziora, P.; Aass, L.; Hildrum, K.I.; Hollung, K. Peroxiredoxin-6—A potential protein marker for meat tenderness in bovine longissimus thoracis muscle. *J. Anim. Sci.* 2009, *87*, 2391–2399.
214. Chelh, I.; Meunier, B.; Picard, B.; Reecy, M.J.; Chevalier, C.; Hocquette, J.F.; Cassar-Malek, I. Molecular profiles of Quadriceps muscle in myostatin-null mice reveal PI3K and apoptotic pathways as myostatin targets. *BMC Genomics*2009, *10*.
215. Raes, K.; De Smet, S.; Demeyer, D. Effect of double-muscling in Belgian Blue young bulls on the intramuscular fatty acid composition with emphasis on conjugated linoleic acid and polyunsaturated fatty acids. *Anim. Sci.* 2001, *73*, 253–260.
216. Aldai, N.; Nájera, A.I.; Dugan, M.E.R.; Celaya, R.; Osoro, K. Characterisation of intramuscular, intermuscular and subcutaneous adipose tissues in yearling bulls of different genetic groups. *Meat Sci.* 2007, *76*, 682–691.
217. Aldai, N.; Dugan, M.E.R.; Nájera, A.I.; Osoro, K. N-6 and n-3 fatty acids in different beef adipose tissues depending on the presence or absence of the gene responsible for double-muscling. *Czech J. Anim. Sci.* 2008, *53*, 515–522.

CHAPTER 3

Novel analytical technologies of Quality in the Sheep & Goat Dairy Sector

Psathas G* and Tzamaloukas O

Department of Biological Sciences - Technological University Limassol, Cyprus

ABSTRACT

France, Greece, Italy, and Spain the major sheep and goat milk producers in Europe, got the initiative through an interim of 30 years (1985 – 2015), to organize - through IDF - 7 symposiums aiming at the upgrade of the Dairy Sector. One major interest was the milk quality. Stepwise, the efforts followed cow milk structural development in organizing the testing laboratory, the definition of quality in composition (fat, protein, lactose, total solids), in Hygienic (Total Bacteria Count, Somatic Cell Count, antibiotics residues and Aflatoxin M1 contamination), adulteration with water and detection of milk species. EU, by Decision 2002/657/EC, defined the rules for the Directive 96/23/EC, enumerated the testing methods, the demands in sampling procedures, the minimum limits of performance, the quality control (ISO 17025) and the interpretation of results. The equivalency of results from different laboratories, through ISO 17025, demanded the application of ISO analytical standards. In a period of 5 years these standards have been reformed in order to cover the needs of sheep and goat milk analysis. Novel development in analytical instrumental methods, like FTIR interferometer working with mid-Infrared region of the spectrum from 3 - 10 μm corresponding to 1000 – 5000 cm $^{-1}$, gives opportunities of collecting data by making measurements for new parameters in complex dairy products, like Urea in milk, FFA profile, casein, pH, etc. Flow Cytometry for Individual Bacteria Count and Somatic Cells Count is opening the road for applications on testing for raw milk in hygienic control of the sheep and goat dairy sector. New developed screening methods are applied to give rapid results for antibiotics or aflatoxin M1 residues. Milk quality in the Central Testing Laboratory of today, is upgraded and promoted through rapid, trustful and traceable systems of LIMS in payments and herd management. Milk species mixtures can today be analysed in milk and dairy products with screening and reference methods. PDO products need today testıng methods for authenticity of animals' breed and geographical region of production.

Keywords: Sheep and goat dairy sector; Milk quality; Testing lab; Novel technology

1. INTRODUCTION

Sheep and goat dairy products have interesting characteristics in their levels of flavor, taste, aromas and leanness as well as the specific composition of fats, proteins, amino and fatty acids. Their quality is very much linked to historical and cultural uniqueness right through the production, marketing and consumption chains. Despite the superiority of their characteristics, they represent only 4% of total world production. More than 50% of this production is produced in Mediterranean basin. Four countries, **Greece, Italy, Spain** and **France** are the main contributors in Europe which by addition of Turkey, in Mediterranean basin, represent 58% worldwide. This extensive production refers to farming systems with dominant extensive grazing situations, specific technologies and conditions for the transformation processes of cheese-making and its ripening; The products are characterized by traditional nutritional habits of the consumers. While their organoleptic properties are very important, the sanitary aspects become more and more influential and tend to modify the accepted definitions of these products and their quality (e.g. non-pasteurized milk, production chains, hygiene, transformation methodologies, etc.). In a rapidly changing world the sustainability of the Sector became a major issue. New technologies through research and development were required to monitor the products in a better position of the international dairy food market. The upgrading of the sector became an utmost necessity and challenge for the interested parties.

1.1 The Forum of Sheep and Goat Dairy Sector

An initiative of Greece - with warm embrace from France, Italy and Spain on 1985-, started for sheep & goat dairy sector under IDF with ICAR, EAAP, CIRVAL, EDA and CIEAM. This initiative created a Forum which in an interim of 30 years – up to 2015 – produced about 800 scientific works, with important messages in the major fields of the sector Raw Material, Processing and Product, Products Characteristics, and Marketing Strategies.

Since Quality was a dominating issue in all above fields, analytical testing became an utmost necessity as a tool to reveal the advantages below of these milk species.

- Revelation of Fatty acids Superior Biological Value in Human Nutrition
- Proteins polymorphism and the connection to milk production and cheese yield
- Genetic character of lactic acid bacteria in fermented products – probiotics and bacteriocins production

At the same time, the Forum in the process of the 7 symposiums disclosed the major key trends governing the sector industry [1]:

- The milk production system cannot survive without governmental support. This support should aim to farm reformation and producers' education. A further development in breeding control, milk collection and farmers' payment based on quality schemes is required.
- To this effort, we should account the issues of animal health and nutrition and the environmental impact, bearing in mind the numbers of more than one billion of dairy sheep and 600 million of dairy goats.
- On the other hand, we cannot ignore the demand of consumers for better quality (flavor, texture, natural ingredients), better biological value and or organic products.
- Finally, food safety away from animal diseases, microbial contaminations, residues of antibiotics and aflatoxins.

1.2 The Milk Central Testing Laboratory (CT Lab)

Organizing the process from Research to Development three major partners is involved: Producers, Processors and the CTLAB. Quality is playing a significant role. This role is revealed and described in detail only by CTLAB. The contribution of CTLAB is focused in 3 major missions:

- Massive analytical data for several parameters for the schemes of payments to producers
- Milk recording activities – Herd management (Nutrition, breeding control)
- Safety controls as demanded by EU REG. 853/2004 for milk hygiene

The principle of organizing the CTLAB is the production of massive data with highly organized controls with automation in sampling, samples registration, automatic analysis, automatic handling of analytical data and rapid briefing of interested customers either producers or processors.

Novel technologies of today are providing solutions to fulfill the below characteristics of the modern CT Lab:

- Massive number of samples
- Automatic sampling procedures by road tankers during milk collection and delivery to processors
- Automatic detection of producers and processors and capture of sampling data by the use of GPS and GSM systems
- Use of RFID system for milk recording in Herd management
- Automatic sample registration and preparation of batches for analysis
- Use of automatic instruments for analysis
- Handle of analytical data produced by the use of Laboratory Information Management System (LIMS)
- Rapid final results up to 48 hrs and interested customers briefing through Internet or cellular phone

1.3 Quality system and quality assurance in the Lab

For a trustful laboratory, it is necessary to provide evidence of its proficiency in providing validated analytical data. This can be succeeded by organized steps in quality assurance for its work. The basic characteristics are listed below:

- Methods selection (Decision 2002/657/EC)
- Novel technology equipment with protocols for standard operation, performance controls and maintenance
- Methods validation
- Standard Operating Procedures – SOPs

- Quality Manual
- Controls (Standard samples, control charts)
- Proficiency testing
- Documentation
- Accreditation (17025)

1.4 Analytical methods

The testing methods for milk composition, during the decade of '70, were semiautomatic nephelometry for fat, amidoblack chromatometric for protein and oxidation iodometric chloramine T for lactose. In the next decade, the new generation turn to Infrared Spectroscopy with automatic modes of MidIR filters technique giving the chance in one test to get results for fat, protein, lactose and total solids. In '90s, there was a further development in analysis, technology with models based on FTIR (Fourier Transformation Infrared Medium Spectrum}. This development opened the door to make measures in an area equivalent to 700 filters which means that in Medium Spectrum through calibration from a reference method with calibration procedures, somebody could create new parameters, like pH, FFA, Casein, Urea, Furosine, etc. Since every parameter needs a reference method, an ISO standard needed to follow in order to calibrate the routine instrument. Until 2004, nobody put in doubt these ISO standards whose collaborative studies were based only on cow milk which means that errors were incorporated in analysis of sheep and goat milk. The following methods were reconsidered through new collaborative studies and adopted with annexes for sheep and goat milk precision figures in a period of 5 years:

- Fat in milk: - Rose Gottlieb (FIL/IDF 1D) – Reference method
- Nitrogen in milk – Kjeldahl (FIL/IDF 20B: 1993) – Reference method
- Freezing Point in milk (FIL/IDF 108:2001) – Reference method
- Urea in milk (FIL/IDF- ISO 14637 Draft) – Reference method
- Somatic cells count in milk (FIL/IDF 148A: 1995) – Reference method
- Microbial inhibitor tests in milk (ISO 13969/IDF 183 Draft)
- Determination of Fat, Protein and Lactose content in milk (FIL/IDF 141A:1990- Mid IR guidance) –Routine method
- Milk - Quantitative determination of bacteriological quality - Guidance on evaluation of routine methods (FIL/IDF 161B)

2. APPLICATION OF NOVEL TECHNOLOGY IN MILK COMPOSITION BY INFRARED SPECTROSCOPY

IR spectrum technology is based on the Lambert Beer law:

$$A = e\,l\,c = \log I_o / I$$

Where A: absorbance, e: coefficient of molar absorptivity, I: cuvette length, c: molar concentration of analytical parameter, lo: Intension of falling beam, l: intension of beam after absorbance

With the previous Mid IR technique of filters (Figure 1), between the light source and the sample in the cuvette, a filter permits a specific wavelength light to pass through. This light is absorbed by the specific bonds which characterize the fat or the protein or the lactose (Figure 2). Fat has two cases of different filters, the filter A for C=O bonds of fatty acids on the triglycerides at 5,7 μm wavelength and the filter B for the C-H of the main skeleton of triglycerides at 3,5 μm wavelength. Protein measure is based on amides bonds $N\text{-}H_2$ at 6,7 μm wavelength and Lactose on -OH bonds of lactose ring at 9,5 μm wavelength.

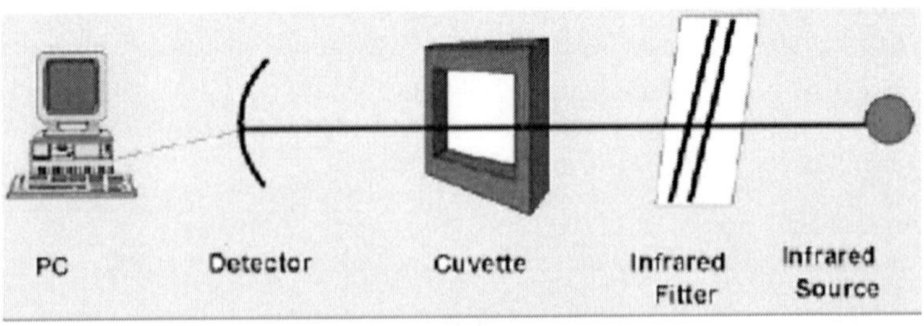

Figure 1: Previous Mid IR Filters technique

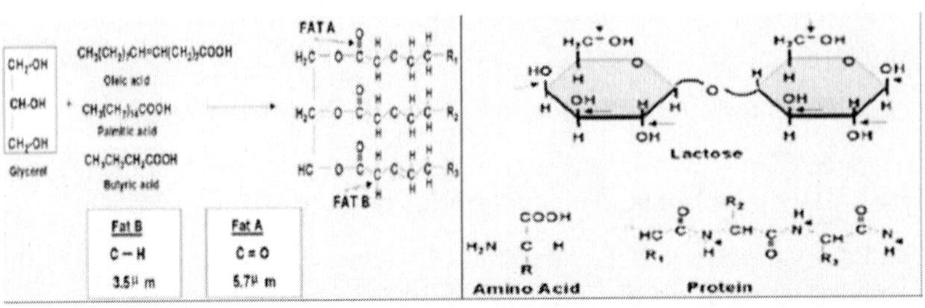

Figure 2: Wavelengths absorbance at specific bonds of fat, protein, lactose

Novel technology uses Fourier Transformation Infrared Spectroscopy (FTIR).

2.2 The infrared optical system

A built in Interferometer FTIR based on the principle of fixed filters is used in the unit. However, the unit detects the FTIR full infrared range, which can be used in innovative applications. The collection of data from the full range ensures at the same time the flexibility and additional features in the IR analysis.

2.3 The basic principles of Interferometry

Interferometry is a very competitive method compared to traditional techniques. All frequencies from the IR source used without any default, which allows the collection of full- spectrum in time less than a second. This is based on the principle of light contributions (interferences), which shape the range of a signal, as a function of the difference path between two interactive sources. An interferometer captures the intensity of light received from the detector as a function of path difference generated by the displacement of a moving mirror. The small displacement measurement of this mirror is achieved, with the help of a laser beam that follows the same path with a IR radiation. However, at this level, the interferogram refers to the location of the moving mirror and not at the wavelength of interest.

The infrared radiation from the source of IRE impinging on sliced-radiation, which reflects half the radiation in a fixed mirror and the other half of the radiation in a mobile mirror. From mirrors, radiation IR reflected and reconnected before reaching the detector. All IR frequencies are transported via interferometer simultaneously, and quick small mirror movements allow simultaneous production of the whole range of IR.

2.4 The transformation by Fourier principle

The principle of transformation in Fourier (FT) is based on the fact, that each function can be divided into a set of sinusoidal functions, where each function sine wave is defined by two values: the frequency (wavelength) and intensity. Fourier transformation is a mathematical process that allows the separation of a fringe effect contribution (interferogram) into a set of sinusoidal functions, each of which represents a given wave. The frequency and intensity of this, is calculated by the data of the interferogram. In a few seconds, the frill contribution (interferogram) collected from the spectrometer, is processed through processing in Fourier transform, and turns into a full spectrum of the sample (Figure 3).

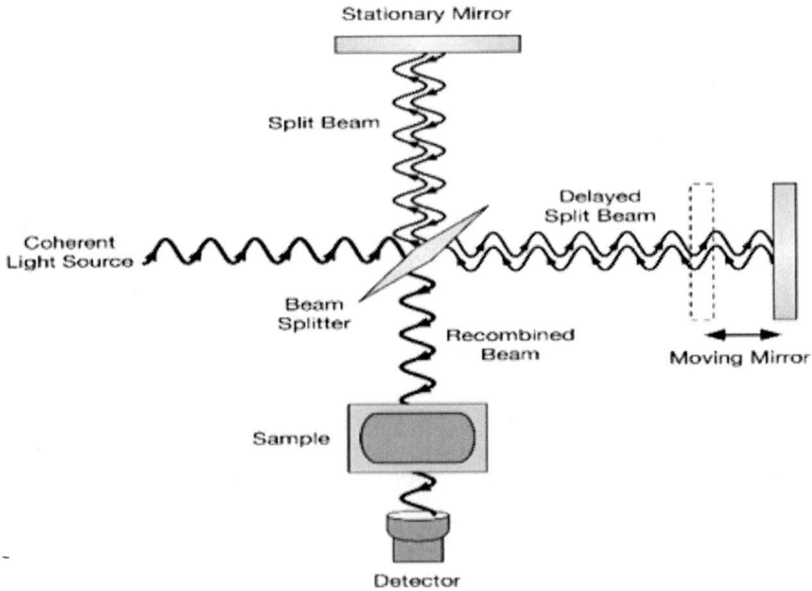

Figure 3: Novel technique of FTIR Interferometer

It is very valuable for the information retrieved. One can see the list of compounds that can be analyzed today (Table 1). Research in sheep and goat has shown superiority of biological value of certain fatty acids (Table 2). This information here is very significant in the hands of nutritionist for the herd management.

2.5 Targeted models

It is remarkable to say that FTIR technology permitted the creation of targeted model for the profile of pure milk, in order to avoid any adulterations with vegetable fats, melamine or cyanuric acid to show higher protein content, or water addition. This below mentioned curve (Figure 4), represents cow milk adulteration in Nederland. Y axis represents absorbance and X axis the wavelength.

Table 1: Basic analytical parameters with FTIR

Component	Measuring range	Performance range	Repeatability	Accuracy bulk	Accuracy single animal
Fat (A)	**0-15%**	**2-15%**	**Cv<0,5%**	**Cv<1,0%**	**Cv<1,5%**
Fat (B)	0-15%	2-15%	**Cv<0,5%**	CV<1,0%	Cv<1,5%
Protein	0-10%	2-10%	**Cv<0,5%**	Cv<0,9%	Cv<1,5%
Lactose	0-10%	2-10%	**Cv<0,5%**	Cv<0,9%	Cv<1,5%
Solids	0-20%	2-20%	**Cv<0,5%**	Cv<1,0%	Cv<1,5%
Urea	10-100 mg/dl	10-100 mg/dl	Sd<1,5mg/dl	Sd<1,5mg/dl	Sd<3,5mg/dl
Citric acid	0,1-0,5%	0,1-0,5%	Sd<0,005%	Sd<0,01%	Sd<0,015%
FPD	400-600m°C	450-550 m°C	Sd<0,5m°C	Sd<4m°C	N/A
Component	**Measuring range**	**Performance range**			
Casein	2,17-3,45%	2,17-3,24%	N/A	N/A	N/A
FFA mmol/10kg	0,072-10,04	0,49-6,63	N/A	N/A	N/A
Ketosis screening (acetone> 0,15 mmol/lt, acetoacetic acid and b hydroxybutyric acid – BHB>0,1 mmol/lt					

Table 2: List of fatty acids analyzed by FTIR [2]

Component	Calibration range	Validation range
Saturated Fatty Acids	0,87 – 5,08%	0,87 – 5,08%
Mono Unsaturated Fatty Acids	0,31 – 2,84%	0,31 – 2,84%
Poly Unsaturated Fatty Acids	0,03 – 0,36%	0,03 – 0,36%
Unsaturated Fatty Acids	0,35 – 5,39%	0,35 – 5,39%
Trans Fatty Acids	0,04 – 0,08%	0,04 – 0,08%
Short Chain Fatty Acids	0,35 – 0,44%	0,35 – 0,44%
Medium Chain Fatty Acids	1,58 – 2,06%	1,58 – 2,06%
Long Chain Fatty Acids	0,38 – 3,48%	0,38 – 3,48%
C:14_0*	0,16 – 1,38%	0,16 – 1,38%
C:16_0*	0,17 – 4,48%	0,17 – 4,48%
C:18_0*	0,11 – 1,25%	0,11 – 1,25%
C:18_1*	0,26 – 4,29%	0,26 – 4,29%
**FA g/100g milk (applies to all fatty acids components). **Calculated component*		

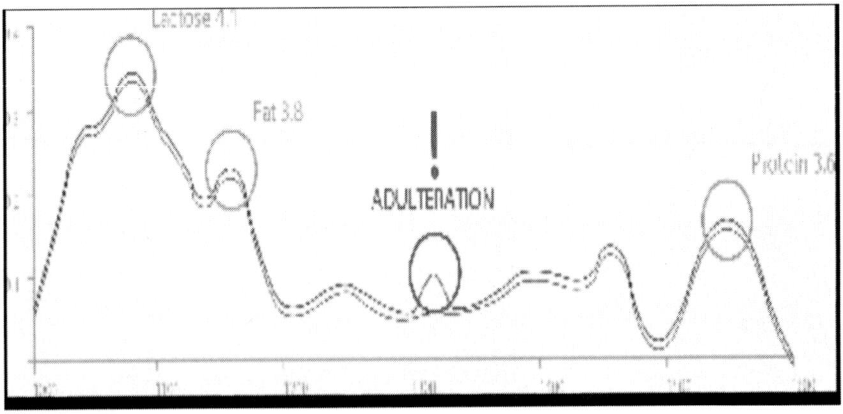

Figure 4: Adulteration predicting curve modeling [2]

3. NOVEL TECHNOLOGIES FOR SOMATIC CELLS COUNTING AND FOR INDIVIDUAL BACTERIA COUNTING

3.1 Somatic Cells Counting

Somatic Cells counting were always a significant indicator for mastitis in milk for all mammal species. 30 years ago, several methods had been tried. Major importance was given to direct microscopic counting. The same path was followed for bacteria counting. Direct epifluorscent filter technique has developed further the microscopical observation. Quantitative measuring was done by Conductivity method. But the most successful routine method was the Fluoro-opto-electronic SCC counters, based on rotating disc method. Somatic cell's DNA was dyed with edithium bromide, emitting fluorescent to an electronic microscope by spraying the sample on the edge of a rotating disc. The light emitted was converted to electronic signal which in turn was interpreted in measuring numbers of somatic cells. The somatic cells counters were quite successful for sheep and goat milk with calibration of instruments with more than 1 million of somatic cells.

3.2 For bacteria counting

Bactoscan technique was used. Through different sample preparation, centrifugation for isolation of alive bacteria, dying the DNA of these bacteria with Acridine orange, counting was made with electronic microscope on the edge of a rotating disc. The Bactoscan technique was not applicable for sheep and goat milk for different reasons. For goat milk a high number of somatic cells fragments – produced during sample centrifugation - were affecting seriously the results turning in inaccuracies. For sheep milk, because of the high content of fat and protein, a new recipe for breaking down the protein micelles was required. It was something that changed totally the performance of the instrument.

After 2000, novel technology came out in the analytical testing market, with the Flow Cytometry. This methology was based on the principle of regulating a standard liquid stream carrying the somatic cells (Figure 5), or the individual bacteria both dyed with their respective fluorescent dye, passing through a standard cuvette permitting the pass of one somatic cell or individual bacteria (Figure 6). In both cases this technique was quite successful for all milk species. Especially for sheep and goat milk it was saving a serious cost of using the reference method of petri dish counting, but also permitted measuring of much more samples and produced results in the same day.

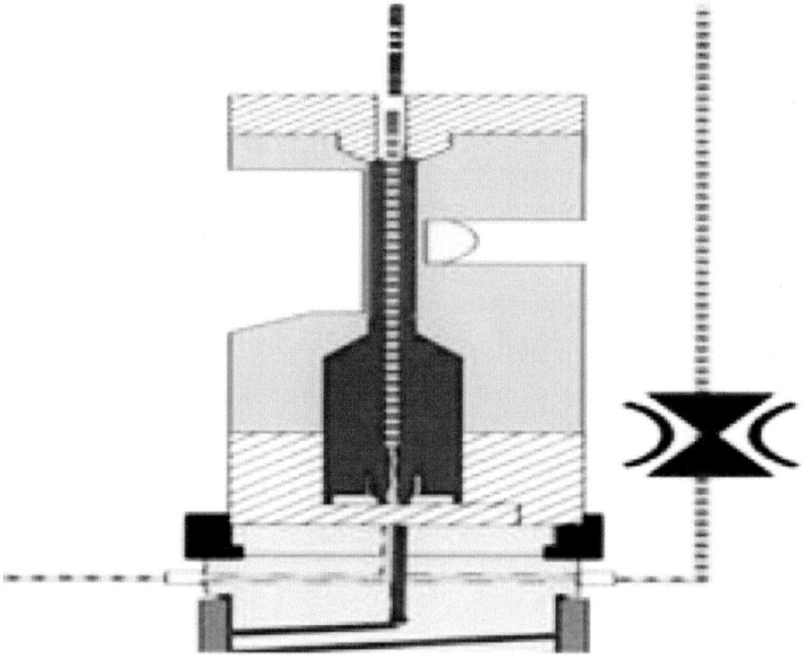

Figure 5: Flow Cytometry for SCC [3]

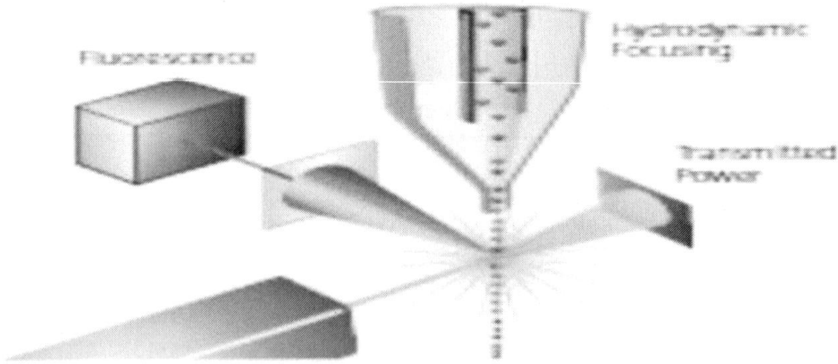

Figure 6: Flow Cytometry for Individual Bacteria Counting [4]

3.3 Screening methods for antibiotics residues detection

The importance of controlling the Hygiene of milk delivered to the processors remains in the first line of testing. Antibiotics residues in milk should be avoided by the milk producers. The Authorities engaged in controls, are seeking in effective methods for vigorous and reliable methods to prevent contaminations in the products processing lines.

Today, ALL COMMERCIAL METHODS ARE AIMING IN MEETING THE MRLs of DIRECTIVE 96/23 EC for Residues of Harmful substancesNovel

- The methods are classified in Qualitative and Quantitative
- Choice of method depends on milk quantity and time limits
- All methods are applied with same practice for all milk species
- Confirmatory testings are required on positive results of 1st screening

Here below, we present major firms in the market (Table 3).

The novel technology characteristic is the automation in processing and reading of results, adopted directly in the system of controlling data matched with other results of the producers. Here below we present major manufacturers' commercial instruments in the market (Figures 7,8,9 and 10).

Table 3: Manufactures and test kits products with their performance

Manufacturer	Name of test	Detection limits	Analysis time	Test method
AIM, Germany	BRT	Betalactam, 2 ppb	2,5 hrs	Qualitative
	*MCR3	Betalactam, <2 ppb	6 min	Quantitative
DSM, Netherlands	Delvotest®Accelerator	Betalactam, 2 ppb, Tetracycline, 40 ppb	1.5-2 hrs	Quantitative
	Delvotest® BLF	Betalactam, 4 ppb	5 min	Qualitative
Charm Science, USA	Charm II	Betalactam, 4 ppb, Tetracycline, 40 ppb	8-10 min 10-15 min	Qualitative
	Rosa Combo	Betalactam, 4 ppb, Tetracycline, 40 ppb	2 min	Qualitative
IDEXX Laboratories, USA	SNAP®	Betalactam, 4 ppb, Tetracycline, 50 ppb	10 min	Qualitative
Neogen, USA	Betastar®	Betalactam, 4 ppb	5 min	Qualitative
	Betastar® Combo	Betalactam, 4 ppb, Tetracycline, 40 ppb	5 min	Qualitative
Unisensor, Belgium	TwinSensor	Betalactam, 4 ppb, Tetracycline, 40 ppb	6 min	Qualitative

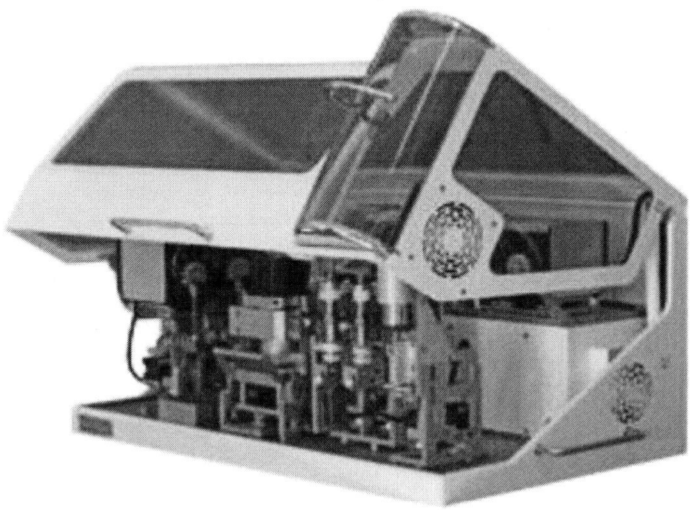

Figure 7: Firm: AIM, Model: MCR3

4. MILK ADULTERATION AND MILK MIXTURES IDENTIFICATION

4.1 Milk adulteration with water

Milk adulteration with water was already mentioned in composition analysis, as one of the new routine parameters of Infrared Spectroscopy. As a reference method, it uses the Freezing Point in milk (FIL/IDF 108:2001). The equipment available in the market, is a completely automated apparatus thermistor cryoscope (Figure 11)

Figure 8: Firm: Charm Science, Model: Charm II

Figure 9: Firm: DSM, Model: Delvotest Accelerator

Figure 10: Firm: Neogen, Model: Betastar

Figure 11: Fiske Thermistor cryoscope

4.2 Milk mixtures identification

Milk mixtures identification was always an issue for testing since milk species have different value in fat and protein content. Several methods were established in bibliography. These methods are evaluated not only for their accuracy for the analyte, but also for their practical performance in the lab, the cost of analysis and the time required for definite result. Here below it is presented a Compendium of methods for milk mixtures (Table 4) In last 15 years, almost 80 papers markers of cow's milk and methods used in the detection have been published. (Table 5)

Table 4: A Compendium of methods for Milk Mixtures Detection [6]

PHYSICOCHEMICAL METHODS					
No	Analyte	Technique	Application field	Advantages	Disadvantages
1.	Fatty Acids	-TLC	-Milk, Cheese - (Cow in sheep- goat)	-Small changes in fat - Research method	-Sensitivity 10% - Not applied in defat-ted products
2.	Triglycerides	-GLC -HPLC	-Milk, Cheese - (Cow in sheep- goat)	-Almost rapid method - Research method	-Very expensive - Experimental stage
3.	Caseins (αs1)	Ion exchange chroma-tography FLPC	-Milk, Cheese (2 months old) - (Cow in sheep- goat)	-Almost rapid method - Research method	-Very expensive -Expertise personnel
4.	-Peptides -Aminoacids -Whey proteins	-HPLC -RP HPLC -GLC	-Various dairy products - (Cow in sheep- goat)	-Almost rapid method - Research method -Very sensitive 0.1%	- Very expensive - Very complicate in sample preparation
ELECTROPHORETIC METHODS					
1.	Caseins (αs1)	PAGE	-Milk, Yogurt - (Cow in sheep- goat)	-Sensitivity 1% -Moderate cost -Recognized on National scale various European countries	- Poor quantitative accuracy - Very complicate in sample preparation
2.	Caseins (αs1)	Electrophoresis on Acetic Cellulose	-Milk, Yogurt - (Cow in sheep- goat)		Poor sensitivity over than 10%
3.	Para-κ-casein	PAGE	- Cheese - (Cow in sheep- goat)	Moderate cost	- Poor sensitivity - Very complicate in sample preparation - Very slow method Expertise personnel
4.	$\gamma_{1,2}$-casein	IEF	-Milk, Cheese -(Cow in sheep- goat)	-EU reference method -Sensitivity 1%	- Very complicate in sample preparation -Very slow method -Expertise personnel
5.	$\gamma_{1,2}$-casein	IEF	-Milk, Cheese -(Goat in sheep)	-Sensitivity 1%	- Poor quantitative accuracy - Very complicate in sample preparation - Very slow method - Expertise personne -Not recognized
6.	Whey proteins (β-lactoglobulin)	PAGE	-Milk, Fresh Cheese -(Cow in sheep- goat)	-Moderate cost - Sensitivity 1-2%	- Complicate in sample preparation - Poor accuracy with heated samples
7.	Whey proteins (β-lactoglobulin)	CZE	-Milk, Cheese -(Mixtures of cow, goat and sheep milk)	-Rapid method -Sensitivity 1-5% -Moderate cost -Good accuracy 2-5%	-Expensive equipment - At experimental stage

ELECTROPHORETIC METHODS					
No	Analyte	Technique	Application field	Advantages	Disadvantages
8.	Whey proteins (β-lactoglobulin)	IEF	- Milk, Cheese - (Cow in sheep- goat) -(Goat in sheep)	-Sensitivity 0.5-3%	-Expensive equipment -At experimental stage - Expertise personnel
9.	Para-κ-casein	CZE	- Milk, Cheese - (Goat in sheep)	-Sensitivity 1% -Accuracy 1-2.6%	-Expensive equipment -At experimental stage - Expertise personnel
10.	Caseins (αs1)	CZE	- Milk, Cheese - (Cow in sheep- goat)	-Sensitivity 1% -Accuracy <5%	-Expensive equipment -At experimental stage -Expertise personnel
IMMUNOLOGICAL, IMMUNOELECTROPHORETIC METHODS					
1.	Immunodiffusion	On immunoglobulins	-Milk, Cheese -(Mixtures of cow, goat and sheep milk)	- Very simple in application - Sensitivity 1-2.5% - Low cost of equipment - Recognized in France	- Very expensive in reagents /sample - Very poor quantitative accuracy especially in the presence of colostrum or with heated samples
2.	Dual Radial Immunodiffusion	On immunoglobulins	-Milk, Cheese -(Mixtures of cow, goat and sheep milk)	- Very simple in application - Sensitivity 1-4% - Low cost of equipment - Recognized in France	- Very expensive in reagents /sample - Better quantitative accuracy but over 20% mixtures is CV>15%
3.	Rocket Immunoelectrophoresis	Antiserum from goats for cow's milk	-Milk, Cheese -(Mixtures of cow, goat and sheep milk)	- Sensitivity 1-5%	-Limited application and recognition
4.	Crossed Immunoelectrophoresis	Antiserum from rabbit for cow's whey proteins	Limited application on blue cheeses	- Sensitivity 0.1-2% -Not affected by heat treatment -Recognized in France	-Limited application - Unknown accuracy, cost, or accuracy
5.	Counter - current	Antiserum from rabbit for cow's whey proteins	-Milk, Cheese -(Mixtures of cow, goat and sheep milk)	- Sensitivity 0.1-2%	-Very poor accuracy -Not applied in heat treated products or with developed proteolysis
6.	Immunodotting technique	Antiserum from rabbit for cow's whey proteins	-Milk, Cheese -(Mixtures of cow, goat and sheep milk)	- Sensitivity 0.1-2% - Not affected by heat treatment	Very poor accuracy
7.	ELISA	Polyclonic antibodies for cow/goat immunoglobulins	-Milk, Cheese -(Mixtures of cow, goat and sheep milk)	-Sensitivity 0.1-1% -Rapid method -Easy application -Cheap equipment -Recognized in Germany	-Poor quantitative accuracy/ Can be characterized as semi quantitative - Expensive in reagents/ High cost per sample

Table 5: Marker and methods [5]

Marker	Methodology	
α_{s1}-CN	PAGE, PAGE-SDS, IEF, CE, para-κ-casein	PAGE, CE, IEF
γ-CN	PAGE-SDS, IEF	
β-Lg	HPLC, FPLC, IEF, CE	
α-La	IEF, CE, MALDI-TOF	
IgG	ELISA	
Casein fractions	CE, HPLC	
CMP	ELISA	

The Commission Decision 93/256/EEC has stated that "Methods based only on chromatographic analysis (or other descriptive techniques) *without use of molecular spectrometric detection are not suitable for use as confirmatory methods*". This official position has encouraged development of mass spectrometry (MS)- based methods

4.3 Authenticity of sheep and goat PDO products

Protected Demographic Origin (PDO) products are remaining as a very valuable asset for the Sheep and Goat Dairy Sector. Since these products are very popular especially in their original place of production and usually are targeted by cow milk processors to copy them, the specific characteristics that prescribe the particular character of these products are listed below:

- Characteristics of Raw Milk Used for Cheese Making: Specific cultures used, or Eco flora isolated in product, milk specie, composition of non-adulterated raw product, animal breed, local vegetation, specific aromatic compounds, trace elements and metals
- Clotting Enzymes: Natural rennet from young animals, artificial preparation enzymes etc.
- Processing Practices: Particular techniques applied in product manufacturing
- Ripening Process for Cheese: Salting process, temperature used, ripening time.
- Character of Cheese: Soft-semi hard- hard cheese – Characteristic peptides, amino acids in hard cheese
- Characteristic Texture: Holes in the cheese, elasticity of cheese...
- Flavor and Aroma: Connected to specific aromatic compounds resulted from ripening process

Many analytical methods are involved, very often in two or three combination to verify specific analyte[5]:

- Electrophoresis - gel (WP, C)
- Capillary electrophoresis
- Chromatographic methods RP-HPLC (WP, C)
- Immunological methods (WP, C)
- Polymerase Chain Reaction (somatic cell's DNA)
- Electronic Nose Aroma Detector

5. CONCLUSIONS

We are already in a new era of testing technology. Many obstacles in the sector of sheep and goat dairy sector have been removed. An enormous number of scientific reports in several subsectors i.e. nutrition, genetics, dairy products technology, have revealed superiority of sheep & goats milk ingredients. Novel technology in analytical testing is to help producers' associations for further improvement of their raw material. On the other hand, the challenge remains in hand of processors to respond to consumers demand for biological products.

References

1. Beheregaray LB, Gibbs JP, Havill N, Fritts TH, Powell JR, et al. (2004) Giant tortoises are not so slow : Rapid diversification and biogeographic consensus in the Galápagos. PNAS 101: 6514–9.
2. Foss (2003) Manual on FTIR instrument FT plus authenticity curve.
3. Foss (2003) Manual on Somatic Cells Counter Leaflet.
4. Manual on Somatic Cells Counter Leaflet
5. Detection of Species Adulteration in Milk and Dairy Products – Key Note presentation by Francesco Addeo- Sardinia 2007 Alghero – Italy
6. A Compendium of methods for Milk Mixtures Detection – By George Psathas CMIO internal report 2000

CHAPTER 4

Synbiotic as Feed Additives Relating to Animal Health and Performance

Hozan Jalil Hamasalim

Department of Animal Production, Faculty of Agricultural Sciences, University of Sulaimani, Sulaimani, Iraq

ABSTRACT

According to the increasing of human population in the world, it reached about seven billion people and it continuously increased. In this background, the food source in both animal and plant origin must be increased accordingly. For these we must use and add some feed additives such as antibiotic, probiotic, prebiotic, postbiotic and synbiotic for the animal feed to increase production (meat, egg, milk and fish) and improve health. In early cases, probiotic as mono or mixed beneficial live microorganism was used as feed additive that plays a significant role in several health conditions and performances. In another way, the scientists use some ingredients indigestible with carbohydrates origin, especially oligosaccharides as a source of energy for beneficial microorganisms in the body which were called prebiotic, and it is indigestible fermented food substrates that stimulate the growth, composition and activity of microorganisms in gastrointestinal and improve host. Most of the scientists urged to use all the above in such way that have more benefits in animal health and performance which were therefore called synbiotic, that was a combination between probiotic and prebiotic which beneficially had significant effects on the host by improving the survival and implantation of live microbial dietary supplements in the gastrointestinal tract, and thus improving animal health and performance. So, it was proposed that the synbiotic in this research increased beneficial microorganisms in the gastrointestinal tract and improved intestinal architect, and then promoted intestine environment. Consequently, it can improve blood indices, and especially decrease bad cholesterol (Low-density lipoprotein), decrease harmful microorganisms and toxins. However, it can also improve ingredient product, increase mineral absorption and nutrient. In conclusion, it can improve animal health and performance.

Keywords: Synbiotic, Animal, Intestinal Tract, Health and Performance

1. INTRODUCTION

People of the world have increased from 3 billion in 1959 to more than 7 billion in March 2012. As the world's people grow, hunger persists in many locations and almost 1 billion people are reported to be malnourished [1]. By 2050, farmers will need to double crop production to meet the demand. In this background, the world needs food products with annual growth of 2.5% for the next 10 years [2]. Many scientists and nutrition specialists believe that animal production can play a role in increasing food production. A huge amount of antibiotics have been used to control diseases, improve performances and increase production in livestock. Additionally, the most commonly used alternatives to antibiotics have been probiotic, prebiotic, postbiotic and synbiotic.

Probiotic is defined as live microbial feed additive that beneficially affects the host animal by improving the intestinal microbial balance [3] - [6]. Also, prebiotic is "indigestible fermented diet substrates that selectively stimulate the composition, growth and activity of microflora in gastrointestinal tract" [7] [8]. Synbiotics refer to nutritional supplements combining probiotic and prebiotic in a form of synergism, hence, synbiotic can enhance their isolated beneficial effects. When two nutritional ingredients or supplements are given together, the resulting positive effect generally follows one of the three patterns: potentiation, synergism and additivity [9] [10]. Many studies have evaluated the effects of different synbiotic preparations [11] [12]. Further, Synbiotic affects the host by improving the survival and establishment of live microbial dietary supplements in the gastrointestinal tract by selectively stimulating the growth by activating the metabolism of one or a limited number of health promoting microorganisms and thus improving the host [7] [10] [13]. Furthermore, synbiotic can exert beneficial effects in the gastrointestinal tract as a result of alteration in whole body, feed consumption, and absorption of nutrient and beneficial changes in intestinal architecture [14]. In fact, intestinal microorganisms play an important role in the immunological, physiological, nutritional and protective functions of the host [15] and can be influenced by the food [16]. The most alternative additives for livestock and poultry feed include probiotic, prebiotic and synbiotic [17]. Though, the use of synbiotics may possibly produce greater benefits rather than the application of individual portions [18].

Synbiotics provide more additive benefits in growth performance, feed conversion ratio, hematological and biochemical parameters than probiotic and prebiotic individual use of these additives [19]. Moreover, synbiotic could increase the digestibility and availability of many nutrient elements such as, vitamins, mineral elements and proteins [20]. After all, limited data are available regarding the application of synbiotic [21] - [24].

The research questions that were asked in this review were as follows: Is the feeding of synbiotic supplement important while used in applied research and commerce? Does the use of synbiotic supplement in animal lead to improved intestinal microbiota and prevent pathogen? Does the use of synbiotic in animal benefit the performance and enhance blood picture and immunity? Hence, the aim of this review study is to update our information regarding the influences of synbiotic. This review focuses on the collection of the most scientific evidences concerning the aspects of synbiotic and its effect on animal growth, production and health, including the immune system, digestive tract, metabolic, intestinal organ and blood.

2. GUT MICROFLORA

The intestinal microbiota (a term that has now replaced the old denomination of "microflora" [25] - [27] is an ecosystem shaped by a diversity of ecological niches, made of several bacterial species and a very large amount of strains [27] [28]. Also, intestine is populated by 100 trillions of microorganisms (called "microbiota") that are important for health [29] [30]. While, the transition from soil and plants to the animal gut, has three areas of genomic adaptation [31]. The intestine's normal microflora is a metabolically active but as yet unexplored area of host defense [32] [33]. Major functions of the gut microbiota include metabolic activities that result in salvage of energy and absorbable nutrients, trophic effects on the intestinal epithelium and protection of the host against invasion by harmful microorganism [34] [35]. The gastrointestinal tract which places a huge microbial ecosystem; the colon alone is estimated to contain over seventy percentage of all the microbes in the body [36] [37]. The gut microbiota or microflora has an important role in disease and well-being [38]. Commensal intestinal flora plays an essential role in the maintenance of the host health. However, reports of shifts in the composition of intestinal microbiota with age, with increased numbers of bacteria (mainly Enterobacter) and a decrease in the number of beneficial organisms such as Lactobacilli and Bifidobacteria [39]. These changes, along with a general decrease in species diversity in most bacterial groups, and changes to diet and digestive physiology such as intestinal transit time, may result in increased putrefaction in the colon and a greater susceptibility to infection and disease [40]. The three major units of the gastrointestinal tract are the stomach, the small intestine, and the large intestine. Every unit has its own distinct microbiota [41] [42]. The amount and composition of microbial species differs along the digestive tract. Families, phyla and genera of the microbiota enriched in each particular niche are listed. Main bacterial phyla are represented in the mammalian gut microbiota; Streptococcus, Lactobscilus, Bacteroides, Clostridium, Streptococci, Lactobacilli, Eubacterium, Peptococcus, Streptococcus Fusobacterium and Bifidobacterium [43]

3. PROBIOTIC

Probiotic was first defined by Metchnikoff in 1908 based on his observations on the longevity of individuals who lived in a certain part of Bulgaria and which he attributed to their ingestion, on a regular basis, of a fermented milk product [44] [45]. Which contain of rod-shaped bacteria (Lactobacillus spp). Therefore, these bacteria affect the gut microflora positively and reduction the microbial toxic activity in intestine [46] - [49]. The expression "probiotic" comes from the Greek word "pro bios" which means "for life" as different to "antibiotics" which means "against life". The history of probiotic began with the consumption of fermented diets by Greek and Romans [46]. Probiotic is defined as mono or mixed cultures of "live microorganisms which, when administered in adequate amounts confer a health benefit on the host" [43] [50] [51]. A new description by the American Academy of Pediatric Committee on Nutrition states that probiotic is "microorganisms that generate small molecular metabolic by-products that exert beneficial regulatory effect on host biological functions and may function as immunomodulators" [52]. Probiotic is mainly lactic acid creating bacilli, mostly Lactobacilli (L. acidophilus DDS-1, Lactobacillus casei (L. casei), L. lactis, L. rhamnosus, L. salviarius) and Bifi dobacteria (B. longum, B. infantis, B. bifi dum), and the yeasts Sacharomyces boulardii (Brewer's yeast) and Sacharomyces cervisiae (Baker's yeast) [53]. Probiotic may play a beneficial role in several health conditions and performance, including Intestinal microbial composition, therapeutic effects, metabolic effects and immunomodulation [43] [54] [55]. The overall performance and well-being benefits of probiotic microorganisms are repre-sented in Figure 1.

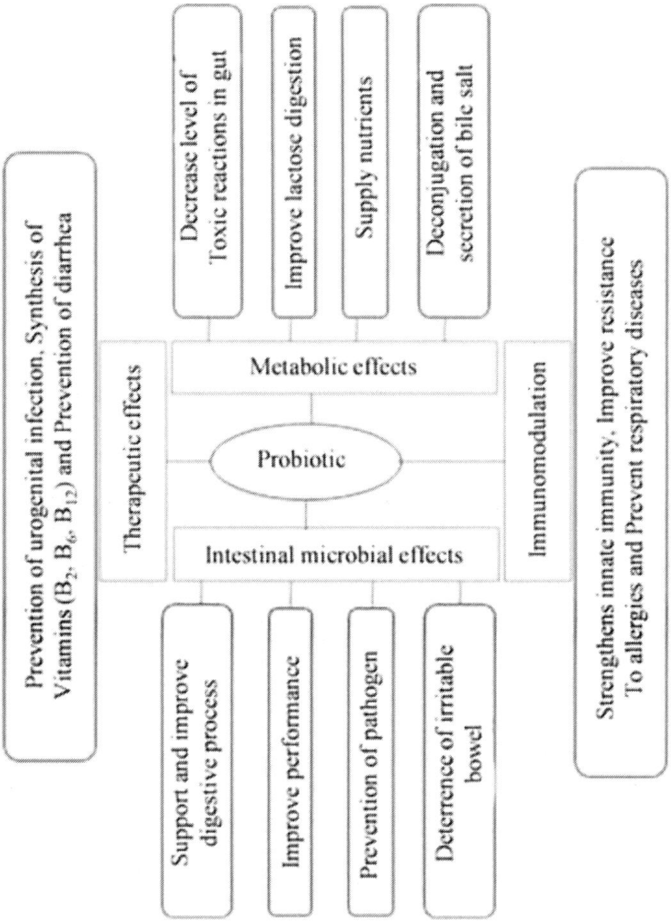

Figure 1. General benefits of probiotic on health and performance.

4. PREBIOTIC

The word prebiotic was first used by Gibson and Roberfroid in 1995 [7]. Dietary fiber is the most generally utilized prebiotic [7] [8]. Also, Prebiotic is "indigestible fermented diet substrates that selectively stimulate the composition, growth, and activity of microflora in gastrointestinal tract and thus improve hosts' health" [56] - [58]. Prebiotic carbohydrates are found naturally in such fruit and vegetables tomatoes, Jerusalem artichoke, as oatmeal, bananas, flaxseed, asparagus, barley, berries, garlic, wheat, onions and chicory, greens and legumes [59]. It is considered important to determine the definite health bonuses associated with prebiotic intake in people and their mechanisms of action [60] [61]. Prebiotic, such as fructooligosaccharides can be used as a dietary supplement for animals like, dogs and cats to maintain gastrointestinal well-being [62]. Prebiotic is not broken down by gastric enzymes, but pass unaltered into the large intestine, where they are then selectively fermented, creating beneficial effects [63] [64]. A study has shown that administration of prebiotic also results in increased numbers of beneficial intestinal flora (especially Bifidobacteria) [65] [66]. Probiotic, prebiotic and synbiotic administration have been documented to increase intestinal levels of beneficial Bifidobacteria, Enterocooci and Lactobacilli, with reduced levels of Enterobacter [67] [68]. Prebiotic have been demonstrated to increase mineral absorption, chiefly that of magnesium and calcium [69] [70]. The main prebiotic functions are shown in Figure 2.

5. SYNBIOTIC

Synbiotic refers to nutritional supplements combining prebiotic and probiotic and in a form of synergism. The main aim for using a synbiotic is that a true probiotic, without its prebiotic food, does not survive well in the digestive system. Synbiotic refers to nutritional supplements combining Probiotic and Prebiotic that are thought to act together; i.e. synergism. It has been suggested that a combination of a probiotic and a prebiotic, i.e. Synbiotics, might be more effect than either a probiotic or prebiotic alone [71] - [75]. Furthermore, synbiotic is a mixture of probiotic and prebiotic which beneficially affect the host by improving the survival and the implantation of live microorganisms dietary supplements in the gastrointestinal tract, and thus improving host health [76]. The United Nations Food and Agriculture Organization (FAO) recommend that the word "synbiotic" be used only if the net health benefit is synergistic [77]. The first attempts should be to combine Probiotic and Prebiotic which have demonstrated individual benefits to determine if there are additive effects, alternatively, a more structured approach would be to determine the specific properties that a prebiotic requires to be beneficial to the probiotic and select the prebiotic accordingly [78]. Synbiotic is designed not only to present beneficial microorganisms populations, but also to promote proliferation of autochthonous-specific strains in the intestinal tract [79]. Studies on the effects of synbiotic on metabolic health still are limited. It is worth mentioning that the health effect will likely depend on the synbiotic combination. Therefore, synbiotics seem promising for the modulation of the gut microbiota composition [80].

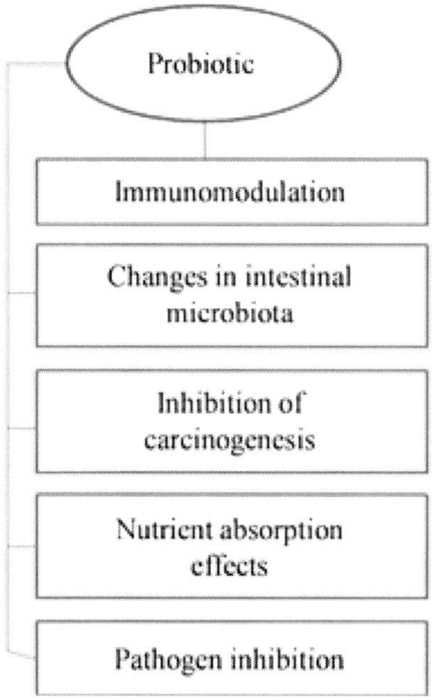

Figure 2. The mechanism action of prebiotic.

6. PROBIOTIC, PREBIOTIC AND SYNBIOTIC AS COMPARED WITH EACH OTHER

Unlike probiotic, prebiotic do not add to an existing colony of bacteria, rather they provide nourishment for existing flora, allowing the colony to grow naturally and flourish. As probiotic is principally active in the small intestine and prebiotic is only effective in the large intestine, the combination of the two may give a synergistic influence. Probiotic is a live microbial food ingredient which is beneficial to host [3] [76] [81] - [83] while prebiotic is a non-digestible food ingredient which beneficially affects the host by selectively stimulating the growth and activity of one or a limited number of microorganisms in the colon having the potential to improve host well-being [76] [84] [85] nevertheless, synbiotic is a mixture of probiotic and prebiotic which beneficially affect the host and thus improving host health and well-being [74] [76] [86]. Examples of list of probiotic, prebiotic and synbiotic applied or studied for application in animal feed are included in Table 1.

7. ACTION MECHANISM OF SYNBIOTIC

The gastrointestinal tract has a compound community of microbiota that provides benefits to its host in numerous different ways, including drug metabolism, nutrient production, and protection against pathogens, detoxification and regulation of the immune system. Animal studies have demonstrated that changes in these gut microbial communities can cause immune dysregulation; improve growth and influence on performance and there are information that support the use of probiotic and prebiotic and especially synbiotic. The synbiotic concepts about mechanism of action: altering the composition of intestinal microbiota by viable benefit organism and non-absorbable organism substrates are shown in Figure 3.

8. SYNBIOTIC AND INTESTINAL FLORA

The gastrointestinal tract is an important immune organ and the largest defense barrier protecting the host from toxins, pathogens and subsequent inflammation while allowing commensal microorganisms to grow [87]. Also, the intestinal tract is a host to a vast ecology of microbes [88]. The notion of modulating bacterial activities directed towards improving gut microbial function has a long history [89]. The importance of the natural gut microflora for decreasing diseases in animals has long been recognized and it is now apparent that the composition of the microflora plays a crucial role both in digestion and in resistance to diseases [90] [91]. When, probiotic and prebiotic are administered at the same time, the combination is called Synbiotic. The prebiotic in the synbiotic mixture improves the survival of the probiotic microorganisms in the intestinal tract, and stimulates the activity of the host's endogenous bacteria [79] [92] [93]. Then, supplemental probiotic from anaerobic microflora with prebiotic improved crude protein and dry matter digestibility as well as reducing noxious gas emission and enter pathogenic bacteria in early-weaning pigs [94].

Smith and Jones [95] reported that supplemental synbiotic increased the production of lactate and antibody, altered intestinal bacteria colonies and reduced harmful bacteria growth in animals. In addition, synbiotic is also considered to reduction harmful bacteria counts and aid the adhesion of beneficial bacteria through the decrease of intestinal pH [96]. Also, Synbiotic supplementation maintaining populations of unprofitable or potential pathogens (E. coli) at relatively low levels (numerically) in the cecal digesta and small intestinal [97]. Further, synbiotic reduced Escherichia coli and total coliform populations in the intestines of broiler chickens. On the contrary, concentrations of the synbiotic higher than the suggested levels in the diet increased the lactic acid bacteria population in the gut of broiler chickens [98]. Furthermore, the addition of synbiotic increased the villus height/crypt depth ratio and villus height in ileum. However, the ileal crypt depth was reduced by dietary supplementation of synbiotic compared with control of broiler Chickens [14].

Table 1. Examples of list of probiotic, prebiotic and synbiotic applied or studied for application in animal feed.

Probiotic	Prebiotic	synbiotic
Lactobacillus sps.	Inulin	*Lactobacilli* + inulin
Bifidobacterium sps.	Galactooligosaccharides (GOS)	*Bifidobacteria* + FOS
Saccharomyces sps.	Fructo-oligosaccharides (FOS)	*Lactobacilli* + FOS
Streptococcus sps.	Lactulose	*Bifidobacteria* and *Lactobacilli* + inulin
Bacillus coagulans	Lactitol	*Bifidobacteria* and *Lactobacilli* + FOS
Propionibacterium	Cereals fibres	*Lactobacilli* + lactitol
Bacillus coagulans	Xylooligosaccharides	*Bifidobacteria* + GOS
Enterococcus faecium	Isomaltooligosaccharides	
Freudenreichii		
Homeostatic Soil		

9. SYNBIOTIC AND PERFORMANCE

Supplemental antibiotic in animal feed improve feed efficiency and growth performance [99]. However, supplemental antibiotic increases the hazard of antibiotic residues and increases bacterial resistance [100], making their use in animal production harmful to human health. Recently, the most commonly used replacements to antibiotic have been synbiotic. Supplemental synbiotic can be expected to improve swine production by improving the feeding environment of early weaning pigs [94]. Further, Supplemental synbiotic with ficus-indica var. saboten could decrease sulfide gas emissions and ammonia of finishing pigs [101]. Synbiotic had significant effect on growth performance in Danio rerio [102]. Conversely, supplemental probiotic from anaerobic microflora with prebiotic did not affect performance in Weaning Pigs [94]. Likewise, the results clarify that using synbiotic in sheep diet did not influence on performance traits such as feed conversion rate, daily gain and dry matter intake. However, Digestibility of dry matter, crude protein and organic matter were not affected with synbiotic [103].

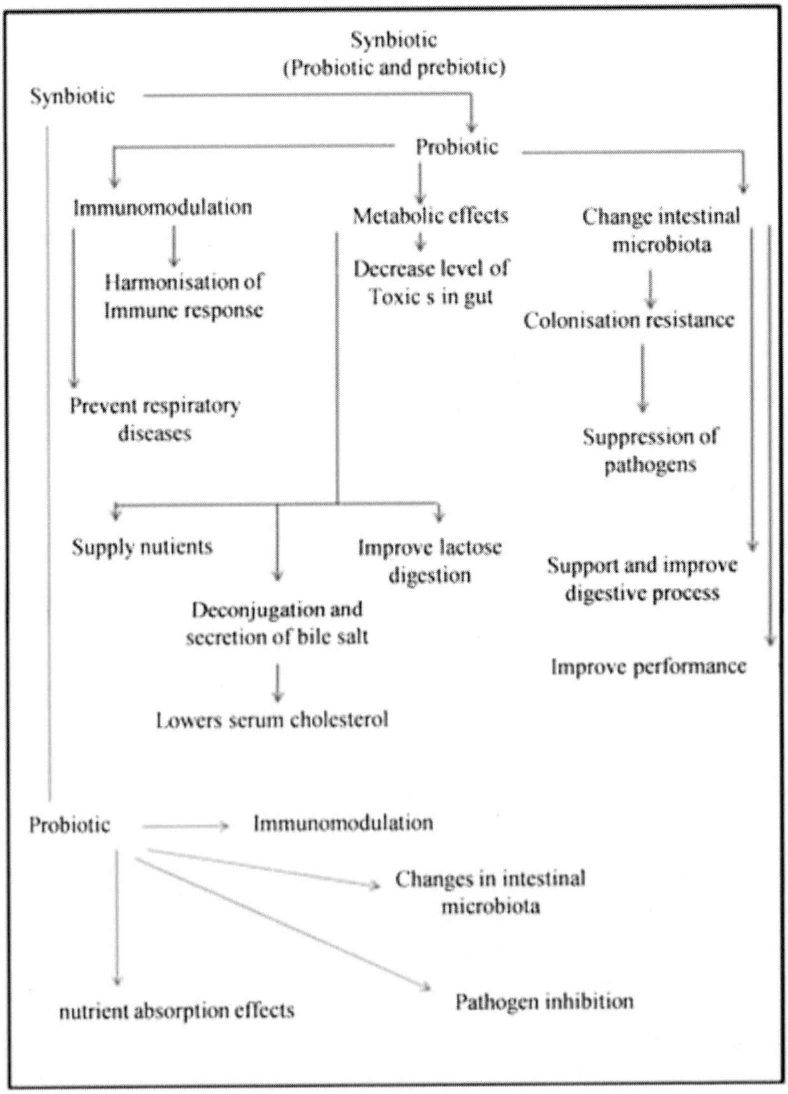

Figure 3. The mechanism action of synbiotic.

Combination of mannan oligosaccharides and Bacillus spp. as synbiotic in European lobster larvae (Hommarus gammarus L.) [104], combination of mannan oligosaccharides and Enterococus faecalis as synbiotic in rainbow trout (Oncorhyncchus mykiss) [21] and combination of Bacillus subtilis and fructooligosaccharides as synbiotic in yellow croaker, Larimichtys crocea [22], synbiotic added to feed was not able to increase the growth and survival rate in grass carp but the best survival rate was obtained in feed added synbiotic [10].

Application of synbiotic treatment combination of probiotic and prebiotic, showed improvement of digestive enzyme activity (lipase, protease and amylase) of Humpback Grouper (Cromileptes altivelis) [105]. Common carp fed dietary synbiotic showed better digestive enzyme activities and significantly higher trypsin and chymotrypsin activities compared with the control treatment [106]. Synbiotic could significantly improve growth parameters (length gain, specific growth rate, percentage weight gain and weight gain) but did not display any effect on survival rate of common carp [106]. Administration of synbiotic (E. faecalis and MOS/PHB) in rainbow trout for periods not affects the survival rate of fish [21]. Japanese flounder feeding B. clausii and MOS/FOS, in which fish maintained active ingestion, exhibited proper growth and survived for all time [107]. In yellow croaker and cobia, administration of B. subtillis/FOS or B. subtillis/chitosan respectively, not affects the survival rate, with no alterations among different dietary treatments [22] [108]. The study of Mehrabi et al. showed that after 60 days groups fed diets containing different levels of synbiotics (0.5, 1.0 and 1.5) improved body weight gain about 50, 59 and 53%, respectively, in comparison with the control group [109]. Body composition has been investigated by Rodriguez-Estrada [21], Ye [107] and Mehrabi [109]. First of these studies, reported no significant differences in crude ash, moisture, crude proteins and crude lipids contents among all the experimental groups. Contrary to our finding, Jung [110] and Erdogan [111] reported that the diet supplemented with synbiotic had no effect on feed intake, body weight, feed conversion efficiency and weight gain, of broilers. The final body weight, feed conversion efciency, weight gain were significantly higher in synbiotic supplemented broilers compared with the control group [97]. Body weight gain was increased significantly in birds' supplemented synbiotic compared with control and probiotic treated group [112]. The results showed that 1 g/kg synbiotic inclusion in the diet significantly improved body weight and feed conversion ratio of the ostrich chicks compared to control group [113]. Also, synbiotic substantially increased blood glucose but reduced cholesterol while Serum total protein and uric acid also decreased in the all dietary levels of synbiotic compared to control group of the ostrich chicks. Synbiotic has the ability to reduction the concentration of cholesteryl esters CE in LDL-cholesterol [114]. Prebiotic that present in the synbiotic mixture has hypocholesterolemic effects thereby decreasing the absorption of lipids in the intestine through binding bile acids, increasing cholesterol elimination and hepatic synthesis of new bile acid [115]. Finally, supplemental Aspergillus oryzae, one kind of prebiotic, increased growth performance and nitrogen retention in pigs, while supplemental Fermacto 500® from Aspergillus oryzae culture increased the digestibility of protein and fat in pigs [116].

10. SYNBIOTIC HEMATOLOGICAL AND BIOCHEMICAL PARAMETERS

Among the hematological and biochemical parameters evaluated in animal after treatments with synbiotic find triglyceride, cholesterol, high-density protein cholesterol, low-density protein cholesterol, albumin, globulin, total serum protein, glucose and hematocrit. Accordingly, a positive contact of prebiotic and probiotic was found concerning the number of leukogram and white blood cells while supplements to feed rations for ewes [117]. Also, Lambs fed on diet contain probiotic and prebiotic plasma was significant decrease of the level of total cholesterol and HDL-cholesterol fraction increase were observed [117]. But, supplementation of synbiotic had no influence on blood urea nitrogen, glucose and albumin concentrations in sheep [103]. Application of synbiotic treatment showed improvement of biochemical plasma (glucose and triglyceride) and haematology parameter (hematocrit, haemoglobin and phagocytic activity) of Humpback Grouper (Cromileptes altivelis) [105]. Rodriguez-Estrada and colleagues [21] found synbiotic influence on hematocrit value and significant higher hematocrit value was recorder in the feed added feed additives while compare control group. Serum biochemical parameters evaluated by Ye et al. [107], reported the following results: The triglyceride level was lower or tended to be lower in fish fed the MOS, FOS and/or B. clausii-containing diets as opposed to the control diet. Dietary administration of commercial synbiotic in rainbow trout resulted in an increase of total serum protein content but other recorded parameters such as albumin/globulin ratio and triglycerides in fish fed with different levels of synbiotic did not show any significant difference in comparison with that from the control group at the end of the experiment [109]. In addition, diets containing synbiotic had no significant effect on total serum protein [118] [119]. Ashayerizadeh [120] and Mokhtari [121] found enhanced serum cholesterol level in broiler chickens in response to synbiotics supplementation and Shams [122] and Sharifi [123] also reported better hematological and cholesterol picture in broiler chickens when synbiotic was given in the diet. Liong and Shah [124] presented that the use of synbiotic consumption in broilers regulates the concentration of the organic acids and decrease cholesterol levels. Also,

Sahin et al. [125] reported that adding a synbiotic supplements in the diet had a significant effect on blood parameters, including total serum protein and cholesterol. Uric acid, creatinine, Packed cell volume and urea concentration were increased significantly in birds' supplemented synbiotic compared with control group [112].

11. SYNBIOTIC AND HEALTH

The gut-associated lymphoid tissue and gut microbiota are fundamental components of the both digestive and immune system function and homeostasis. Microbes of the GIT can be generally separated into potentially pathogenic or beneficial groups. Harmful microorganisms may be involved in localized or systemic infections, toxin formation, and intestinal putrefaction. Some intestinal organisms may have useful effects such as vitamin production, stimulation of the immune system through nonpathogenic mechanisms, and inhibition of the growth and establishment of harmful microbial groups [126]. Antimicrobial action of Probiotic (Bifidobacterium sp), Prebiotics (chicory roots) and Inulin and synbiotic: (Bifidobacterium sp + chicory) and (Bifidobacterium sp + Inulin) against Pathogenic bacteria was tested by using agar diffusion assay [127]. The synergistic inhibitory influence of Synbiotic (Bifidobacterium sp + inulin) and (Bifidobacterium sp + chicory) on Pathogenic bacteria was higher than the effect of Bifidobacterium sp alone, chicory alone and inulin alone [128]. The microbiota of the GIT of mammals can be considered a metabolically active organ with its wide biodiversity in term of species and the high number of cells [129] - [131]. The positive effects of both probiotic and prebiotic on immune system in different animal strains have been confirmed [91] [132] [133]. Supplementation of synbiotic had influence blood metabolites, both non-esterifies fatty acids and total immunoglobulin of in sheep [103]. Lysozyme activity: Lysozyme is one of the significant bactericidal enzymes of innate immunity, and constitutes an essential defense mechanism against pathogens in fish [134]. As synbiotic therapy may offer a suitable alternative for controlling pathogens, the effectiveness of synbiotic in terms of protection against infectious agents could be evaluated by a challenge test. To date, challenge test carried out in fish following to symbiotic administration, have employed Vibrio sp species as pathogens, specifically V. anguillarum and V. harveyi [100] [107] [108]. Today, specific health effects are being investigated and documented, including alleviation of chronic intestinal inflammatory diseases [135], and prevention and treatment of pathogen induced diarrhea [136], urogenital infections [137], and atopic diseases [138]. The combination of a probiotic and prebiotic as single administration, is called synbiotic which is characterized by antimicrobial, anticarcinogenic, immune stimulating actions and antiallergic. It also improves the absorption of minerals, protects from diarrhea and optimizes nutrient digestion processes [139]. Supplemental synbiotic from Lactobacillus sp reduced diarrhea, as well as increased feed efficiency and performance in initial weaning pigs [140].

12. CONCLUSION AND FUTURE PERSPECTIVE

From the research, we could conclude that the intestine was a variable niche of beneficial and harmful microorganisms. And in a different way they benefit or harm their host. The gastrointestinal tract has a compound community of microbiota that provides benefits to its host in many different methods, including nutrient production, protection against pathogens, detoxification, drug metabolism and regulation of the immune system. Animal studies have demonstrated changes in these gut microbial communities, and there are documents funding the use of probiotic, prebiotic and synbiotic. Though, the practice of synbiotic may possibly produce greater benefits rather than the application of individual portions. In recent years, synbiotic has become an integral part of the animal practices for improving the growth, enhancing blood picture, stimulating intestinal architecture, improving performance and competing against the pathogenic microbes. But limited information is available regarding the application of synbiotic in animal science. Experts recommend further research to substantiate preventive and therapeutic health benefits, mechanism of action, optimal intake, duration of treatment and selection of the best synbiotic (specific probiotic strains and type of prebiotic) for a targeted outcome in agriculture's animals.

REFERENCES

1. Alexander, R. (2013) Does a Child Die of Hunger Every 10 Seconds? BBC News Magazine, Accessed 15 October 2013. http://www.bbc.co.uk/news/magazine-22935692
2. Mack, M. (2009) Role of Technology Is Crucial in Improving Food Security. Syngenta CEO in USDA Outlook Forum, Washington DC and Syngenta International AG, Basel.
3. Fuller, R. (1989) A Review: Probiotics in Man and Animals. Journal of Applied Microbiology, 66, 365-378.http://dx.doi.org/10.1111/j.1365-2672.1989.tb05105.x
4. Kelly, D. (1998) Probiotics in Young and Newborn Animals. Journal of Animal and Feed Sciences, 7, 15-23.
5. Ko, S.Y. and Yang, C.J. (2008) Effect of Green Tea Probiotics on the Growth Performance, Meat Quality and Immune Response in Finishing Pigs. Asian-Australasian Journal of Animal Sciences, 21, 1339-1347.http://dx.doi.org/10.5713/ajas.2008.70597
6. Ko, S.Y., Bae, I.H., Yee, S.T., Lee, S.S., Uuganbayar, D.J., Oh, I. and Yang, C.J. (2008) Comparison of the Effect of Green Tea By-Product and Green Tea Probiotics on the Growth Performance, Meat Quality, and Immune Response of Finishing Pigs. Asian-

Australasian Journal of Animal Sciences, 21, 1486-1494. http://dx.doi.org/10.5713/ajas.2008.70604
7. Gibson, G.R. and Roberfroid, M.B. (1995) Dietary Modulation of the Human Colonic Microbiota: Introducing the Concept of Prebiotics. Journal of Nutrition, 125, 1401-1412.
8. Hamasalim, H.J. (2015) Oligosaccharides as Prebiotic. Health Journal, 1, 4-9.
9. Chou, T.-C., Rideout, D., Chou, J. and Bertino, J.R. (1991) Chemotherapeutic Synergism, Potentiation and Antagonism. In: Dulbecco, R., Ed., Encyclopedia of Human Biology, Vol. 2, Academic Press, San Diego, 371-379.
10. Nekoubin, H. and Sudagar, M. (2012) Assessment of the Effects of Synbiotic (Biomin Imbo) via Supplementation with Artificial Diet (with Different Protein Levels) on Growth Performance and Survival Rate in Grass Carp (Ctenopharyngodon Idella). World Journal of Zoology, 7, 236-240.
11. Haghighi, H.R., Gong, J., Gyles, C.L., Hayes, M.A., Sanei, B., Parvizi, P., Gisavi, H., Chambers, J.R. and Sharif, S. (2005) Modulation of Antibody-Mediate Immune Response by Probiotics in Chicken. Clinical and Diagnostic Laboratory Immunology, 12, 1387-1392.
12. Metzler, B., Bauer, E. and Mosenthin, R. (2005) Microflora Management in the Gastrointestinal Tract of Piglets. Asian-Australasian Journal of Animal Sciences, 18, 1353-1362. http://dx.doi.org/10.5713/ajas.2005.1353
13. Awad, W.A., Gharee, K. and Böhm, J. (2011) Evaluation of the Chicory Inulin Efficacy on Ameliorating the Intestinal Morphology and Modulating the Intestinal Electrophysiological Properties in Broiler Chickens. Journal of Animal Physiology and Animal Nutrition, 95, 65-72. http://dx.doi.org/10.1111/j.1439-0396.2010.00999.x
14. Awad, W., Ghareeb, K. and Böhm, J. (2008) Intestinal Structure and Function of Broiler Chickens on Diets Supplemented with a Synbiotic Containing Enterococcus faecium and Oligosaccharides. International Journal of Molecular Sciences, 9, 2205-2216. http://dx.doi.org/10.3390/ijms9112205
15. Vispo, C. and Karasov, W.H. (1997) The Interaction of Avian Gut Microbes and Their Host: An Exclusive Symbiosis. In: Mackie, R.J., White, B.A. and Issacson, R.E., Eds., Gastrointestinal Microbiology: Gastrointestinal Microbes and Host Interactions, Chapman and Hall, New York, 116-155. http://dx.doi.org/10.1007/978-1-4615-4111-0_5
16. Rehman, H., Vahjen, W., Awad, W.A. and Zentek, J. (2007) Indigenous Bacteria and Bacterial Metabolic Products in the Gastrointestinal Tract of Broilers. Archives of Animal Nutrition, 61, 319-335. http://dx.doi.org/10.1080/17450390701556817
17. Doley, S., Gupta, J.J. and Reddy, P.B. (2009) Effect of Supplementation of Ginger, Garlic and Turmeric in Broiler Chicken. Indian Veterinary Journal, 86, 644-645.
18. Merrifield, D.L., Dimitroglou, A., Foey, A., Davies, S.J., Baker, R.T.M. and Bogwald, J. (2010) The Current Status and Future Focus of Probiotic and Prebiotic Applications for Salmonids. Aquaculture, 302, 1-18. http://dx.doi.org/10.1016/j.aquaculture.2010.02.007
19. Abdel-Fattah, F.A.I. and Fararh, K.M. (2009) Effect of Dietary Supplementation of Probiotic, Prebiotic and Synbiotic on Performance, Carcass Characteristics, Blood Picture and Some Biochemical Parameters in Broiler Chickens. Benha Veterinary Medical Journal, 20, 9-23.
20. Naji, S.A.H., Al-kassie, G.A., Al-Hadithi, M.F., Al-Hillali, A.H. and Jameel, Y.J. (2009) Poultry Health Management. Brochure No. 27. Iraqi Poultry Producer Association. (In Arabic)
21. Rodriguez-Estrada, U., Satoh, S., Haga, Y., Fushimi, H. and Sweetman, J. (2009) Effect of Single and Combined Supplementation of Enterococcus faecalis, Mannan Oligosaccharide and Polyhydrobutyric Acid on Growth Performance and Immune Response of Rainbow Trout Oncorhynchus mykiss. Aquaculture Science, 57, 609-617.
22. Ai, Q., Xu, H., Mai, K., Xu, W., Wang, J. and Zhang, W. (2011) Effects of Dietary Supplementation of Bacillus subtilis and Fructooligosaccharide on Growth Performance, Survival, Non-Specific Immune Response and Disease Resistance of Juvenile Large Yellow Croaker, Larimichthys crocea. Aquaculture, 317, 155-161. http://dx.doi.org/10.1016/j.aquaculture.2011.04.036
23. Zhang, Q., Tan, B., Mai, K., Zhang, W., Ma, H. and Ai, Q. (2011) Dietary Administration of Bacillus (B. licheniformis and B. subtilis) and Isomaltooligosaccharide Influences the Intestinal Microflora, Immunological Parameters and Resistance against Vibrio

alginolyticus in Shrimp, Penaeus japonicus (Decapoda: Penaeidae). Aquaculture Research, 42, 943-952. http://dx.doi.org/10.1111/j.1365-2109.2010.02677.x

24. Daniels, C., Merrifield, D., Boothroyd, D., Davies, S., Factor, J. and Arnold, K. (2010) Effect of Dietary Bacillus spp. and Mannan Oligosaccharides (MOS) on European Lobster (Homarus gammarus L.) Larvae Growth Performance, Gut Morphology and Gut Microbiota. Aquaculture, 304, 49-57. http://dx.doi.org/10.1016/ j.aquaculture. 2010.03.018
25. Quigley, E.M. (2010) Prebiotics and Probiotics; Modifying and Mining the Microbiota. Pharmacological Research, 61, 213-218. http://dx.doi.org/10.1016/j.phrs.2010.01.004
26. Laparra, J.M. and Sanz, Y. (2010) Interactions of Gut Microbiota with Functional Food Components and Nutraceuticals. Pharmacological Research, 61, 219-225. http://dx.doi.org/10.1016/j.phrs.2009.11.001
27. Aureli, P., Capursob, L., Castellazzi, A.M., Clerici, M., Giovanninie, M., Morellif, L., Poli, A., et al. (2011) Probiotics and Health: An Evidence-Based Review. Pharmacological Research, 63, 366-376. http://dx.doi.org/10.1016/j.phrs.2011.02.006
28. ones, B.V. and Marchesi, J.R. (2007) Accessing the Mobile Metagenome of the Human Gut Microbiota. Molecular BioSystems, 3, 749-758. http://dx.doi.org/10.1039/b705657e
29. Hooper, L.V. and Macpherson, A.J. (2010) Immune Adaptations That Maintain Homeostasis with the Intestinal Microbiota. Nature Reviews Immunology, 10, 159-169. http://dx.doi.org/10.1038/nri2710
30. Kamada, N., Seo, S.U., Chen, G.Y. and Nú&ntildeez, G. (2013) Role of the Gut Microbiota in Immunity and Inflammatory Disease. Nature Reviews Immunology, 13, 321-335. http://dx.doi.org/10.1038/nri3430
31. Lebeer, S., Vanderleyden, J. and De Keersmaecker, S.C. (2008) Genes and Molecules of Lactobacilli Supporting Probiotic Action. Microbiology and Molecular Biology Reviews, 72, 28-764. http://dx.doi.org/10.1128/MMBR.00017-08
32. Berg, R.D. (1996) The Indigenous Gastrointestinal Microflora. Trends in Microbiology, 4, 430-435. http://dx.doi.org/10.1016/0966-842X(96)10057-3
33. Cummings, J.H., Gibson, G.R. and Macfarlane, G.T. (1989) Quantitative Estimates of Fermentation in the Hind Gut of Man. Acta Veterinaria Scandinavica. Supplementum, 86, 76-82.
34. Guarner, F. and Malagelada, J.R. (2003) Gut Flora in Health and Disease. Lancet, 361, 512-519. http://dx.doi.org/10.1016/S0140-6736(03)12489-0
35. Backhed, F., Ding, H., Wang, T., Hooper, L.V., Koh, G.Y., Nagy, A., et al. (2004) The Gut Microbiota as an Environmental Factor That Regulates Fat Storage. Proceedings of the National Academy of Sciences of the United States of America, 101, 15718-15723. http://dx.doi.org/10.1073/pnas.0407076101
36. Ley, R.E., Peterson, D.A. and Gordon, J.I. (2006) Ecological and Evolutionary Forces Shaping Microbial Diversity in the Human Intestine. Cell, 124, 837-848. http://dx.doi.org/10.1016/j.cell.2006.02.017
37. Whitman, W.B., Coleman, D.C. and Wiebe, W.J. (1998) Prokaryotes: The Unseen Majority. Proceedings of the National Academy of Sciences of the United States of America, 95, 6578-6583. http://dx.doi.org/10.1073/pnas.95.12.6578
38. Vyas, U. and Ranganathan, N. (2012) Probiotics, Prebiotics, and Synbiotics: Gut and Beyond. Gastroenterology Research and Practice, 2012, Article ID: 872716. http://dx.doi.org/10.1155/2012/872716
39. H&eacutebuterne, X. (2003) Gut Changes Attributed to Ageing: Effects on Intestinal Microflora. Current Opinion in Clinical Nutrition & Metabolic Care, 6, 49-54. http://dx.doi.org/10.1097/00075197-200301000-00008
40. Woodmansey, E.J. (2007) Intestinal Bacteria and Ageing. Journal of Applied Microbiology, 102, 1178-1186. http://dx.doi.org/10.1111/j.1365-2672.2007.03400.x Dethlefsen, L., Eckburg, P.B., Bik, E.M. and Relman, D.A. (2006) Assembly of the Human Intestinal Microbiota. Trends in Ecology & Evolution, 21, 517-523. http://dx.doi.org/10. 1016/ j.tree.2006.06.013
41. Savage, D. (1977) Microbiology of the Gastrointestinal Tract. Annual Review of Microbiology, 31, 107-133. http://dx.doi.org/10.1146/annurev.mi.31.100177.000543

42. Hamasalim, H.J. (2015) The Impact of Some Widely Probiotic (Iraqi Probiotic) on Health and Performance. Journal of Biosciences and Medicines, 3, 25-36. http://dx.doi.org/10.4236/jbm.2015.38003
43. Chauhan, S.V. and Chorawala, M.R. (2012) Probiotics, Prebiotics and Synbiotics. International Journal of Pharmaceutical Sciences and Research, 3, 711-726.
44. Eamonn, M. and Quigley, M. (2010) Prebiotics and Probiotics; Modifying and Mining the Microbiota. Pharmacological Research, 61, 21218. http://dx.doi.org/10.1016/j.phrs.2010.01.004
45. Gismondo, M.R., Drago, L. and Lombardi, A. (1999) Review of Probiotics Available to Modify Gastrointestinal Flora. International Journal of Antimicrobial Agents, 12, 287-292. http://dx.doi.org/10.1016/S0924-8579(99)00050-3
46. Guarner, F., Perdigon, G., Corthier, G., Salminen, S., Koletzko, B. and Morelli, L. (2005) Should Yoghurt Cultures Be Considered Probiotic? British Journal of Nutrition, 93, 783-786. http://dx.doi.org/10.1079/BJN20051428
47. Akir, I.C. (2003) Determination of Some Probiotic Properties on Lactobacilli and Bifidobacteria. PhD Thesis, Ankara University, Ankara.
48. Chuayana Jr., E.L., Ponce, C.V., Rivera, M.R.B. and Cabrera, E.C. (2003) Antimicrobial Activity of Probiotics from Milk Products. The Philippine Journal of Microbiology and Infectious Diseases, 32, 71-74.
49. Sekhon, B.S. and Jairath, S. (2010) Prebiotics, Probiotics and Synbiotics: An Overview. Journal of Pharmaceutical Education and Research, 1, 13-36.
50. FAO and WHO (Food and Agriculture Organization of the United Nations and World Health Organization Report) (2007) Probiotic. http://www.fao.org/es/ESN/ probio/probio.html
51. Hickey, L., Jacobs, S.E. and Garland, S.M. (2012) Probiotics in Neonatology. Journal of Paediatrics and Child Health, 48, 777-783. http://dx.doi.org/10.1111/j.1440-1754.2012.02508.x
52. Kumar, A.V. (2013) Probiotics: Nature's Medicine. International Journal of Nutrition, Pharmacology, Neurological Diseases, 3, 219-228. http://www.ijnpnd.com
53. Brown, A.C. and Valiere, A. (2004) Probiotics and Medical Nutrition Therapy. Nutrition in Clinical Care, 7, 56-68.
54. Anandharaj, M., Sivasankari, B. and Rani, R.P. (2014) Effects of Probiotics, Prebiotics, and Synbiotics on Hypercholesterolemia: A Review. Chinese Journal of Biology, 2014, Article ID: 7572754.
55. Roberfroid, M.B. (2000) Prebiotics and Probiotics: Are They Functional Foods? American Journal of Clinical Nutrition, 71, 1682S-1687S.
56. Gibson, G.R., Beatty, E.R., Wang, X. and Cummings, J.H. (1995) Selective Stimulation of Bifidobacteria in the Human Colon by Oligofructose and Inulin. Gastroenterology, 108, 975-82. http://dx.doi.org/10.1016/0016-5085(95)90192-2
57. Patel, P.J., Singh, S.K., Panaich, S. and Cardozo, L. (2012) The Aging Gut and the Role of Prebiotics, Probiotics, and Synbiotics: A Review. Journal of Clinical Gerontology & Geriatrics, 5, 3-6. http://dx.doi.org/10.1016/j.jcgg.2013.08.003
58. Crittenden, R. and Payne, M.J. (2008) Nutrition News. Facts and Functions of Prebiotics, Probiotics and Synbiotics. May 2008, pp. 1-2. Department of Human Nutrition, K-State Research and Extension, Kansas State University; Prebiotics. In: Lee, Y.K. and Salminen, S., Eds., Handbook of Probiotics and Prebiotics, 2nd Edition, Chap. 4, Wiley-Interscience, Hoboken, 535-582.
59. Hamasalim, H.J. (2012) Inulin. Zanisti Sardam, 47, 175-178.
60. Venter, C.S. (2007) Prebiotics: An Update. Journal of Family Ecology and Consumer Sciences, 35, 17-25.
61. Lynn, S.S. (2010) Why "Biotics"? Using Prebiotics and Probiotics in Your Practice. Practical Applications for Achieving Gastintestinal Wellness Conference Proceeding 2010, Supplement to Compendium: Continuing Education for Veterinarians.
62. Williams, N.T. (2010) Probiotics. American Journal of Health-System Pharmacy, 67, 449-458. http://dx.doi.org/10.2146/ajhp090168

63. Hamasalim, H.J. (2009) The Effect of Different Levels of Feeding on Karadi Lambs Response to Local Iraqi Probiotics. Master's Thesis, University of Sulaimani, Sulaimani.
64. Vulevic, J., Drakoularakou, A., Yaqoob, P., Tzortzis, G., Gibson, G.R. (2008) Modulation of the Fecal Microflora Profile and Immune Function by a Novel Transgalactooligosaccharide Mixture (B-GOS) in Healthy Elderly Volunteers. The American Journal of Clinical Nutrition, 88, 1438-1446.
65. Schiffrin, E.J., Kumar, V.B., Brown, C., Hager, C., Van't Hof, M.A., Morley, J.E., et al. (2007) Systemic Inflammatory Markers in Older Persons: The Effect of Oral Nutritional Supplementation with Prebiotics. The Journal of Nutrition Health and Aging, 11, 475-479.
66. Ahmed, M., Prasad, J., Gill, H., Stevenson, L. and Gopal, P. (2007) Impact of Consumption of Different Levels of Bifidobacterium lactis HN019 on the Intestinal Microflora of Elderly Human Subjects. The Journal of Nutrition Health and Aging, 11, 26-31.
67. Bartosch, S., Woodmansey, E.J., Paterson, J.C., McMurdo, M.E. and Macfarlane, G.T. (2005) Microbiological Effects of Consuming a Synbiotic Containing Bifidobacterium bifidum, Bifidobacterium lactis, and Oligofructose in Elderly Persons, Determined by Real-Time Polymerase Chain Reaction and Counting of Viable Bacteria. Clinical Infectious Diseases, 40, 28-37. http://dx.doi.org/10.1086/426027
68. De Preter, V., Hamer, H.M., Windey, K. and Verbeke, K. (2011) The Impact of Pre- and/or Probiotics on Human Colonic Metabolism: Does It Affect Human Health? Molecular Nutrition & Food Research, 55, 46-57. http://dx.doi.org/10.1002/mnfr.201000451
69. Demigné, C., Jacobs, H., Moundras, C., Davicco, M.J., Horcajada, M.N., Bernalier, A., et al. (2008) Comparison of Native or Reformulated Chicory Fructans, or Non-Purified Chicory, on Rat Cecal Fermentation and Mineral Metabolism. European Journal of Nutrition, 47, 366-374. http://dx.doi.org/10.1007/s00394-008-0736-5
70. Bengmark, S. (2005) Bioecological Control of the Gastrointestinal Tract: The Role of Flora and Supplemented Probiotics and Synbiotics. Gastroenterology Clinics of North America, 34, 413-436.
71. Bengmark, S. and Martindale, R. (2005) Prebiotics and Synbiotics in Clinical Medicine. Nutrition in Clinical Practice, 20, 244-261. http://dx.doi.org/10.1177/ 0115426505020002244
72. Rioux, K.P., Madsen, K.L. and Fedorak, R.N. (2005) The Role of Enteric Microflora in Inflammatory Bowel Disease: Human and Animal Studies with Probiotics and Prebiotics. Gastroenterology Clinics of North America, 34, 465-482. http://dx.doi.org/10.1016/j.gtc.2005.05.005
73. Panesar, P.S., Kaur, G., Panesar, R. and Bera, M.B. (2009) Synbiotics: Potential Dietary Supplements in Functional Foods. Food Science Central. http://scialert.net/fulltext/?doi=crds.2012.17.23&org=12
74. De Vrese, M. and Schrezenmeir, J. (2008) Probiotics, Prebiotics, and Synbiotics. In: Stahl, U., Donalies, U.E.B. and Nevoigt, E., Eds., Food Biotechnology, Advances in Biochemical Engineering/Biotechnology, Vol. 111, Springer, Berlin, 1-66. http://dx.doi.org/10.1007/10_2008_097
75. Harish, K. and Varghese, T. (2006) Probiotics in Humans—Evidence Based Review. Calicut Medical Journal, 4, e3.
76. Cecic, A. and Chingwaru, W. (2010) The Role of Functional Foods, Nutraceuticals, and Food Supplements in Intestinal Health. Nutrients, 2, 611-625. http://dx.doi.org/10.3390/nu2060611
77. Geier, M.S., Butler, R.N. and Howarth, G.S. (2007) Inflammatory Bowel Disease: Current Insights into Pathogenesis and New Therapeutic Options; Probiotics, Prebiotics and Synbiotics. International Journal of Food Microbiology, 115, 1-11. http://dx.doi.org/10.1016/j.ijfoodmicro.2006.10.006
78. Gourbeyre, P., Denery, S. and Bodinier, M. (2011) Probiotics, Prebiotics, and Synbiotics: Impact on the Gut Immune System and Allergic Reactions. Journal of Leukocyte Biology, 89, 685-695. http://dx.doi.org/10.1189/jlb.1109753
79. Scavuzzi, B.M., Henrique, F.C., Miglioranza, L.H.S., Sim ão, A.N.C. and Dichi, I. (2014) Impact of Prebiotics, Probiotics and Synbiotics on Components of the Metabolic Syndrome. Annals of Nutritional Disorders & Therapy, 1, 1009.

80. Food and Agriculture Organization of the United Nations; World Health Organization (FAO/WHO) (2001) Evaluation of Health and Nutritional Properties of Probiotics in Food Including Powder Milk with Live Lactic Acid Bacteria. Report of a Joint FAO/WHO Expert Consultation, Córdoba, Argentina.ftp://ftp.fao.org/docrep/fao/009/a0512e/ a0512 e00.pdf
81. Vershuere, L., Rombaut, G., Sorgeloos, P. and Verstraete, W. (2000) Probiotic Bacteria as Biological Control Agents in Aquaculture. Microbiology and Molecular Biology Reviews, 64, 655-671. http://dx.doi.org/10.1128/MMBR.64.4.655-671.2000
82. Fuller, R. (Ed.) (1992) Probiotics: The Scientific Basis. Chapman & Hall, London. http://dx.doi.org/10.1007/978-94-011-2364-8
83. Duggan, C., Gannon, J. and Walker, W.A. (2002) Protective Nutrients and Functional Foods for the Gastrointestinal Tract. The American Journal of Clinical Nutrition, 75, 789-808.
84. Roberfroid, M. (2007) Prebiotics: The Concept Revisited. Journal of Nutrition, 137, 830-837.
85. Batvani, R. (2009) Synbiotics, the Combination of Probiotics, Prebiotics and Immune System Stimulants. Farda Etouk Research Group. http://www.roavs.com/pdf-files/Issue_1_2012/10-13.pdf
86. Medzhitov, R. and Janeway, C. (2000) Innate Immunity. The New England Journal of Medicine, 343, 338-344. http://dx.doi.org/10.1056/NEJM200008033430506
87. Huilan, S., Zhen, L.G., Mathan, M.M., Mathew, M.M., Olarte, J., Espejo, R., Khin Maung, U., Ghafoor, M.A., Khan, M.A., Sami, Z. and Sutton, R.G. (1991) Etiology of Acute Diarrhoea among Children in Developing Countries: A Multicentre Study in Five Countries. Bulletin of the World Health Organization, 69, 549-555.
88. Naidu, A.S., Bidlack, W.R. and Clemens, R.A. (1999) Probiotic Spectra of Lactic Acid Bacteria (LAB). Critical Reviews in Food Science and Nutrition, 39, 13-126. http://dx.doi.org/10.1080/10408699991279187
89. Awad, W.A., Ghareeb, K., Abdel-Raheem, S. and Böhm, J. (2009) Effects of Dietary Inclusion of Probiotic and Symbiotic on Growth Performance, Organ Weights, and Intestinal Histomorphology of Broiler Chickens. The Journal of Poultry Science, 88, 49-55. http://dx.doi.org/10.3382/ps.2008-00244
90. Provenza, F.D. and Villalba, J.J. (2010) The Role of Natural Plant Products in Modulating the Immune System: An Adaptable Approach for Combating Disease in Grazing Animals. Small Ruminant Research, 89, 131-139. http://dx.doi.org/10.1016/ j.smallrumres. 2009.12.035
91. Vandenplas, Y., De Greef, E., Devreker, T., Veereman-Wauters, G. and Hauser, B. (2013) Probiotics and Prebiotics in Infants and Children. Current Infectious Disease Reports, 15, 251-262. http://dx.doi.org/10.1007/s11908-013-0334-4
92. Boirivant, M. and Strober, W. (2007) The Mechanism of Action of Probiotics. Current Opinion in Gastroenterology, 23, 670-692. http://dx.doi.org/10.1097/ MOG.0b013e3282 f0cffc
93. Lee, S.J., Shin, N.H., Ok, J.U., Jung, H.S., Chu1, G.M., Kim, J.D., et al. (2009) Effects of Dietary Synbiotics from Anaerobic Microflora on Growth Performance, Noxious Gas Emission and Fecal Pathogenic Bacteria Population in Weaning Pigs. Asian-Australasian Journal of Animal Sciences, 22, 1202-1208. http://dx.doi.org/10.5713/ajas.2009.90045
94. Smith, H.W. and Jones, J.E.T. (1963) Observations on the Alimentary Tract and Its Bacterial Flora in Healthy and Diseased Pigs. The Journal of Pathology and Bacteriology, 86, 387-412. http://dx.doi.org/10.1002/path.1700860214
95. Underdahl, N.R., Torres-Median, A. and Doster, A.R. (1982) Effect of Streptococcus faecium C-68 in the Control of Escherichia coli-Induced Diarrhea in Genotociotic Pigs. American Journal of Veterinary Research, 43, 2227-2232.
96. Abdel-Raheem, S.M., Abd-Allah, S.M.S. and Hassanein, K.M.A. (2012) The Effects of Prebiotic, Probiotic and Synbiotic Supplementation on Intestinal Microbial Ecology and Histomorphology of Broiler Chickens. International Journal for Agro Veterinary and Medical Sciences, 6, 277-289.

97. Dibaji, S.M., Seidavi, A., Asadpour, L. and da Silva, F.M. (2012) Effect of a Synbiotic on the Intestinal Microflora of Chickens. The Journal of Applied Poultry Research, 23, 1-6. http://dx.doi.org/10.3382/japr.2012-00709
98. Hays, V.W. (1977) Effectiveness of Feed Additive Usage of Antimicrobial Agents in Swine and Poultry Production. Office of Technology Assessment, Washington DC. Witte, W. (2000) Selective Pressure by Antibiotic Use in Livestock. International Journal of Antimicrobial Agents, 16, S19-S24. http://dx.doi.org/10.1016/s0924-8579(00)00301-0
99. Ra, J.C., Han, H.J. and Song, J.E. (2004) Effect of Probiotics on Production and Improvement of Environment in Pigs and Broilers. Korean Journal of Veterinary Public Health, 28, 157-167.
100. Nekoubin, H., Gharedaashi, E., Imanpour, M., Nowferesti, H. and Asgharimoghadam, A. (2012) The Influence of Synbiotic (Biomin imbo) on Growth Factors and Survival Rate of Zebrafish (Danio rerio) Larvae via Supplementation with Biomar. Global Veterinaria, 8, 503-506.
101. Kazemi-Bonchenar, M., Ghasemi, H.A., Khodaei-Motlagh, M., Khaltabadi-Farahani, A.H. and Ilani, M. (2013) Influence of Feeding Synbiotic Containing Enterococcus faecium and Inulin on Blood Metabolites, Nutrient Digestibility and Growth Performance in Sheep Fed Alfalfa-Based Diet. Scientific Research and Essays, 8, 853-857.
102. Daniels, C.L., Merrifield, D.L., Ringo, E. and Davies, S.J. (2013) Probiotic, Prebiotic, Synbiotic Applications for the Improvement of Larval European Lobster (Homarus gammarus) Culture. Aquaculture, 416-417, 396-406. http://dx.doi.org/10.1016/j.aquaculture.2013.08.001
103. Marlida, R., Suprayudi, M.A., Widanarni and Harris, E. (2014) Growth, Digestive Enzyme Activity and Health Status of Humpback Grouper (Cromileptes altivelis) Fed with Synbiotic. Pakistan Journal of Nutrition, 13, 319-326. http://dx.doi.org/10.3923/pjn.2014.319.326
104. Dehaghani, P.G., Baboli, M.J., Moghadam, A.T., Ziaei-Nejad, S. and Pourfarhadi, M. (2015) Effect of Synbiotic Dietary Supplementation on Survival, Growth Performance, and Digestive Enzyme Activittraies of Common Carp (Cyprinus carpio) Fingerlings. Czech Journal of Animal Science, 60, 224-232. http://dx.doi.org/10.17221/8172-CJAS
105. Ye, J.D., Wang, K., Li, F.D. and Sun, Y.Z. (2011) Single or Combined Effects of Fructo- and Mannan Oligosaccharide Supplements and Bacillus clausii on the Growth, Feed Utilization, Body Composition, Digestive Enzyme Activity, Innate Immune Response and Lipid Metabolism of the Japanese Flounder Paralichthys olivaceus. Aquaculture Nutrition, 17, 902-911. http://dx.doi.org/10.1111/j.1365-2095.2011.00863.x
106. Geng, X., Dong, X.H., Tan, B.P., Yang, Q.H., Chi, S.Y., et al. (2011) Effects of Dietary Chitosan and Bacillus subtilis on the Growth Performance, Non-Specific Immunity and Disease Resistance of Cobia, Rachycentron canadum. Fish and Shellfish Immunology, 31, 400-406. http://dx.doi.org/10.1016/j.fsi.2011.06.006
107. Mehrabi, Z., Firouzbakhsh, F. and Jafarpour, A. (2012) Effects of Dietary Supplementation of Synbiotic on Growth Performance, Serum Biochemical Parameters and Carcass Composition in Rainbow Trout (Oncorhynchus mykiss) Fingerlings. Journal of Animal Physiology and Animal Nutrition, 96, 474-481. http://dx.doi.org/10.1111/j.1439-0396.2011.01167.x
108. Jung, S.J., Houde, R., Baurhoo, B., Zhao, X. and Lee, B.H. (2008) Effects of Galacto-Oligosaccharides and a Bifidobacteria lactis-Based Probiotic Strain on the Growth Performance and Fecal Microflora of Broiler Chickens. Poultry Science, 87, 1694-1699. http://dx.doi.org/10.3382/ps.2007-00489
109. Erdogan, Z., Erdogan, S., Aslantas, O. and Celik, S. (2010) Effects of Dietary Supplementation of Synbiotics and Phytobiotics on Performance, Caecal Coliform Population and Some Oxidant/Antioxidant Parameters of Broilers. Journal of Animal Physiology and Animal Nutrition, 94, e40-e48. http://dx.doi.org/10.1111/j.1439-0396.2009.00973.x
110. Ahmed, K.S., Hasan, M., Asaduzzaman, M., Khatun, A. and Islam, K. (2015) Effects of Probiotics and Synbiotics on Growth Performance and Haemato-Biochemical Parameters in Broiler Chickens. Journal of Science, 5, 926-929.

111. Fallah, R., Fosoul S.S.A.S. and Rezaei, H. (2014) Effect of Synbiotic on Performance and Serum Biochemical Parameters of Ostrich Chicks. Journal of Farm Animal Nutrition and Physiology, 9/1, 51-56.
112. Grimes, J.L., Maurice, D.V., Lightsey, S.F. and Lopez, J.G. (1997) The Effect of Dietary Fermacto on Layer Hen Performance. The Journal of Applied Poultry Research, 6, 399-403. http://dx.doi.org/10.1093/japr/6.4.399
113. klebaniuk, R. and Czech, A. (2007) The Influence of Synbiotic Participation in Feed Ration for Ewes on Selected Lamb Blood Parameters. Annales, 2, 1-5.
114. Al-Kassi, G.A.M. (2009) Influence of Plant Exteracts Derived from Thyme and Cinnamon on Broiler Performance. Pakistan Veterinary Journal, 29, 169-173.
115. Farinu, G.O., Ademola, S.G., Ajayi Obe, A.O. and Babatunde, G.M. (2004) Growth, Haematological and Biochemical Studies on Garlic and Ginger-Fed Broiler Chickens. Moor Journal of Agricultural Research, 5, 122-128
116. Ashayerzadeh, A., Dabiri, N., Mirzadeh, K.H. and Ashayerzadeh, A. (2010) The Comparison between Antibiotics, Probiotics and Prebiotics on Growth Response and Some Blood Parameters in 42 Days Old Broiler. Proceedings of 4th Animal Science Congress, Karaj, 786-789.
117. Mokhtari, R., Yazdani, A.R., Rezaee, M. and Ghorbani, B. (2009) The Effect of Synbiotics and Phytobiotic on Broiler Blood Parameters. Proceedings of 4th Animal Science Congress, Karaj, 142-145.
118. Shams, M., Azadeghan Mehr, M., Dastar, B. and Hosseini, S. (2008) Effect of Different Levels of Protein and Probiotics on Production Traits and Some Blood Parameters in Broiler. Journal of Agriculture Science, 14, 15.
119. Sharifi, M.R., Shams, M., Dastar, B. and Hosseini, S. (2009) Effect of Different Levels of Synbiotics Supplementation on Performance and Some Blood Parameters in Japanese Quail. Proceedings of 4th Animal Science Congress, Karaj, 1-4.
120. Liong, M.T. and Shah, N.P. (2006) Effects of a Lactobacillus casei Symbiotic on Serum Lipoprotein, Intestinal Microflora and Organic Acid in Rats. Journal of Dairy Science, 89, 1390-1399. http://dx.doi.org/10.3168/jds.S0022-0302(06)72207-X
121. Sahin, T., Kaya, I., Unal, Y. and Emali, D.A. (2008) Dietary Supplementation of Probiotics and Prebiotics Combination on Performance, Carcass Quality and Blood Parameters in Growing Quail. Journal of Animal and Veterinary, 7, 1370-1373.
122. Min-Tze, L., Dunshea, F.R. and Shah, N.P. (2007) Effect of Synbiotic Containing Lactobacillus acidophilus ATCC 4962 on Plasma Lipid Profiles and Morphology of Erythrocytes in Hypercholestrolaemic Pigs on High- and Low-Fat Diets. British Journal of Nutrition, 98, 736-744.
123. Zhang, W.F., Li, D.F., Lu, W.Q. and Yi, G.F. (2003) Effect of Isomaltooligosaccharides on Broiler Performance and Intestinal Micro Flora. Poultry Science, 82, 657-663. http://dx.doi.org/10.1093/ps/82.4.657
124. Jeurissen, S.H., Lewis, F., Van der Klis, J.D., Mroz, Z., Rebel, J.M. and Ter Huurne, A.A. (2002) Parameters and Techniques to Determine Intestinal Health of Poultry as Constituted by Immunity, Integrity, and Functionality. Current Issues in Intestinal Microbiology, 3, 1-14.
125. Ibrahim, S.A. and Salameh, M.M. (2001) Simple Rapid Method for Screening Antimicrobial Activities of Bifidobacterium sp of Human Isolates. Rapid Methods and Automation in Microbiology, 9, 52-63.
126. Salman, J.A.S. (2009) Synbiotic Effect of Probiotic (Bifidobacterium sp) and Prebiotics (Chicory and Inulin) against Some Pathogenic Bacteria. Um-Salama Science Journal, 6, 355-360.
127. Murphy, W.J., Larkin, D.M., van der Wind, A.E., Bourque, G., Tesler, G., Auvil, L., Beever, J.E., Chowdhary, B.P., Galibert, F., Gatzke, L., Hitte, C., Meyers, S.N., Milan, D., Ostrander, E.A., Pape, G., Parker, H.G., Raudsepp, T., Rogatcheva, M.B., Schook, L.B., Skow, L.C., Welge, M., Womack, J.E., O'brien, S.J., Pevzner, P.A., Lewin, H.A. (2009) Dynamics of Mammalian Chromosome Evolution Inferred from Multispecies Comparative Maps. Science, 309, 613-618. http://dx.doi.org/10.1126/science.1111387

128. Backhed, F., Ley, R.E., Sonnenburg, J.L., Peterson, D.A. and Gordon, J.I. (2005) Host-Bacterial Mutualism in the Human Intestine. Science, 307, 1915-1920. http://dx.doi.org/10.1126/science.1104816
129. Macfarlane, S., Macfarlane, G.T. (2004) Bacterial Diversity in the Human Gut. Advances in Applied Microbiology, 54, 261-289. http://dx.doi.org/10.1016/S0065-2164(04)54010-8
130. Benyacoub, J., Maulden, G.L.C., Cavadini, C., Sauthier, T., Anderson, R.E., Schiffrin, E.J. and Weid, T.V.D. (2003) Supplementation of Food with Enterococcus faecium (SF68) Stimulates Immune Functions in Young Dogs. Journal of Nutrition, 133, 1158-1162.
131. Newman, K. (1994) Mannan-Oligosaccharides: Natural Polymers with Significant Impact on the Gastrointestinal Microflora and the Immune System. In: Lyons, T.P. and Jacques, K.A., Eds., Biotechnology in the Feed Industry. Proceeding of Alltech's Tenth Annual Symposium, Nottingham University Press, Nottingham, 167-175.
132. Lindsay, G.J.H. (1986) The Significance of Chitinolytic Enzymes and Lysozyme in Rainbow Trout (Salmo gairdneri) Defence. Aquaculture, 51, 169-173. http://dx.doi.org/10.1016/0044-8486(86)90306-6
133. Mach, T. (2006) Clinical Usefulness of Probiotics in Inflammatory Bowel Diseases. Journal of Physiology and Pharmacology, 57, 23-33.
134. Yan, F. and Polk, D.B. (2006) Probiotics as Functional Food in the Treatment of Diarrhea. Current Opinion in Clinical Nutrition & Metabolic Care, 9, 717-721. http://dx.doi.org/10.1097/01.mco.0000247477.02650.51
135. Reid, G. (2008) Probiotic Lactobacilli for Urogenital Health in Women. Journal of Clinical Gastroenterology, 42, S234-S236. http://dx.doi.org/10.1097/mcg.0b013e31817f1298
136. Vanderhoof, J.A. (2008) Probiotics in Allergy Management. Journal of Pediatric Gastroenterology and Nutrition, 47, S38-S40. http://dx.doi.org/10.1097/ 01.mpg.0000338810.74933.c1
137. Gruzauskas, R., Lekavicius, R., Raceviciute-Stupeliene, A., Sasyte, V. and Guarner, F. (2004) Inulin and Oligofructose: Impact on Intestinal Diseases and Disorders. British Journal of Nutrition, 93, 61-65.
138. Pollman, D.S. (1986) Additives, Flavors, Enzymes and Probiotics in Animal Feeds. Proceedings of the 22nd Annual Nutrition Conference, University of Guelph.

CHAPTER 5

Prevention and Control of Production Limiting Goat Diseases on Small Farms

Marshall RW*, Gebrelul S and Adams C

Southern University Agricultural Research and Extension Center, College of Agriculture, Baton Rouge, LA 70813, Louisiana, United States

ABSTRACT

This article was written in support of development of new strategies to control economically important diseases of goats, and advance the body of knowledge of agricultural professionals and small farmers on disease surveillance and biosecurity for farm operations. The goat industry has been looking for more effective ways to alleviate health problems for a long time. There is an evolving need for scientists, agricultural professionals with expertise in small ruminants, along with the input from goat producers, to develop strategies tailored for specific disease prevalence that affects sustainable small ruminant productions.

Keywords: Goat production; Disease surveillance; Biosecurity; Investigation; Health

1. INTRODUCTION

The US livestock industry generates $100 billion annually. Serious economic impacts are expected if animal diseases are introduced, suspected or prevalent. Each small ruminant herd's productivity is directly impacted by the health of the animals in the operation. Not only can diseases reduce efficiency, illness outbreaks are detrimental to animal welfare and consumer confidence. Control measures largely depend on the knowledge base of small ruminant producers for their success. New approaches and enhancements are needed to improve goat herd health and productivity. We at the Southern University Ag Center have learned a great deal and we understand what is needed to enhance our investigative tools and incorporate new approaches to make small livestock systems work more efficiently.

Traditional veterinary medicine consisting of healing, curing and treating of single animals or entire herds has lost its importance and will more and more be replaced by preventive measures and health management. The key to success in handling animal diseases is early detection. If a disease can be detected in an early phase of its development, the possibility exists that it can be arrested and eliminated before it actually inflicts damage. Early detection presupposes that there is a surveillance system in place that will bring infection to light when it is first seen. Veterinarians, agricultural professionals and producers are then able to manage the problem before it becomes uncontrollable, thus protecting the local livestock industry and ensuring food security for those closely dependent upon livestock. Early detection enables early warning and an early reaction.

2. GOAT INDUSTRY

The USDA defines a small farm with $250000 or less in farm sales. In Louisiana 84.5% of the sales is $50,000 or less and only 6.7% of the farms have $500,000 or more in sales. Until recently, the meat goat industry was the fastest growing class of livestock in the U.S.A. The number of meat goat farms almost doubled from 63,000 in 1997 to about 123,000 in 2007 but declined to 100,910 in 2012 [1-3]. There were about 2,550,000 meat goats in 2009, a 23% increase since 2002 [4]. More than 71% of the U.S. meat goat population is in the South [5]. Most of the recent growth in the goat industry has occurred in Hawaii, Oklahoma, Louisiana, New Mexico and Idaho [1,2,4,6]. According to the Louisiana Agricultural Summary (2008) for the period between 1998 and 2008, the number of does increased from about 4,000 to 16,700, and goat producers increased from 400 to over 1,200 [7].

As an indicator of goat meat consumption, more than 10,763,842 million kg of frozen goat meat was imported from Australia and New Zealand into the U.S. in 2008. Concurrently, the domestic harvest of goats in federal and state plants in the US has also increased with more than 670,709 head in federally inspected and 195,100 head in other (state-inspected) plants [5].

For the period between1997 to about 2007, the number of goats sold increased from 533,000 to 1,235,000, an increase of 232%. Income from goat sales as a fraction of gross animal farm income in Louisiana averaged $1.35/$1,000.

Parasites in goats or lack of control therein, are forcing some goat producers to re-think about possible diversification to or inclusion of lamb production in their operations. As of Jan. 1, 2013, there were 5.3 million head of sheep in the United States. Forty-five percent (2.4 million) of these sheep are located in Texas, California, Colorado, Wyoming and Utah. In 2012, there were 80,000 sheep operations in the United States. The five states with most sheep operations are Texas (8,700), Arizona (5,000), California (4,100), Pennsylvania (3,800) and Iowa (3,500). The number of sheep producers in Louisiana ranged from high of 330 in 2010 to a low of 208 in 2014 [8]. The corresponding number for goats was 876 and 596, respectively. The number of breeding ewes fluctuated from 2449 to 3835 between 2010 and 2014. Does numbers decreased from 14,242 in 2010 to 6578 in 2014. Much of the decline could be contributed to parasite and foot rot problems. Total farm value for sheep was $2.6 million as compared to $3.3 million for goats. Sheep and goats account for only 0.10% of all income derived from livestock in Louisiana.

Economic evaluations have shown that major losses due to parasitism have been associated with decreased animal production, cost of prevention and treatment, and ultimately death of infected animals [9,10]. In Ghana, the average producer loses up to 34% of their small ruminant herd to diseases per year [11]. For a small farmer with an average size herd of 50 animals, this represents a significant loss of more than 15 animals per year. Considerable costs may be incurred in controlling animal diseases, though the control measures paid for may not always be effective. This is particularly true for small farmers, who often lack information and have limited diagnostic data to make disease-control and treatment decisions. The US livestock industry generates $100 billion annually. Serious economic impacts are expected if animal diseases are introduced, prevalent or suspected.

3. IMPORTANCE OF DISEASE SURVEILLANCE

Disease surveillance will help determine the most important diseases in farm operations and will present clinical approaches that can improve detection, diagnosis, and treatment of herd-based problems [12]. Surveillance is the primary key to effective disease management. For small farmers, gastrointestinal nematodes, foot rot, and caseous lymphadenitis (CL) are a huge production limiting problem.

Because of warm, humid conditions during the summer months, GINs are the greatest health and production challenge for goats in southeastern states [12,13]. Of the GINs, *H. contortus* is the most detrimental because of substantial economic losses. *H. contortus* or barber pole worm thrives in warm, humid climates and is a voracious blood feeder that can cause anemia and death to the animal if left untreated. Over the past few decades, chemical anthelmintics were used to control GINs. Recently, GINs have developed resistance to most anthelmintics, and now reliance on anthelmintics is nearly impossible.

Foot rot is a costly disease for both sheep and goat producers due to animal weight loss, low fleece weight, decreased milk and reproductive capabilities, premature culling, and increased labor and treatment costs. This infection is prevalent in the Southern regions where there are prolonged periods of warmth and wetness [14]. When pastures have been consistently wet with no dry spells there can be a higher incidence. It can be prevented through concentrated management [14].

Transmission of CL can be spread in high density stocking areas including barns and dry lots [14]. This organism can survive for extended periods in dark, damp areas, soil, and manure [15-17]. Because CL is not considered a reportable disease in the United States there is very little information available about its prevalence.

A study by van Leuken, et al., reported that hotspots (areas most likely to contain the actual source) could be identified during the early outbreak stages based on the earliest 2-10% of the case notifications in Q fever outbreaks in the Netherlands [18]. Distances between the hotspots and suspected goat farms varied from 300-1500 m. In regional likelihood rankings that included all large dairy farms, the suspected goat farms consistently ranked first. The model they developed identified the most likely location of sources in an airborne pathogen outbreak area, even at early stages. This model can help to reduce the number of potential sources to be investigated by microbial testing, and allow rapid implementation of interventions to limit the number of human infections and reduce the risk of source-to-source transmission.

4. HERD INVESTIGATION

A systematic herd investigation strategy for disease management is pivotal to define the problems, understand important risk factors, develop a plan, and make accurate recommendations for disease management. An initial survey of the goat operation should be conducted to evaluate the herd health practices in place and the existing knowledge of the farmer concerning the control of diseases.

A review of records, feeding routines, housing, routine procedures, vaccination, and treatment protocols begins the investigation and determines which diagnostic procedures and testing strategies are most useful for an operation. The degree of prevalence of gastrointestinal nematodes, foot rot, and CL on goat farms should be determined. This is determined by the collection of on farm samples and analysis. Farm samples should include fecal, blood, aspirates of any intact abscesses, forage and soil samples, along with equipment.

Disease management is most effective when the problem source is well defined and the exposure can be limited. Screening examinations performed regularly or done at strategic times improves detection of disease and can be used to monitor treatment outcomes, and can avoid disease outbreaks.

5. BIOSECURITY

National and state agencies and institutions have identified bio-security related to animal agriculture as a matter of high priority [19]. To keep illness out of an operation, goat producers are encouraged to maintain biosecurity standards in their operations. Biosecurity (biological safety and well-being) is the set of management practices that prevent infectious diseases from being carried into a herd. Effective biosecurity management practices prevent the spread of disease by minimizing movement of biological organisms and their vectors onto and within the premises. These management practices are based on the principle that it is easier to prevent disease than it is to treat or react to a problem caused by disease. The advantages of adopting a biosecurity program are numerous. An effective program can improve the cost-efficiency of the farm, improve the reputation of the producer, and allow the producer to better maintain the health status of the herd [20].

6. CONCLUSION

Small scale farms face the greatest challenge to generate sufficient income. Diseases affecting livestock can have a significant impact on animal productivity and production, on trade in live animals, meat and other animal products, on human health zoonosis, and, consequently, on the overall process of economic development. Effective approaches to mitigate production limiting diseases will enable these goat farms to thrive and improve their niche markets by helping with disease surveillance and implementation of farm specific herd health management strategies. Quantification of disease populations on these farms through herd investigation and early detection can help with risk assessment and development of risk management strategies needed to control the colonization of these pathogens within goat operations.

REFERENCES

1. National Agricultural Statistics Service (NASS) (2009b) Sheep and Goats. Washington, DC: United States Department of Agriculture.
2. National Agricultural Statistics Service (NASS) (2014a) 2012 Census of Agriculture, United States, Summary and State Data, Geographic Area Series, Part 51. Washington, DC: United States Department of Agriculture.
3. National Agricultural Statistics Service (NASS) (2007) Overview of the U.S. sheep and goat industry. Mt An 8-7 (9-07) released September 28, 2007, Washington, DC: United States Department of Agriculture.
4. National Agricultural Statistics Service (NASS) (2009a) Census of Agriculture. United States Summary and State Data. Volume 1 Geographical Area series Part 51. Washington, DC: United States Department of Agriculture.
5. Pinkerton F, McMillin K (2009b) Goat Industry Update Part 2. Goat Rancher, 13-14, 16, 18-19.
6. National Agricultural Statistics Service (NASS) (2014b) Sheep and Goats Final Estimates 2009-2013. Washington, DC: United States Department of Agriculture.
7. LA Agricultural Summary (2008) Louisiana Summary: Agriculture and Natural Resources. LSU AgCenter, Baton Rouge, Louisiana.
8. Jones KG (2014) Trends in the US sheep industry. USDA. ERS. Agric. Info Bul # 787.
9. McLeod RS (1995) Costs of major parasites to the Australian livestock industries. International J Parasitol 25: 1363-7.
10. Perry BD, Randolph TF (1999) Improving the assessment of the economic impact of parasitic diseases and of their control in production animals. Vet Parasitol 84: 145-68.
11. Baah J, Tuah AK, Addah W, Tait RM (2012) Small ruminant production characteristics in urban households in Ghana. Livestock Research for Rural Development 24 (5).
12. Marshall RW (2011) Herd-Health Programs for Limited-Resource Farmers: Prevention Versus Treatment. J Extension 49: 5COM2.
13. Kaplan RM, Burke JM, Terrill TH, Miller JE, Getz WR, et al. (2004) Validation of the FAMACHA eye color chart for detecting clinical anemia in sheep and goats on farms in the southern United States. Vet Parasitol 123: 105-20.
14. Pugh DG (2002) Sheep and goat medicine. Bacterial Diseases: Caseous Lymphandenitis. Diseases of the foot: Infectious foot rot.

15. Smith MC, Sherman DM (1994) Subcutaneous swellings. Pages 45–61 in Goat Medicine. Lippincott Williams & Wilkins, Philadelphia, PA.
16. Williams LH (2001) Caseous lymphadenitis in small ruminants. Vet Clin North Am. Food Anim Pract 17: 359-71.
17. Anderson DE, Rings DM, Pugh DG (2002) Diseases of the integumentary system. Pages 197–222 in Sheep & Goat Medicine. W. B. Saunders Co., Philadelphia, PA.
18. van Leuken JP, Havelaar AH, van der Hoek W, Ladbury GA, Hackert VH, et al. (2013) A model for the early identification of sources of airborne pathogens in an outdoor environment. PLoS One 8: e80412.
19. Smith MH, Meehan CL (2012) Assessment of Bio-Security Risks Associated with 4-H Animal Science Exhibition Practices in California. J Extension 50: 3FEA10.
20. Mobley R, Lyttle-N'guessan C (2009) The Herd Health Handbook for Goat Producers: Biosecurity at the Farm Level. Florida A&M University.

CHAPTER 6

Comparison of Weight Gain, Milk Production, and Milk Composition of Iranian Mamasani Goat and its Cross with Saanen

Seyed Mahdi Hosseini[1], Li Guo Yang[1], Sayed Haidar Abbas Raza[*2], Rajwali Khan[2], Majid Kalantar[3], Shahid Faraz Syed[5], Mohib Ullah Kakar[4] and Robina Manzoor[4]

[1]*Key Laboratory of Agricultural Animal Genetics, Breeding and Reproduction of the Ministry of Education, College of Animal Science and Technology, Huazhong Agricultural University, Wuhan, China*

[2]*College of Animal Science and Technology, Northwest A&F University, Yangling, Shaanxi, China*

[3]*Animal Science Department, Agricultural Research Center of Qom, Qom, Iran*

[4]*Lasbela University of Agriculture, Water and Marine Sciences(LUAWMS), Uthal, Baluchistan*

[5]*Faculty of veterinary and Animal Sciences, Lasbella University of Agriculture, Water and Marine Sciences,Uthal, Baluchistan, Pakistan*

ABSTRACT

Crossbreeding can be a breeding strategy to facilitate genetic improvement in native goat production. For this purpose, the performance of weight gain, milk production and milk composition of native Mamasani goat with Mamasani goat × Saanen (F_1) as crossed goat breed during 2012 to 2014 years were compared. According to the results, effect of genetic groups on birth weight (BW) was not significant but Weaning weight (WW) was significant (P <0.05). The average birth weight and weaning weight for native Mamasani and crossed F_1 goats were 2.85 ± 0.23; 2.98 ± 0.19 and 9.52 ± 0.28; 11.61 ± 0.28 respectively. Milk yield in crossed F_1 goats differed from Mamasani goats significantly, however milk composition in term of milk fat and milk protein content decreased significantly (P <0.05). The average daily milk records were 0.65 ± 0.05 kg for Mamasani goats, and 1.31 ± 0.12 kg for crossed F_1 goats respectively. These findings indicated that crossbreeding of native goats with Saanen had positive effects on performance traits such as weight gain and milk yield but could not improve the milk composition traits including fat and protein content.

Keywords: Goat; Mamasani Saanen; Milk production; Milk composition

1. INTRODUCTION

The performance of native Iranian goats is said to be adequate in terms of meat and hair production, although poor with regard to milk yields, probably because of unfavorable conditions [1]. Cross-breeding with Saanen goats promises improvements in milk production and growth rates, but the rearing performance and meat production ability of the native goats needs further evaluation [2]. There is a great potential for development of milk production from goats and recently projects for this purpose have been established in Iran [3]. These research projects are mainly concerned with evaluating milk production from native and crossbred goats [1]. Crossbreeding has the benefit of heterosis, as an initial stage of transition in establishing a breed which means "grading up" for the development of a new breed. The use of established dairy breeds for upgrading low producing or nondescript breeds for milk production is particularly relevant in unfavorable environments [2,4,5]. Whether up-grading or the development of a new breed is the best policy will depend on the environment and level of management [4]. The option of crossbreeding to introduce suitable genetic material for milk production is a much more rapid method than that of attempting to improve yield of native goat breeds by selection [6,7]. Crossbreeding results in heterosis for milk production and weight gain as it is apparent when the average performance of crossbred progeny is superior to the average performance of the two parents [8,9]. Mamasani goat is a native goat in Iran located in central regions of Iran. This breed of goat was kept to produce of fluff and milk, hair [1]. In this regard we described in the present study the futures of native Mamasani goat and its crosses with Saanen breed and evaluate the present and potential milk production capacity and weight gain of the existing native and imported goats stock.

2. MATERIALS AND METHODS

In this study all animals were fed hay, wheat straw and concentrate with 12% protein. Number of 94 Mamasani goats and 37 Mamasani × Saanen (F_1) were selected at a local farm in south of Qom province. Weights of experimental animals were recorded and weight gain data were adjusted regards to birth date and birth type according to birth date and birth type using Least Square Means (LSM) calculating procedures included Birth Weight (BW) and Weaning Weight (WW) of related kids. Raw milk samples (n = 140) were obtained hygienically during 5 month lactation period and stored at 5 oC before analysis. Milk composition included fat and protein were analyzed by Milk Scope Julli C2 auto-analyzer in a laboratory of animal hospital in Qom city. Data were analyzed statistically using T-student procedure. P-values less than 0.05 were statistically significant. Estrus Synchronization and artificial insemination technology were used for Mamasani native goats and Saanen × Mamasani goats (F_1).

3. RESULTS AND DISCUSSION

Birth body weights were similar between Mamasani and crossed F_1, when compared with those of previous reports, this finding indicated that there is no similarity with the findings of other researchers [4]. Hoseini *et al.* (2012) reported differences in birth weights of black Lori goats and their crosses with Saanen. The results depicted in Table 1 related to weaning weight indicated that weaning weight of kids increased ($P<0.05$) the P-value indicates the difference is significant. These finding were in line with other results [1,5,6]. The average of daily milk records were 0.65 ± 0.05 kg and 1.31 ± 0.12 kg for native and crossed F_1 respectively. These significant results were similar to the latest research about native black Lori goat × Saanen in Qom province. When the results of this study are compared with the results of previous research, it is found that they are very close to related findings, which reported significant differences between native with crossed goats, and confirmed the positive effects of crossbreeding on milk yield traits directly [4,5,7]. In this study the milk yield increased significantly, however milk composition in term of milk fat and protein decreased through crossbreeding (Table 2). Although reduction in milk composition of crossbred offspring is a normal phenomenon due to negative correlation between quantitative and qualitative traits through animal breeding [2,7]. In this study, the reduced milk quality maybe due to higher quantity of the milk produced by crossbred goats. Guzler *et al.* (2010) also reported significant reduction in milk composition in Turkish native goat (Kilis) when crossed with Saanen [10].

Table 1: Average of body weight of Mamasani goat and their crosses with Saanen

Body weight (kg)	Mamasani Goat	Crossed F_1
Birth Weight	2.85 ± 0.23	2.98 ± 0.19
Weaning Weight	9.52 ± 0.28^{b}	11.61 ± 0.28^{a}

^{ab}Superscript letters within the same row are significantly different (P<0.05)

Table 2: Average of milk yield (L/day) and Composition of Mamasani goat and their crosses with Saanen

	Mamasani Goat	Crossed F_1
Milk yield	0.65 ± 0.05^{a}	1.31 ± 0.12^{b}
Fat%	4.85 ± 0.13^{a}	3.91 ± 0.11^{b}
Protein%	4.12 ± 0.03^{a}	3.6 ± 0.15^{b}

^{ab}Superscript letters within the same row are significantly different (P<0.05)

Other reports indicated that crossbreeding of native goats with genetically modified goats resulted in improved performance traits such as milk and gain traits [5,10,11]. Moameni *et al.* (2012) reported that crossbreeding between Sahelian goats with Anglo-Nubian goats improved their growth rate and milk production.

4. CONCLUSION

According to the results of present study we can demonstrate that weight gain and milk production traits of crossed goats were improved after crossbreeding of native Mamasani goat × Saanen. Also milk yield in crossed F_1 goats increased significantly compare to Mamasani goats, however milk composition (fat and protein percentage) decreased significantly. These findings indicated that crossbreeding of native goats with Saanen had positive effects on performance traits such as weight gain and milk yield.

REFERENCES

1. Heidari M, Khashui A (2008) Comparison of growth traits and milk composition on seven goat genetic groups in Yazd province. Proceeding of 3th Animal Science Congress, Ferdowsi University of Mashad, Iran.
2. Assan N (2013) Crossbreeding as a strategy to increase productivity in resource poor goat keepers in the rural areas of Zimbabwe. Int J Sci Knowled Vol 2: 52-6.
3. Hoseini SM (2012) Comparison the genetic potential of some growth and milk traits of crossed Saanen with native goats versus native goats in Qom province, M.Sc Thesis Azad University of Saveh, Iran.
4. Donkin EF (1997) Productivity and Disease of saanen, Indigenous and Crossbred goats on zero Grazing, Medical University of Southern Africa.
5. Dinh VB, Do TV, Pham TB (1998) Study to find out the performance of F1 between Indian and local Co in Vietnam raising condition, Proceeding of Workshop of Animal Production Science.
6. Karua SK, Banda JW (1989) The performance of the small East African goats and their saanen crosses in Malawi. M.Sc Thesis, University of Malawi, Malawi.
7. Serradilla JM (2001) Use of high yielding goat breeds for milk production. Livestock Product Sci 71: 59-73.
8. Koch RM, Dikerson GE, Cundif LV, Gregory KE (1985) Retained heterosis in advanced generations among Angus and Herford crossbred cattle. J Anim Sci 60: 1117-32.
9. Momani, MS, Sanogo S, Coulibaly D, Al-Olofi S, Alkhewani T (2012) Growth performance and milk yields in Sahelian × Anglo-Nubian goats following crossbreeding in the semi-arid zone of Mali. Agricultural Tropica et subtropica 45: 117-25.
10. Guzler N, Say D, Kacar A (2010) Compositional changes of Saanen × Kilis goats, milk during lactation. GIDA 35: 325-30.
11. Gebrelul S, Sartin LS, Iheanacho M (1994) Genetic and non-genetic effects on the growth and mortality of Alpine, Nubian and crossbred kids. Small Ruminant Res 13: 169-76.

CHAPTER 7

Hanwoo cattle: origin, domestication, breeding strategies and genomic selection

Seung-Hwan Lee[1†], Byoung-Ho Park[2†], Aditi Sharma[1†], Chang-Gwon Dang[1], Seung-Soo Lee[2], Tae-Jeong Choi[2], Yeon-Ho Choy[2], Hyeong-Cheol Kim[1], Ki-Jun Jeon[1], Si-Dong Kim[2], Seong-Heum Yeon[1*], Soo-Bong Park [2]and Hee-Seol Kang[1]

[1]*Hanwoo Experiment Station, National Institute of Animal Science, RDA, Pyeong-Chang 232-950, Korea.*

[2]*Animal Genetic and Breeding Division, National Institute of Animal Science, Cheon-An, Korea.*

ABSTRACT

Hanwoo (Korean cattle) is the native, taurine type of cattle breed of Korea and its history as a draft animal dates back to 5000 Years. In earlier times Hanwoo was used extensively for farming, transportation. Over the period of time, Hanwoo has changed to be meat type cattle. Full-scale production of Hanwoo as meat-type cattle has occurred since 1960s with the rapid growth of the Korean economy. Hanwoo is one of the most economically important species in Korea as it is a significant source of nutrition to the Korean people. Hanwoo beef is the most cherished food of Korea. One of the main goals of researchers is to increase the meat quality, quantity and taste of the beef. In this review we describe the origin, domestication of Hanwoo cattle and breeding program initiated from 1980's. Moreover the advent of technological advancement had provided us a platform to perform genome wide selection on economic traits and its implementation into traditional breeding programs.

Keywords: Hanwoo ,Origin, Domestication and breeding program

1. INTRODUCTION

Hanwoo (Korean cattle) is the native, taurine type, small sized cattle breed of Korea and its history as a draft animal dates back to 5000 Years. In earlier times Hanwoo was used extensively for farming, transportation and for religious sacrifice. Over the period of time Hanwoo cattle has evolved to be meat type cattle. Full-scale production of Hanwoo as meat-type cattle has occurred since 1960s with the rapid growth of the Korean economy. Despite its high price, beef from Hanwoo cattle is very popular among both Koreans and foreigners. Hanwoo beef is known for its marbled fat, tenderness, juiciness and characteristic flavor. There are four breeds of Hanwoo in Korea viz. brown Hanwoo, brindle Hanwoo, black Hanwoo and jeju black (Figure 1). Among all these colour variants brown is the most common one (Figure 1). National Institute of Animal Science along with the small stakeholders has taken an initiative for a better management and improvement of Hanwoo cattle. Hanwoo Experiment Station (A Research Station of National Institute of Animal Science) in Pyeongchang-gun, Gangwon-do, Korea is dedicated exclusively for research on Hanwoo cattle. Along with Hanwoo experiment station there are other institutes and universities as well which are working towards the betterment of this native cattle breed.

Korea along with Hanwoo has a market for imported beef from Australia, USA, New Zealand, Canada and Mexico as well (KMTA, [1]). When Hanwoo beef was compared with the Australian Angus, it showed characteristic compositional and quality differences that should result from genetic and environmental differences between them [2]. Hanwoo is lower in cholesterol compared to other beef. It also has a higher omega 3 fatty acid count which makes it healthier than beef from other cattle. Taste-wise, it's very soft, juicy, and delicious. The marbling in Hanwoo steak is excellent as well and it has the right balance of meat and fat. Korean consumers decided their overall acceptability of Hanwoo beef are as follows: weights of tenderness 55%, juiciness 18%, and flavor-likeness 27% [3]. No wonder Korean consumers, despite the high price prefer Hanwoo beef to the imported one. In order to produce high quality Hanwoo beef along with specially designed breeding programs special care and feed strategies are deployed.

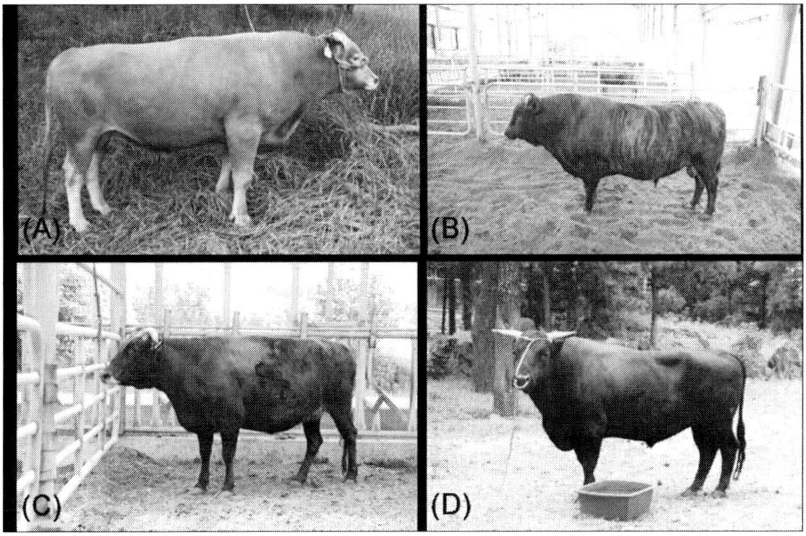

Figure 1 Pictures of four Hanwoo cattle. (A) Brown Hanwoo, (B) Brindle Hanwoo, (C) Black Hanwoo and (D) Jeju black Hanwoo.

The major goal of researchers is to increase both the quality (Marbling, tenderness and flavor) and the quantity (Carcass weight) of the meat to benefit the Hanwoo beef industry. Therefore, the current selection index used in Korea Proven Bulls (KPN) program is based on 1) carcass weight on marbling score 2) the comparison on carcass weight. In order to support the beef industry along with the conventional selection program, the genomic selection is also being incorporated in selection and breeding programs. In the present review we discussed the origin, domestication, breeding strategies and genomic selection in Hanwoo cattle.

2. ORIGIN AND DOMESTICATION

Origin and domestication of not just Hanwoo but all north-east asian cattle is a topic of discussion among researchers. Researchers are of divided view when it comes to ancestry and domestication of Hanwoo. There are some studies on the origin and ancestry of this cattle breed. Han S. W., [4] suggested Hanwoo to be a cross between zebu and taurine cattle, which migrated to Korea from Mongolia via north china while Lee & Pollack [5] proposed Hanwoo to have originated as a hybrid from auroch and zebu cattle. Yoon et al. [6] suggested an independent domestication event for Hanwoo. Mannen et al. [7] also suggested the independent mitochondrial origin in Asian cattle. Their study included Mongolian, Korean and Japanese cattle. In a recent study McTavish et al. [8] suggested that the asian cattle (Hanwoo and Japanese Black) to be of hybrid taurine–indicine origin.We carried out cattle diversity analysis which showed a clear differentiation of asian taurine from western taurine cattle (Figure 2B-2C). Based on 50 K SNP chip data we computed a neighbor joining phylogenetic tree where we observed a clear partitioning for African taurine, European taurine, Asian taurine and Zebu cattle. The tree clearly clustered Korean cattle breeds on a separate node than the European taurine cattle breeds. This might suggest a separate domestication event for north-east asian cattle. Also, mitochondria based studies have identified a haplotype T4 that is observed in north east asian cattle breeds but not in other breeds (Mannen et al., [7]). The PCA plot also clearly separated Hanwoo and Wagyu (Japanese cattle) from Hereford and angus cattle thus supporting the inference from neighbor joining tree.

3. POPULATION GENETIC PARAMETERS IN HANWOO

Linkage disequilibrium (LD) and effective population size (Ne) are important genetic parameters to understand population structure for optimal breeding program design. Lee et al., [9] performed a LD analysis and studied Ne for the entire Korean Hanwoo cattle genome, which was the first LD map and effective population size estimate ever calculated for this breed. A panel of 4,525 markers was used in the final LD analysis. The pairwise *r2* statistic of SNPs up to 50 Mb apart across the genome was estimated. A mean value of $r2 = 0.23$ was observed in pairwise distances of <25 kb and dropped to 0.1 at 40 to 60 kb, which is similar to the average intermarker distance used in this study. The proportion of SNPs in useful LD ($r^2 \geq 0.25$) was 20% for the distance of 10 and 20 kb between SNPs. Analyses of past effective population size estimates based on direct estimates of recombination rates from SNP data demonstrated a decline in effective population size to $Ne = 98.1$ up to three generations ago (Figure 3). In this population genetic parameter (LD and Ne) study, the Ne in Hanwoo was larger overall than in North American Holstein cattle. In particular, the Ne of Hanwoo dramatically decreased 25 generations ago, this is when official Hanwoo breeding programs were initiated. Effect of selection pressure on Ne in the Hanwoo population was evident from this study.

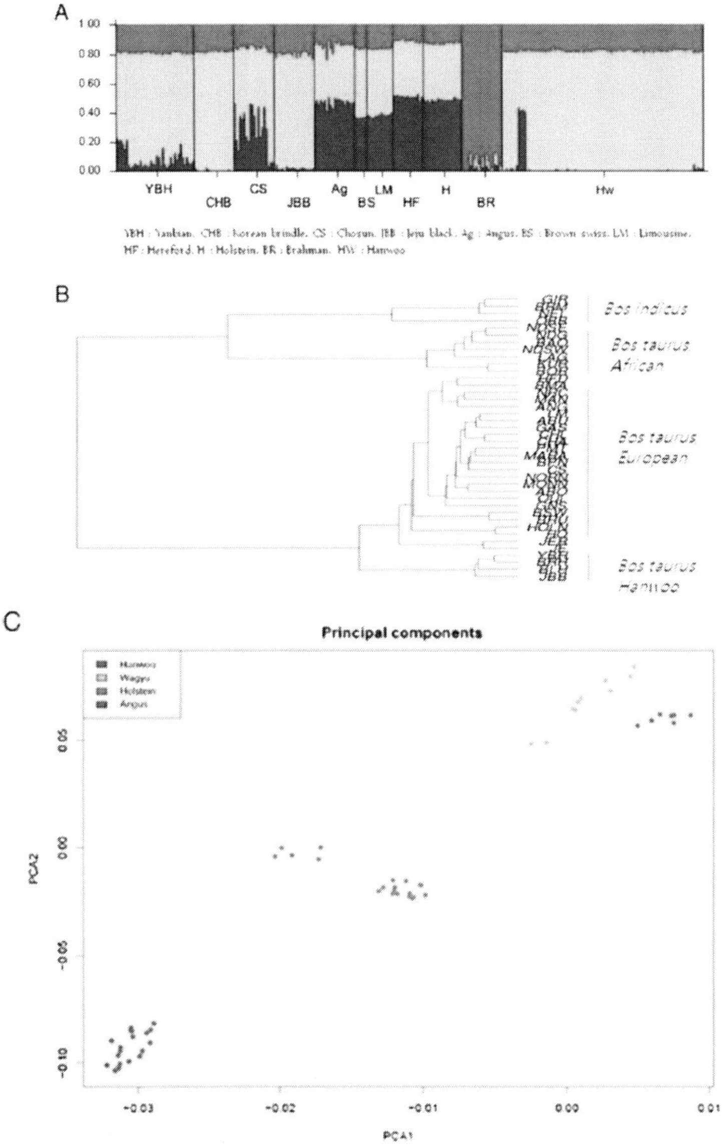

Figure 2 Comparison of Hanwoo compared to other taurine cattle. (A) Model-based population assignment for individuals based on the 50K SNP panel using STRUCTURE (K=3), YBH: Yanbian, CHB: Korean brindle, CS: Chosun, JBB: Jeju black, Ag: Angus, BS: Brown swiss, LM: Limousine, HF: Hereford, H: Holstein, BR: Brahman, HW: Hanwoo (B) A neighbor joining tree based on Pairwise Fst values, (C) Principal component analysis capturing a clear difference of north-east asian cattle.

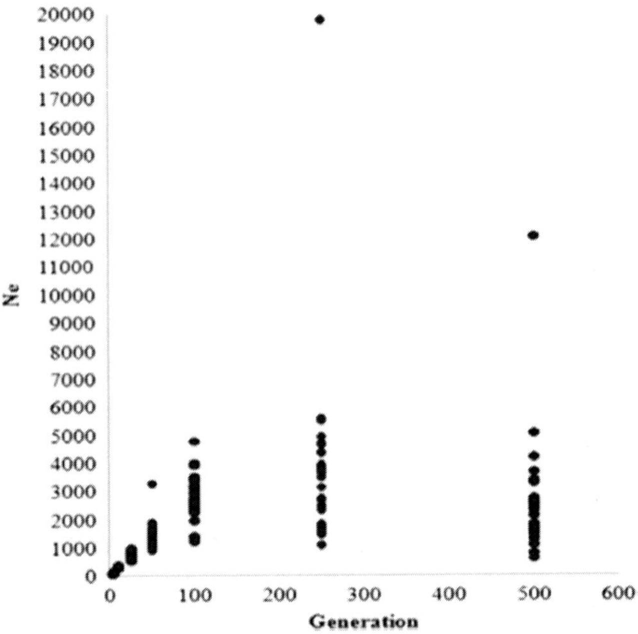

Figure 3 Effective population size from chromosome of the past Hanwoo (Korean cattle) population. Effective population size was estimated in each autosomal chromosome. The generation in the past population was calculated as 1/2c. N_e is plotted against the past generation up to 500 generation. (Extracted from [10]).

4. HANWOO BREEDING PROGRAM

Marbling score (MS), carcass weight (CWT), eye muscle area (EMA) and back fat thickness (BF) are the traits to impact Korean beef industry the most. Therefore the selection index for Korea Proven Bulls uses eye muscle area on marbling score and their comparison on CWT.

The first genetic breeding program called "Hanwoo-Gaeryang-Danji (HGD)" was initiated by Ministry of Agriculture and Forestry (MAF) in 1979. However, due to lack of awareness in the farmers to record the data and scarcity of the trained emulators, the program didn't meet its objective. In 1999, a new program called "Hanwoo-Gaeryang-Nongga (HGN)" was introduced. Whereas HGD focused on a particular province HGN concentrated on individual farms. The cows from HGN program are used as major breeding stock for "Hanwoo Performance and Progeny Test (HPPT)" program (MAF, [11]). Prior to the HPPT program, bulls to be used for artificial insemination were selected solely on the basis of phenotypes. It was in 1987 that the first proven bull was produced using HPPT program. As shown in Figure 4, the HPPT program was a two stage selection program. The first stage was a performance test for the young bulls and the second stage was progeny test of selected young bulls. Young bull calves of current proven bulls were harvested at the age of 6 months from HGN belonging to HGD based on their phenotypic values and underwent a performance test up to age of 12 months. At this stage young bulls were selected based on a combination of their breeding values of weight at 12 months and average daily gain estimated from its performance test records. In the progeny test, cows from the HGN program were inseminated with the semen from young bulls and the male calves were harvested from farms and raised for the station progeny test. These steers were raised in a group until slaughter at 24 months. Carcass data comprised EMA, BF, MAR and CWT. Based on the progeny test results 20 bulls are selected for the AI program. Selection is based on a selection index which used weighted breeding values for BF, MAR and EMA.

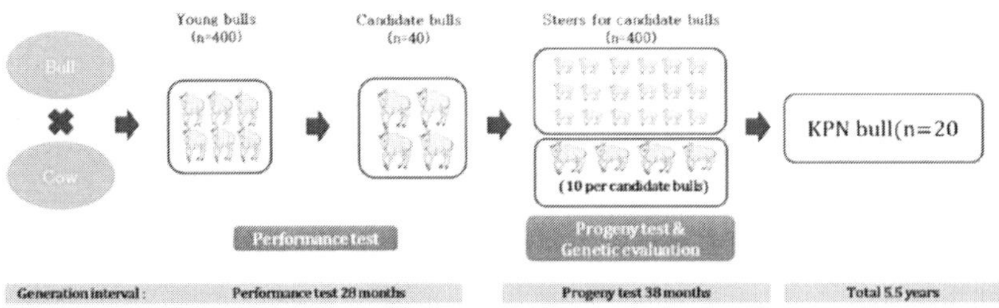

Figure 4 Pictorial representation of Hanwoo performances and progeny test (HPPT) program

Park et al. [12] carried out a study to determine the trend of improvement as well as the estimation of genetic parameters of the traits being used for seedstock selection based on the data collected from the past. In his study he concluded that the performance and progeny test based selection system of Korea is good enough to accommodate circumstances where fewer sires are used on many more cows. Although progeny tests take longer and cost more, they seem to be appropriate under the circumstances of the domestic market with its higher requirement for better meat quality. The study suggested accumulative data collection, genetic evaluation model development, revision of selection indices, as well as cooperation among farms, associations, National Agricultural Cooperative Federation, universities, research institutes and government agencies must be applied to the Hanwoo selection program. According to the Park's study, the current progeny tested evaluation runs every 6 months to select KPN in Korea. The breeding program has achieved a substantial genetic improvement for CWT and EMA. However there has been just a slightly negative genetic response for MS with an -0.036 of genetic response per year [12].

The heritability and genetic correlation are important genetic parameters in breeding program. In Hanwoo population, heritability and genetic correlation were estimated using carcass data from National Hanwoo progeny tested population (Figure 5A). A negative genetic correlation was observed between BF and EMA (-0.2) and MS and BF (-0.02) while positive correlation was observed in all other traits. A high positive correlation was observed between EMA and CWT (0.45) indicating that a greater EMA is associated with a higher production of CWT. The value of correlation between EMA and CWT in our study is more than that reported by and Koch et al. [13] and less as reported by Baik et al. [14] and Cundiff et al. [15]. High estimated heritabilities for BF (0.5), EMA (0.41), MS (0.4) and moderate for CWT (0.33) were observed. This indicates that in Hanwoo cattle a considerable genetic variability still exits which may be used for the improvement of carcass traits. Estimation of heritability, from genome-wide SNPs has recently attracted interest and offers several advantages over traditional pedigree-based methods. A comparison of pedigree and genotype based heritability for carcass traits was thus drawn (Figure 5B). The amount of variation that could be accounted for by SNP genotypes was concordant with pedigree-based heritabilities and varied from very low for CWT to medium-high for BF.

We know conventional breeding methods rely on physical characteristics or phenotypes to calculate the breeding values (BV) of animals. The traditional methods are often not so accurate, inefficient and time consuming as most traits like meat quality are hard to measure and evident only when animal reaches maturity leading to a delay in verifying breeding results. With the advent of genome wide SNP panels we can overcome the drawbacks of the conventional breeding methods. Genome wide SNP panels will allow us to accurately and cost & time effectively determine the genomic estimated breeding values (GEBV) even in young animals and thus select animals at young age.

5. GENETIC ARCHITECTURE FOR CARCASS TRAIT IN HANWOO

Korean beef industry lay emphasis on meat yield and quality traits like carcass weight, marbling, intramuscular fat, eye muscle area. To identify the significant associations of SNPs with these traits can directly affect the selection of animals for a breeding program.

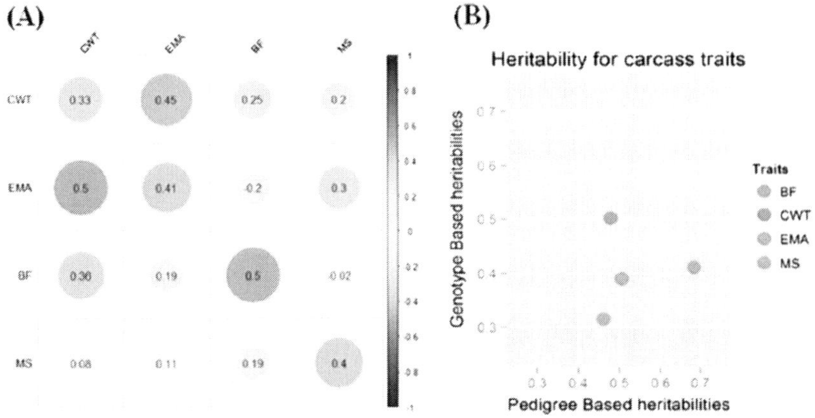

Figure 5 Heritability and genetic correlation for carcass trait in Hanwoo. Data are taken from a report for Hanwoo genetic improvement meeting (2013). (A) Along the diagonal (From upper left) are heritability values, upper triangle shows variance for genetic correlation and lower triangle shows residual variance. (B) Estimate of heritability from a pedigree and genotype for carcass traits in Hanwoo.

Genome-wide association study (GWAS) is a process for inspection and screening of detectable common genetic variants (single-nucleotide polymorphisms) in individuals to identify the variant(s) associated with the trait under study. There are several GWAS studies that are successfully conducted in livestock species. A recent study on Hanwoo cattle conducted by lee et al. (2013) identified 6 highly significant SNPs on chromosome 14 to be associated with carcass weight in Hanwoo cattle (Figure 6). This genome-wide association study was conducted to identify major loci that are significantly associated with carcass weight, and their effects, so as to provide an insight into the genetic architecture of carcass weight in Hanwoo. This genome-wide association study identified a major chromosome region ranging from 23 Mb to 25 Mb on chromosome 14 as being associated with carcass weight in Hanwoo. The most significant SNP detected was *BTB-01280026* ($P = 4.02 \times 10^{-11}$), located in the 25 Mb region on Bos taurus autosome 14 (BTA14). The most significant SNPs accounted for 6.73% to 10.55% of additive genetic variance, which is quite a large proportion of the total additive genetic variance. The most significant SNP (*BTB-01280026*; $P = 4.02 \times 10^{-11}$) had 16.96 kg of allele substitution effect, and the second most significant SNP (*Hapmap27934-BTC-065223; $P = 4.04 \times 10^{-11}$*) had 18.06 kg of effect on carcass weight, which correspond to 44% and 47%, respectively, of the phenotypic standard deviation for carcass weight in Hanwoo cattle. Our results demonstrated that carcass weight was affected by a major Quantitative Trait Locus (QTL) with a large effect and by many SNPs with small effects that are normally distributed. As evident from manhattan plots of GWAS for four carcass traits shown in Figure 6 a highly significant association was observed for carcass weight on BTA14 while for other three traits there was no such association observed which shows the effect of many genes on a trait across all chromosomes.

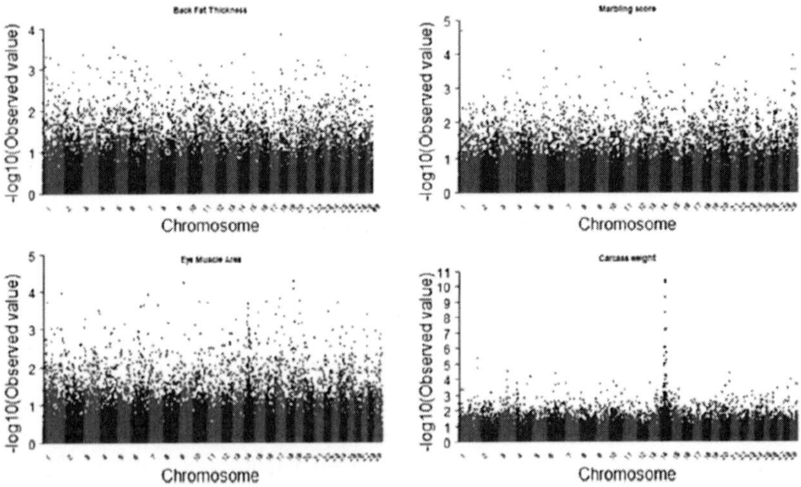

Figure 6 Association of SNPs with the carcass traits in the Hanwoo shown as a Manhattan plot.

In order to determine if polygenic inheritance is a general phenomenon for carcass traits in Hanwoo population, we directly estimated the proportion of phenotypic variance explained by the common SNPs on each chromosomes. Proportion of variance attibuted to each chromosome averaged across four carcass traits against chromosome length (Figure 7) was calculated using genotype relationship matrix (GRM) on each chromosomes which supported the results shown in manhattan plots of GWAS. While for all other traits we can see most of the chromosomes having marginal effect but for CWT where Chromosome 14 is clearly showing a high effect on the trait. By calculating the proportion of the genome represented by each chromosome (not including the length of sex chromosomes), we tested for a correlation between the variance explained by each chromosome relative to its size. It was observed that several of the smaller chromosomes contributed less to the overall variance than several of the larger chromosomes, however trend was not significant.

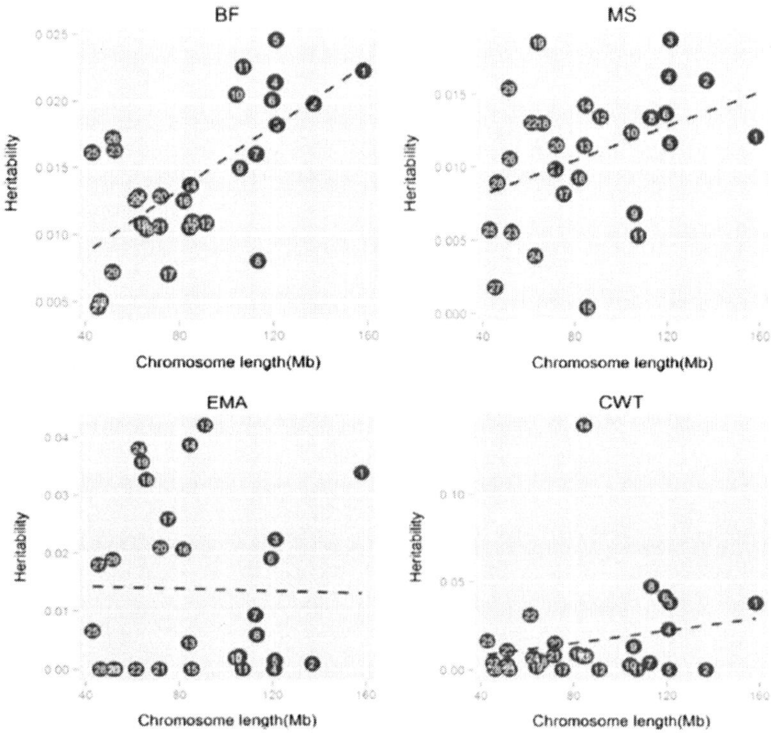

Figure 7 Proportion of variance attibuted to each chromosome averaged across four carcass traits against chromosome length.

Also there are reports on GWAS studies on other traits of economical importance such as MS, IMF, EMA and sensory panel. Yi et al. [16] carried out a GWAS study to identify QTL for growth and carcass quality traits using high-density SNP panels. The data set comprised of 61 sires, their 486 steers, and the 54,001 SNP markers on 29 bovine autosomal chromosomes. Traits to be analyzed in this study were six growth and carcass quality traits including weaning weight (WWT), 365-d yearling weight (YWT), CWT after slaughter, BF, EMA, and MS. A total of 16(0), 18(4), 20(13), 11(23), 10(13) and 19(1) SNPs were detected at the 5% chromosome (genome)-wise level for the traits, WWT, YWT, CWT, BF, EMA and MS, respectively. Among the 148 SNPs, 91 SNPs had dominance effects, suggesting that dominance inheritance mode be considered in genetic improvement for growth and caracass quality in Hanwoo. Thirty five QTL regions on 17 Bos taurus chromosomes (i.e. BTA 3, 4, 5, 6, 7, 11, 12, 13, 14, 15, 16, 17, 18, 20, 23, 26, and 28) were detected. Strong evidence for the QTL influencing CWT were detected on BTA14. Also, the QTL for WWT, YWT, BF, and EMA were detected on BTA20. The GWAS studies could thus greatly help in developing the selection strategies for breeding programs.

6. GENOMIC STRATEGY FOR BREEDING SCHEME

Genomic selection (GS) employs selection of an individual based on the genomic breeding value assessed through evaluating all the genetic markers located throughout the genome of that individual. Thus, the underlying principle of GS is to exploit the linkage disequilibrium of the QTL with one or more genetic marker(s) [17]. Molecular markers can be used to predict GBV of breeding animals by exploiting population-wide linkage disequilibrium between QTL and genetic markers spanned over the genome. The key–factor behind the success of genomic selection is to utilize the next generation sequencing approaches and the associated bioinformatics tools to identify the SNPs. Simulation studies in some domesticated species like, beef cattle, swine and chicken (Meuwissen et al., [18]) have suggested that the breeding values can be predicted with high accuracy using genetic markers alone but its validation is required especially in samples of the population different from that in which the effect of the markers was estimated. In genomic selection, the estimation of genomic breeding values is predicted to sum up all loci that are estimated based on phenotypes and genotypes in training population. This is particularly useful for traits that are very difficult to measure, such as marbling. The accuracy of these genomic predictions depends on the genetic architecture of the complex traits. For example, number of loci affecting the trait and distribution of their effects. In case of Hanwoo, MAS (Marker Associated Selection) and GS will provide a way to predict phenotype of marbling score as a preselection method that can be used in performance test [19].

7. STRATEGIES FOR GENOMIC SELECTION PROGRAM IN HANWOO

Application of genomic selection in Hanwoo preselection process such as performance test and then in progeny test in the breeding program might raise the genetic gain for marbling. A comparison of accuracy of GEBV and traditional EBV is tabulated in Table 1. In order to estimate the genomic breeding value, well organized reference population (Training data set) is to be build. The two most important features of reference population are 1) to representation of the entire Hanwoo population (all the variants) and the sample size. For the same, current progeny tested steers, selected proven bulls and candidate bulls from the progeny test program will be a reasonable population. Reference population must contain minimum of 5,000 animals. A genetic gain of 30% can be observed if we have a large reference population.

Table 1 Accuracy of GEBV and traditional EBV estimated by 50 K SNP chip for Hanwoo cattle

Traits	BLUP	GBLUP	Difference
Eye muscle area (cm^2)	0.11(0.08)	0.29(0.07)	0.18
Back fat thickness (mm)	0.11(0.08)	0.30(0.11)	0.19
Marbling score (1 ~ 9)	0.11(0.08)	0.27(0.12)	0.16

8. CONCLUSION

Recent advances in molecular biotechnology facilitate not only detection of genes that contribute to genetic variation of quantitative traits but also incorporation of genomic information into a conventional animal breeding program. The incorporated molecular information into GEBV may achieve an improvement of EBV and selection accuracy in cattle populations. In conclusion, we suggest incorporating molecular information into the conventional breeding programs to achieve better results in comparatively shorter time. Including molecular information will be a step further to achieve higher standards of Hanwoo beef.

ACKNOWLEDGEMENTS

This study was supported by awards from the AGENDA project (Grant no. PJ907008062013) and Molecular Breeding program (PJ0081882013) of Next Generation BIOGREEN21 project in the National Institute of Animal Science, RDA.

REFERENCES

1. Korea Meat Trade Association (KMTA): Information and Data of Agricultural Statistics Of Korea. 2011, Accessed on January 12, 2011. http://kmta.or.kr/html/sub6-1.html?scode=6.
2. Jo C, Cho SH, Chang J, Nam KC: Keys to production and processing of Hanwoo beef: a perspective of tradition and science. Anim Front. 2012, 2 (4): 32-38. 10.2527/af.2012-0060.
3. Cho SH, Kim J, Park BY, Seong PN, Kang GH, Kim JH, Jung SG, Im SK, Kim DH: Assessment of meat quality properties and development of a palatability prediction model for Korean Hanwoo steer beef. Meat Sci. 2010, 86: 236-242.
4. Han SW: The Breed of Cattles. 1996, 148-160. Sun-Jin publishing.
5. Lee C, Pollak EJ: Genetic antagonism between body weight and milk production in beef cattle. J Anim Sci. 2002, 80: 316-321.
6. Yoon DH, Park EW, Lee SH, Lee HK, Oh SJ, Cheong IC, Hong KC: Assessment of genetic diversity and relationships between Korean cattle and other cattle breeds by microsatellite loci. J Anim Sci Technol (Kor). 2005, 47 (3): 341-354.
7. Mannen H, Kohno M, Nagata Y, Tsuji S, Bradley DG, Yeo JS, Nyamsamba D, Zagdsuren Y, Yokohama M, Nomura K, Amano T: Independent mitochondrial origin and historical genetic differentiation in North Eastern Asian cattle. Mol Phylogenet Evol. 2004, 32: 539-544.
8. McTavish EJ, Decker JE, Schnabel RD, Taylor JF, Hillis DM: New World Cattle Show Ancestry From Multiple Independent Domestication Events. 2013, PNAS Early Edition
9. Lee SH, Cho YM, Lim D, Kim HC, Choi BH, Park HS, Kim OH, Kim S, Kim TH, Yoon D, Hong SK: Linkage disequilibrium and effective population size in Hanwoo Korean Cattle. Asian Australas J Anim Sci. 2011, 24 (12): 1660-1665. 10.5713/ajas.2011.11165.
10. Lee SH, Julius VW, Lee SH, Lim DJ, Park EW, Gondro C, Yoon D, Oh SJ, Kim OH, Gibson J, Thompson J: Genome wide QTL mapping to identify candidate genes for carcass traits in Hanwoo (Korean Cattle). Genes Genom. 2012, 34 (1): 43-49. 10.1007/s13258-011-0081-6.
11. Ministry of Agriculture and Fishery (MAF): Code of Practice For Development Programs In Agriculture And Forestry. 1999, 40
12. Park B, Choi T, Kim S, Oh S-H: National genetic evaluation (system) of Hanwoo (Korean Native Cattle). Asian-Aust J Anim Sci. 2013, 26 (2): 151-156. 10.5713/ajas.2012.12439.
13. Koch RM, Cundiff LV, Gregory EK: Heritabilitied and genetic, environmental and phenotypic correlations of carcass traits in a population of diverse biological types and their implications in selection programs. J Anim Sci. 1982, 55: 1319-1324.
14. Baik DH, Hoque1 MA, Choe HS: Estimation of genetic and environmental parameters of carcass traits in hanwoo (korean native cattle) populations. Asian-Aust J Anim Sci. 2002, 15 (11): 1523-1526.
15. Cundiff LV, Gregory EK, Koch RM, Dickerson GE: Genetic relationships among growth and carcass traits of beef cattle. J Anim Sci. 1971, 33: 550-555.
16. Yi L: Genome-Wide Association Study to Identify Qtl For Growth And Carcass Quality Traits In Korean Native Cattle, Hanwoo. 2012
17. Goddard ME, Hayes BJ: Mapping genes for complex traits in domestic animals and their use in breeding programmes. Nat Rev Genet. 2009, 10: 381-391.
18. Meuwissen T, Hayes B, Goddard M: Prediction of total genetic value using genome-wide dense marker maps. Genetics. 2001, 157: 1819-1829.
19. Lee SH, van der Werf JHJ, Hayes BJ, Goddard ME, Visscher PM: Predicting unobserved phenotypes for complex traits from whole-genome SNP data. PLoS Genet. 2008, 4: e1000231.

CHAPTER 8

Nutritional Influences on Skatole Formation and Skatole Metabolism in the Pig

Raffael Wesoly and Ulrike Weiler *

Institute of Animal Husbandry and Animal Breeding, Behavioral Physiology of Farm Animals (470a), University of Hohenheim, Garbenstr. 17, 70593 Stuttgart, Germany

SIMPLE SUMMARY

Skatole is a tryptophan metabolite with fecal odor. Skatole and the testicular steroid androstenone are regarded as the main compounds leading to 'boar taint,' a sex-specific odor from pork taken from entire males, as elevated concentrations of both substances may be found in adipose boars tissue. High skatole concentrations in adipose tissue are the result of a complex process, which includes microbial formation in the colon, absorption, metabolism and accretion in fat. Several of these steps leading to high skatole concentrations are influenced by feed components and additives. The present paper discusses the mechanisms by which effective feeding strategies and feed additives exert their influence in the prevention of high skatole concentrations in adipose pig tissue.

ABSTRACT

Skatole is a tryptophan (TRP) metabolite with fecal odor. Together with the testicular steroid androstenone it is regarded as a main determinant of boar taint, even if elevated concentrations of skatole occur occasionally in gilts and barrows. Skatole concentrations in adipose tissue result from a complex process, which includes the availability of TRP and the presence of specialized bacteria in the gut in need of TRP for energy production, as well as absorption, transport and accumulation of skatole in adipose tissue. Several steps of this process are influenced by diet and specific feed compounds. In the present paper the current knowledge about physiological mechanisms of skatole dynamics is summarized. Additionally mechanisms are discussed, by which effective feeding strategies and feed additives exert their influence in the prevention of high skatole concentrations in adipose pig tissue. It was concluded that the most effective measures are those which influence several steps of skatole formation. Despite the numerous studies carried out in the field of skatole physiology, interesting aspects still need clarification, such as the effect of adipose tissue turnover. Reliable control of skatole accretion in fat of boars is one of the main prerequisites for pork production with entire males.

Keywords: skatole; pig; feeding; skatole physiology; inulin; raw potato starch; boar taint

1. INTRODUCTION

Boar taint is one of the major problems occurring in meat production of entire male pigs [1]. The two main compounds contributing to boar taint are androstenone (5α-androst-16-ene-3one, [2,3]), a steroidal pheromone produced in the Leydig cells of the testis, and skatole (3-methyl-indole, [4]), a product of microbial degradation of tryptophan (TRP) with fecal odor [5]. Skatole can be found in several species including ruminants, where it is mainly formed in the rumen, and monogastrics, such as the pig, where it is mainly formed in the colon [6]. In the pig, skatole has gained considerable scientific interest and is discussed as one of the principal compounds contributing to the off-odor of meat from entire males, but elevated concentrations occur occasionally in females and barrows [7]. Whereas the physiology of androstenone formation has been studied extensively and environmental and genetic influences are known [8,9], the physiological mechanisms of skatole formation, absorption and its metabolism within the pig are less clear. While androstenone concentration is only moderately influenced by nutrition [9,10,11,12,13], various feeding strategies are known to influence the concentrations of skatole in adipose tissue. This topic has been the subject of several scientific reviews during the last decade [9,14,15]. As a lot of new studies are available, it is the aim of the present paper to describe current knowledge about the mechanisms leading to skatole formation and accretion in adipose tissue of pigs. Based on these mechanisms, an overview on possible feeding strategies or nutritional manipulations of skatole formation is summarized and their efficiency discussed.

2. PHYSIOLOGY OF SKATOLE FORMATION AND DEPOSITION IN ADIPOSE TISSUE

2.1. Biochemical Pathway and Microbial Activity

Skatole results from a multistep degradation of TRP by microbial activity, mainly in the hind gut of the pigs [16]. Analogously, many other metabolites such as indole are formed which may also contribute to off odor in pork [17].

The reaction mechanism of TRP degradation is limited under the anaerobic conditions of the intestinal tract to reductive processes at the 3-position of the TRP indole ring structure, yielding the terminal products skatole and indole [16]. Whereas many bacteria are able to metabolize tryptophan to indole and indole acetic acid (IAA), the key precursor of skatole, only a few specialized gut bacteria, mainly from the Clostridium and Bacteroides genera, can catalyze the steps from IAA to skatole [18,19,20]. It was estimated that these bacteria represent less than 0.01% of the total intestinal flora [21]. In case of skatole, intermediate products are indole-3-lactic acid and indole-3-pyruvic acid, resulting from the deamination of TRP, which is decarboxylated to IAA, and then further decarboxylated to skatole (see Figure 1, upper part) [16,20]. This metabolic pathway has been identified in Lactobacilllus sp. strain 11201 [22] but also in C. drakeii and C. skatologenesis using deuterium labeled TRP or IAA through in vitro studies with bacteria from swine manure [20]. In this study, adding TRP together with glucose reduced the amount of skatole and increased IAA with both types of Clostridia. Similarly, the addition of TRP to cell cultures rich in C. scatologenes did not necessarily increase skatole formation [23]. Such data suggest that the microbial skatole forming activity is reduced, if alternative energy substrates are available, as further discussed below (see Section 3.3).

In two more recent studies which compared the microbial community in the colon of Jinhua and Landrace pigs, C. aminophilum was also suggested to affect skatole production in the Jinhua breed [24]. The same authors published an in vitro study with porcine gut microbes, which points to a contribution of C. disporicum to skatole formation [25]. A summary of formation and metabolism of skatole and indole in the pig is given in Figure 1.

2.2. Anatomical Sites of Skatole Formation

TRP degradation to indole and skatole starts in the proximal part of the colon. In contrast to earlier studies [26,27], which already revealed elevated concentrations of indole in the distal part of the small intestine and the cecum, the majority of more recent studies point to a later onset of TRP degradation to indole and skatole [24,28,29]. In one of the feeding studies [28], only traces of free tryptophan were detectable (<100 μmol/kg digesta) in the stomach but increased in the small intestine. Dividing the small intestine into three parts, elevated concentrations above 200 μmol TRP/kg digesta were already found in the proximal part, increasing to maximum concentrations (about two-fold) in the middle part. In the distal part, TRP concentrations decreased to values comparable to those of the proximal part. No conversion of TRP to indole or skatole was measurable from the stomach to the distal part of the small intestine. Low concentrations of skatole and indole were found in the cecum, but maximum concentrations of IAA and IPA (indolic pyruvic acid) were observed, which decrease continuously along the colon [28]. Correspondingly, almost linearly increasing concentrations of skatole and indole are observed. Maximum concentrations of both compounds are found either in the distal colon or the rectum [14,24,28,29].

Absorption of skatole and indole occurs along the colon and both substances are transferred to the liver via the portal vein. Total daily absorption rates of skatole were estimated between 820 and 365 μmol skatole, depending on the diet [28]. The daily absorption of indole was reported to be about threefold higher and ranged between 2,999 and 929 μmol [28]. The concentrations of skatole in portal blood, peripheral blood and feces is highly correlated within individuals, but that of indole is not, suggesting that the amount of skatole absorbed is proportional to the amount produced [27,28]. Such correlations were less obvious in one study if several individuals were included or between animal correlations were calculated [27,28], as individual differences in metabolism of skatole seem to exist [30].

The direct infusion of skatole into the cecum of pigs via a cannula in the ileo-cecal junction in 4 h intervals for three days increased skatole concentration in feces only slightly and excretion rates via feces were below 5 %. An absorption rate above 90% of the amount of skatole, which is already available in the proximal part of the colon, was discussed [31]. Under physiological conditions, however, only a small amount of the total skatole production is formed in the proximal part of the colon, as shown above [28].

Thus, in a similar experimental approach using TRP infusion into the cecum, an appearance rate in the portal vein of 70% of the skatole formed along the colon was calculated. In contrast to a former study [32], this percentage seemed to be independent on the total amount of skatole synthesized [14].

2.3. Origin of Tryptophan as a Precursor for Skatole Synthesis

The origin of TRP needed for the microbial synthesis of skatole is controversial [24,27]. Whereas in ruminants, the addition of TRP to the diet regularly leads to increased skatole formation [33,34], the effect of TRP in the diet of the pig is less clear [35,36,37]. Diets with low prececal protein digestibility were reported to increase skatole production [5,14,38], whereas feed supplements with synthetic L-TRP below or above the requirements (1.0 g/kg DM to 1.91 g/kg DM) were not effective in systematically changing skatole concentrations in the feces of growing pigs [37]. It was postulated that free TRP is mainly absorbed in the small intestine and is therefore not available for microbial metabolism in the colon. Studies aiming to influence the rate of mitosis and apoptosis in the small intestine experimentally showed that gut cell debris is a major source of tryptophan for the microbial skatole formation [39,40].

Changes in diet may lead to a reorganization of intestinal mucosa and thereby increase the amount of cell debris, as most obvious after weaning in piglets [41]. A clear increase in skatole formation after weaning was clearly shown for piglets of different sex and weaning ages, which, however, was accompanied by an age dependent decrease in SULT1A1 (see Section 2.4; [30,42]). Feed additives (antibiotics and Chinese herbs), which reduced the number of pathogenic bacteria in the intestine were effective in reducing atrophy of villi in piglets after weaning [41]

The direct infusion of 4.9 mmol TRP into the cecum of pigs was effective in increasing skatole concentrations in portal blood. The increase started after 2 h, reached maximum values up to five-fold concentrations after 6 h to 10 h and decreased thereafter [14]. The total conversion of infused TRP to skatole was estimated, based on portal blood concentrations. It ranged between 26% for animals given a low fiber diet and 6% for animals given a high fiber diet. Most of the infused TRP was converted to indole (69% and 35% respectively). In animals fed with high fiber diet a higher amount of indole-propionic acid (IPA) was measured, compared with animals on a low fiber diet (16 vs. 7%) [14].

Thus a generally lower proportion of degradation products of TRP was observed if more fiber was fed simultaneously. Under such conditions, only

57% of the cecally-infused TRP was metabolized to one of the three compounds (skatole, indole, IPA), whereas in animals on a low fiber diet the total amount of infused TRP was degraded to one of the three substances. It was concluded that in the case of a high fiber diet, more TRP is used for the microbial synthesis of protein instead of microbial energy production via TRP degradation [14]. Other factors such as colon motility, fecal transit time, secretory rates, osmolality in the colon which are influenced by the physiological and psychological status of the host may also have a significant role [16].

The transfer of skatole and indole from gut to blood is probably due to passive diffusion, as there is no skatole carrier protein known [43].

2.4. Metabolism of Skatole in Liver and Kidney

Skatole and indole are transported by the portal vein (V. porta) to the liver, where most of the indole derivatives are metabolized by specific enzymes. Small amounts of indoles, which are absorbed in the distal colon or rectum, can bypass the liver and are transferred via the vena cava caudalis directly into the peripheral bloodstream [10,27]. The liver metabolism is highly effective and skatole concentrations of the portal vein may be reduced severely in the liver (up to 90%), as was obvious from parallel measurements in the hepatic vein [32,44]. Based on these data, a half-life for skatole in blood of one hour was calculated [44].

Even if skatole and indole are mainly degraded in the liver [45,46,47], a contribution of other organs, such as the kidneys, as similarly shown for androstenone degradation, may be assumed [48]. The knowledge on hepatic metabolism of skatole and indole has been reviewed recently [8,9]. In brief, hepatic degradation of indoles can be divided in two distinct steps: an oxidative step, phase 1 metabolism, and a conjugative step, phase 2 metabolism. Responsible enzymes are various cytochrome P450 isozymes, which are known to play a predominant role in drug and xenobiotic metabolism. Two specific enzymes, CYP2E1 and CYP2A were identified as major enzymes of the phase 1 metabolism of skatole [47,49]. A minor contribution of other P450 isozymes to the skatole phase 1 metabolism was discussed, but is less clear [50,51,52,53,54]. During phase 1 metabolism skatole is degraded to seven intermediate products (see Figure 1, lower part). The main enzymes of phase 2 metabolism, SULT1A1 (sulfotransferase) and UGT (uridine-di-phosphate-glucuronosyltransferase) further modify these seven compounds mainly by increasing their hydrophilic properties, adding either a sulfate or glucuronyl group to the molecules [44]. Phase 2 metabolism thus results in a variety of terminal products, where 6-sulfatoxy-skatole, sulfated or glucuronic conjugates of 5-

hydroxy-3-methylindole and 3-hydroxy-3-methyloxindole are predominant [55,56,57]. During phase 2 metabolism, the water solubility of the skatole metabolites is increased, facilitating excretion via urine [11,58]. Enzymes of phase 2 metabolism are mainly located in the liver, but can also be found other tissues, such as kidneys and lungs [48]. The main products of phase 1 and phase 2 metabolism are shown in Figure 1 (lower part).

The activity of the phase 1 enzymes is modulated by several gonadal steroids. It was shown that the addition of physiological concentrations of androstenone led to a significant reduction in the activity of CYP2E1 and CYP2A in hepatic microsomes from pigs [47,53], while the addition of 17ß- and 17a-estradiol only revealed such an inhibiting effect in supraphysiological concentrations [59]. Similarly, a higher enzyme activity of these two enzymes in the hepatic tissue of barrows was reported [60], compared to activity in tissue of boars. It was concluded that the lower skatole concentrations in adipose tissue of barrows were due to these differences. Higher levels of enzyme expression were reported for barrows and immunocastrates when compared to boars, but this high expression level did not always result in high enzyme activity, suggesting the influence of additional factors [50]. This hypothesis was further supported by an in vitro study [53], where gender specific effects of a pre-incubation with androstenone on the activity of CYP2E1 and CYP2A in liver microsomes were reported. In this study, enzymes derived from liver microsomes of female pigs were not inhibited by the pre-incubation with androstenone, up to 15 ng/mL, whereas the same enzymes from male liver samples exhibited a decreasing activity after the addition of 15 ng androstenone/mL or 0.5 ng estradiol/mL to the in vitro system. The addition of testosterone at a dose of 5 ng/mL was without effect in both genders. A slight increase (10%) in CYP2A activity was induced by adding 0.5 ng 17ß-estradiol/mL to liver microsomes from female pigs [53].

This effect of sex hormones on hepatic skatole degrading enzyme activity offers an explanation for the regularly reported higher concentrations of skatole in blood and tissue of boars compared to sows, gilts and barrows [7,45,61].

2.5. Accretion of Skatole in Adipose Tissue

As a lipophilic substance, skatole is accumulated in adipose tissue if blood levels of skatole are elevated for a prolonged period. Similarly, skatole concentrations in adipose tissue drop within days, if skatole formation in the colon is reduced due to feeding measures, such as inulin supplementation as described below [10]. These changes occur more rapidly than in the case of the even more lipophilic steroid androstenone [10].

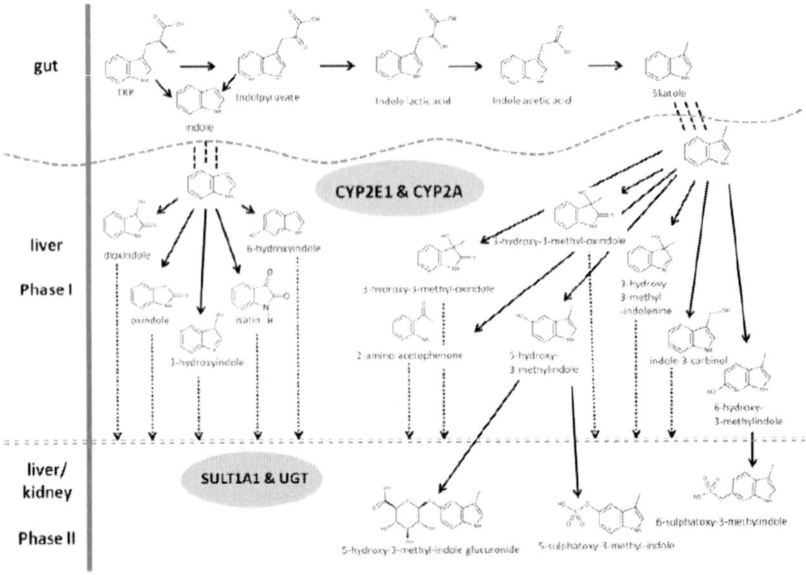

Figure 1. Formation of skatole (3-methyl-indole) and indole from TRP in the gut and further metabolism via phase 1 and phase 2 enzymes, (black arrows: known pathway; broken arrows: assumed pathway) [46,54, 55,57,62].

Adipose tissue differs within and between animals in its fatty acid turnover and in its stability according to its fatty acid composition. Main factors influencing fatty acid composition are the anatomical site (e.g., organ fat vs. subcutaneous or intramuscular fat), the overall degree of fatness of an individual or breed and the diet in monogastrics [63,64]. It is well established that leaner pigs have a lower ability to synthesize fatty acids combined with a greater mobilization, which results in adipose depots with more unsaturated lipids [65,66].

A relationship between skatole deposition and the turnover rate of adipose tissue has not yet been proven, but breed comparisons in skatole concentrations suggest such an influence. Lean breeds, such as the Pietrain, revealed generally lower skatole and androstenone concentrations than the fatter Large White breed [67]. Similarly, the distribution of skatole within the carcass is in accordance with the amount of saturated fatty acids of the specific adipose tissue. Both are higher in flare than in belly fat, and higher in belly fat than in adipose tissue from the neck [68,69,70], suggesting again that adipose tissue with a higher amount of SFA and lower turnover accumulates more skatole. Specific studies on fat turnover and skatole dynamics however need to be conducted.

2.6. Summary of the Physiological Mechanisms Leading to Elevated Skatole in Adipose Tissue

The mechanisms leading to elevated concentrations of skatole in adipose tissue of pigs are summarized in the Figure 2. High concentrations of skatole require (A) a high amount of TRP with low prececal digestibility or cell debris for microbial degradation in the colon, (B) specialized microbes for skatole synthesis, (C) insufficient alternative energy sources for microbial activity, so that the metabolism of TRP to skatole occurs instead of the synthesis of bacterial protein, (D) a high absorption rate, such as in the case of a long transient time of digesta, (E) a reduced degradation of skatole in phase 1 metabolism of liver and phase 2 metabolism in liver and kidney and (F) the deposition in adipose tissue, which requires continuously high concentrations of skatole in peripheral blood and a low turnover of adipose tissue.

This model allows a discussion of the mode of action of feed components on the complex mechanisms leading to a change of skatole concentrations in adipose tissue.

3. EFFECTS OF DIET AND FEED ADDITIVES ON SKATOLE FORMATION, ABSORPTION, METABOLISM AND ACCRETION IN ADIPOSE TISSUE OF PIGS

As shown in Figure 2, high or low concentrations of skatole in adipose tissue are the result of several independent steps, such as the formation, absorption, metabolism and accretion in adipose tissue. In the following section, feed components and feed additives which influence skatole dynamics are summarized and related to the physiological steps (Figure 2), contributing to skatole accumulation in adipose tissue. Feeding techniques, such as liquid feeding vs. dry feeding, exerted only minor effects on skatole formation [71,72]. Additionally, effects of feed intake behavior (feed intake rate, duration of meals) may explain some variability in skatole formation [73] but these aspects are not further discussed in this review.

3.1. Effects of Diet on TRP Availability in the Colon and Consequences for Skatole Formation (Step A, Figure 2)

The effect of TRP availability was investigated by feeding diets with high and low prececal digestibility. It was obvious, as stated earlier, that diets with free TRP supplementation had only a low effect on skatole formation [37], whereas feeding animals using products with a 20% lower prececal digestibility of TRP (blood, meat and bone meal) than the control diet led to increasing skatole concentrations in adipose tissue [38].

In a series of studies, Claus and co-workers changed the TRP availability in the colon by influencing the mitosis and apoptosis rate in the small and large intestine [29,39,40,70,74,75]. These studies utilized an increased supply of energy and purines to increase mitosis and consequently apoptosis [39,40], and also the application of either coated butyrate [74] or butyrate-producing feed components, such as raw potato starch [29,70,75]. The increase of mitosis—as confirmed by histological evaluations—resulted in the postulated increase in skatole in feces, blood and adipose tissue, whereas the inhibition of apoptotic processes in the colon, as in a part of the studies confirmed by histochemical evaluation of the mucosa, led to low skatole concentrations in all compartments analyzed [29,70,74,75]. In contrast, the addition of coated butyrate in low amounts from 1.2 to 1.5% was without effect on the skatole formation in the colon and its accretion in adipose tissue as well [76,77].

3.2. Effects of Diet or Feed Supplements on Microbial Population in the Gut and Consequences for Skatole Production (Step B, Figure 2)

As skatole is a product of microbial degradation, several strategies are documented to influence the microbial ecosystem in the gut of pigs to reduce skatole forming microbes. Such strategies include antimicrobial compounds or feed additives leading to changes in intestinal pH. It has been shown earlier that skatole-producing bacteria are favored under acidic conditions (pH 5.0), whereas indole-forming bacteria predominate if the pH was increased to 8.0 [5,18].

The most radical approaches to influence the microbial population within the gut are the applications of antibiotics such as Tylosin, Virginiamycin and Bacitracin and are mainly limited to studies during the late 90s [18,71,78,79]. Such rude treatments of the intestinal microbial ecosystem resulted in decreasing skatole concentrations in adipose tissue and/or feces if increasing concentrations of these substances were applied [18,71,78], whereas low doses were without effect on skatole in adipose tissue [79].

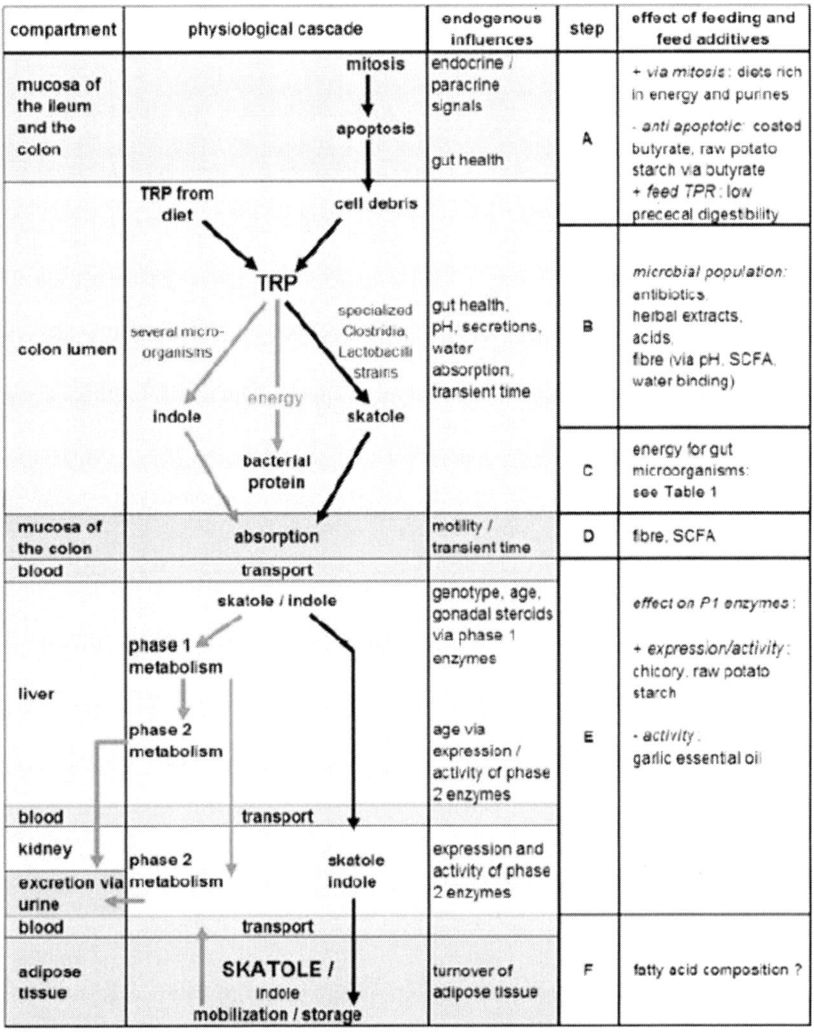

Figure 2. Cascade of physiological events leading to skatole formation, further metabolism and accumulation of skatole in adipose tissue. Steps correspond to Sections 3.1 (A) to 3.6 (F). The right column shows where distinct feeding influences exert their effects. Black arrows: steps leading to high skatole concentrations; brown arrows: skatole reducing or neutral conditions.

More recently, studies were published in which gut microbes were influenced via feeding of organic acids with weakly antibacterial properties (e.g., 1.0% formic acid, 0.85 % benzoic acid, 0.85% sorbic acid, corresponding to 0.85% of pure acid) or components, which are metabolized by gut microbes to terminal SCFA [76,77]. Digesta samples for microbiological examinations were taken from proximal jejunum, colon descendens, and rectum. Pigs fed organic acids had significantly lower levels of coliforms, enterococci, and lactic acid-producing bacteria in all sampling sites of the gastro-intestinal tract. Formic acid had a stronger antibacterial effect on coliforms than benzoic or sorbic acids in the small intestine. It was concluded that supplementing diets with different organic acids reduces the number of coliforms, enterococci, and lactic acid-producing bacteria in the gastrointestinal tract. Even if there were no differences in levels of skatole or indole in colon descendens among pigs fed organic acid-supplemented diets, compared to a control diet, plasma skatole levels were significantly reduced in pigs fed diets containing formic acid or benzoic acid [76]. The levels of skatole in entire male pigs however were not reduced in a similar study [77].

The antimicrobial activity of selected plant extracts, such as Chinese medicinal herbs or feed additive derived from Sanguinaria canadensis, essential oil components from herbs and spices or tannin-rich preparations, may offer a strategy to reduce skatole formation [41,80,81,82]. The addition of tannin-rich extracts of several plants inhibits either microbial activity directly or by reducing the availability of proteins for bacterial metabolism. Such extracts were effective to reduce skatole *in vitro* [82], and *in vivo* in ewes [83]. A study with essential oil components from herbs and spices revealed that carvacrol, thymol, eugenol and trans-cinnamaldehyde influence the microbial ecosystem and fermentation pattern in the gastrointestinal tract of pigs [81]. Under conditions of the jejunum, carvacrol and thymol showed very similar and non-selective antimicrobial properties. In cecal simulations, carvacrol, thymol and trans-cinnamaldehyde were equally effective while eugenol had an effect only on coliforms, but the effects of skatole, however, were not investigated. The addition of Chinese medicinal herbs to the diets of piglets after weaning had a clear effect on the bacterial population in the gut, reduced the occurrence of diarrhea and led to an increased villus height in the small intestine, suggesting reduced availability of cell debris. The effect on skatole formation was not measured [41].

A feed additive derived from Sanguinaria canadensis contains large amounts of sanguinarine, an alkaloid with known *in vitro* antimicrobial properties [84]. It was postulated that this alkaloid inhibits the activity of the amino acid, decarboxylase, which is involved in the synthesis of indole and skatole [85] and its application to prevent skatole formation has been claimed by a patent (EP 0581926, US 20030190344). Convincing studies with growing pigs, however, were not published. Approaches to control skatole formation by the addition of plant extracts may be attractive, as they reveal antimicrobial properties at very low concentrations [41,81] but need more systematic investigations.

In contrast, the addition of different sources of fermentable carbohydrates to the diets of growing pigs led to reduced skatole formation or accretion in adipose tissue in several studies. These studies are summarized in Table 1. Easily fermentable carbohydrates are thought to exert two main effects: One is the change of intraluminal pH within the colon, which is assumed to influence the microbial ecosystem of the colon. The other is related to their function as a microbial energy source and further discussed in Section 3.3.

Thus, *in vitro* results after the addition of 0.5–1.5% fructooligosaccharides (FOS), which revealed an increase in indole formation but not in skatole, were discussed as a result of a changed microbial ecosystem via feed-induced pH changes [86]. Similarly, differences in the intraluminal pH were reported after feeding oat-based diets versus barley-based diets [87]. Oat-based diets led to a higher pH and concomitantly higher indole concentrations when compared to the barley-based diets. As skatole concentrations were not changed by the treatments, it was postulated that indole-producing bacteria favored the higher pH, as similarly shown in *in-vitro* studies [5,18].

Even if the mechanisms were not clarified in detail, differences in the microbial populations were reported after the supplementation of diets of growing entire male pigs during the last weeks before slaughter with either chicory inulin (see Section 3.3) or potato starch in increasing amounts (0 to 15% in pellets). A reduction of enterobacteria counts in colon descendens and rectum was shown only for the inulin-containing diets with a tendency towards reduced levels of Enterococcus spp. in colon descendens and rectum. The concentrations of skatole in the digesta were also reduced in the case of ≥6% of inulin supplementation, but not in the potato starch groups as further discussed below. The addition of raw potato starch in feed pellets was without effect on the microbial population [88].

The effect of a high protein diet may be associated with higher prevalence of the Clostridiaceae family and the Clostridium genus, which are predominantly proteolytic microbes. Furthermore, a reduced prevalence of the order Clostridiales, particularly the predominately saccharolytic microbes, especially butyrate producers, have to be expected (for review see [89]). Consequences for skatole formation, however, were not studied in detail but may favor skatole production.

3.3. Dietary Manipulations Resulting in Higher Energy Availability for Microbial Activity and Their Consequences for Skatole Formation (Step C, Figure 2)

Most of the studies which were effective in reducing skatole formation and deposition in adipose tissue increased energy availability and shifted microbial metabolism from proteolytic to saccharolytic. This feeding strategy was suggested in an early study in which inulin supplementation and bicarbonate were effective in significantly reducing skatole in pig feces. The decrease in adipose tissue was confirmed by two consecutive biopsies one week apart [10]. The importance of fiber as such an energy source was finally proven by the elegant experiment described in part already in Section 2.3 of this review, reporting differences in the conversion rate of TRP infused into the cecum to skatole, if animals were either on a diet with a low or high amount of fiber [14]. Such a diet influenced both the production of skatole in the hindgut and the absorption of skatole to the portal blood. With the low fiber diet, the hindgut bacteria transformed 26% of the infused tryptophan into skatole, resulting in a significant increase of skatole concentration in the portal blood, compared to 6% in the case of a high fiber diet. In the case of the high fiber diet, only 57% of the infused TRP was converted into indoles, whereas in animals on low fiber diets, the total amount was degraded to indoles [14]. It might be assumed that in the case of the high fiber diet, the remaining TRP was used for the synthesis of bacterial protein as similarly discussed after the application of fructooligosaccharides (FOS) to reduce skatole concentrations [86].

Feed supplementation for two weeks with 10% of either raw potato starch, FOS, such as inulin, lupines and barley hull meal was significantly effective in reducing skatole concentrations in daily blood samples of pigs, when compared to a control diet (barley, soya). The addition of either palm cake, coconut cake or sugar beet pulp at the same concentration revealed no significant effect [14]. Since then, numerous studies have been carried out to evaluate the minimum dosage of such fiber additive to reduce skatole formation and thus the risk of skatole tainted pork. The most prominent fiber sources are either chicory inulin or raw potato starch (RPS) added to the diet during the last weeks before slaughter. Such studies are summarized in Table 1.

Diets supplemented with RPS proved to be effective in reducing skatole concentrations in hind-gut and adipose tissue, if at least 20% was added to the diet [29,75,90]. Reducing this amount of RPS to 10% of the diet, in contrast to blood, resulted in no significant reduction of skatole in adipose tissue being observed [18,91].

Table 1. Feed components used to influence skatole formation and accretion via fiber supplementation in recent studies and their effectiveness as influenced by the amount in the diet and duration of application. Amounts of feed components and inulin are given as percentage of the fresh matter (DM: dry matter; n.d.: not determined; +: significant decrease in skatole concentration ($p < 0.05$); (+): tendency for decrease in skatole concentration ($p < 0.10$); - no significant change in skatole concentration).

feed component	% in diet	% effective ingredient in diet	feeding duration (d)	decrease of skatole ($p < 0.05$) in			[ref.]
				adipose tissue	blood	digesta/feces	
feed components with main effective ingredient inulin							
dried chicory roots	10	n.d.	16	-	-	n.d	[11]
	10-13.3	3.6–4.7	7–14	-	-	n.d	[94]
	25	13.95	7	+	+	n.d	[12]
			14	+	+	n.d	
			42	+	+	n.d	
crude chicory roots (25% DM)	56.3	6.9	28	+	+	n.d	
			42	+	+	n.d	
			63	+	+	n.d	
dried chicory root extract	3	1.8	14	-	n.d	n.d	[13]
		2.1	28	-	n.d	-	[93]
		2.1	30	-	n.d	-	[88]
	6	3.6	14	+	n.d	n.d	[13]
		4.2	28	+	n.d	+	[93]
		4.2	30	+	n.d	+	[88]
	9	5.4	14	+	n.d	n.d	[13]
		6.3	28	+	n.d	+	[93]
		6.3	30	+	n.d	+	[88]

Inulin	16.3	15.5	42	+	+	n.d	[12]
Jerusalem artichoke	8.1	4.2	7	-	n.d	+	[95]
	12.2	6.3	7	(+)	n.d	+	
other feed components							
raw potato starch	10	n.d.	28–42	-	n.d	n.d	[91]
	10	n.d.	14	n.d.	+	n.d.	[18]
	20	n.d.	14	+	n.d	n.d	[90]
		n.d.	14–21	+	n.d	+	[75]
	30	n.d.	14–21	+	n.d	+	
lupin seeds	10	n.d.	28–42	-	n.d	n.d	[91]
	25	n.d.	30	+	n.d	+	[88]
sugar beet pulp	10	n.d	28	n.d	+	-	[28]
	15	n.d	fattening period	-	n.d	n.d	[96]
	20	n.d.	30	+	n.d	n.d	[97]

In contrast to the effect of raw potato starch powder, the addition of the same substance before pelleting did not influence skatole concentrations [88], since the process of gelatinisation at high temperatures increased prececal digestibility [92]. Therefore this application type was chosen as a control, when specific effects of raw potato starch on skatole formation were investigated [29,75].

Similar effects to those of RPS on skatole concentrations in fat or feces could be achieved by adding at least 6% inulin via the addition of chicory root extract to the diet [88,93]. However, adding 5% of a chicory extract, containing 66% inulin, had no effect [91]. It has to be kept in mind that these additives may exert their effect also via other mechanisms of skatole formation, e.g., via butyrate formation, as described previously for TRP availability (Section 3.1), or microbial population (Section 3.2). Additional effects on hepatic skatole metabolism are discussed later (Section 3.5).

Four types of carbohydrate sources, sugar beet pulp (SBP), rye grass hay (RYE), alfalfa hay (ALF) and FOS were compared in their effects on the *in vitro* metabolism of L-tryptophan to skatole and indole by a mixed bacterial population from the large intestines of pigs [25]. Addition of SBP or FOS showed a significant inhibitory effect on skatole formation and relative production rate.

In contrast, the addition of both types of hay significantly increased skatole concentrations and the relative rate of skatole production after 24 h. The authors postulated that in the presence of fermentable carbohydrates these are used preferentially as an energy source by the intestinal bacteria, leading to a reduced protein catabolism [25,98]. In the case of a relatively higher fermentable RYE and ALF, the carbohydrate source is rapidly depleted (within 15 h), which in turn leads to a switch in the metabolism to proteolysis thereafter [25]. The availability of the slow fermentable SBP may provide a carbohydrate source for the bacteria over the whole investigation period of 24 h, preventing a metabolic switch to proteolysis of the skatole producing microbes. These results suggested that reduced concentration of skatole observed in the presence of SBP and FOS may be caused by an overall decreased rate of tryptophan degradation [25].

3.4. Effect of Feeding on the Absorption Rate of Skatole (Step D, Figure 2)

The effect of transient time on skatole absorption was more discussed than investigated. Positive correlations between dry matter content of feces and skatole concentrations in feces (μg/g DM) were reported for growing pigs [37]. It was discussed for other species that the formation of short-chain fatty acids, methane, CO_2 and other products from the bacterial metabolism of inulin may also contribute to increased peristalsis in the whole gastro intestinal tract, leading to changed conditions for bacterial fermentation and intestinal absorption in general [99].

Studies with the addition of complex clay minerals to the diet of pigs to bind skatole to prevent the absorption were published, but were not successful in all studies [91,100]. Similar approaches with more advanced additives, which should bind skatole and thus reduce the absorption rate, were carried out as in vitro studies, were not proven to be effective in vivo, but may be promising [101].

3.5. Effects of Feeding on Degradation of Skatole in the Liver and Kidney (Step E, Figure 2)

Various components of feed may either increase or inhibit the activity of skatole degrading enzymes of the phase 1 metabolism. The most impressive effect was obtained by adding garlic essential oil at a dose from zero to 2.15 g/kg to the feed. Concentrations of skatole and indole in adipose tissue increased after 57 days of feeding 2.15 g/kg garlic essential oil from 39.6 ng/g skatole and 35.8 ng/g indole in the control group to 1,001.5 ng/g skatole and 972.5 ng/g indole in the garlic supplemented group. It was suggested that the sulfur containing compounds (allyl di- and tri-sulfides) had an inhibitory effect on the CYP-dependent skatole metabolism, thereby decreasing clearance of skatole and indole from the blood, since similar effects have been observed in other species [38,83,102].

Stimulatory effects of feed components on enzyme expression, as characterized by measuring either transcription through mRNAs or the enzyme activity, as characterized via enzymatic conversion rates, were reported for enzymes of the phase 1 metabolism in the liver [52,97]. For example, activity or expression of phase 1 enzymes were increased after the addition of dried chicory [52], sugar beet pulp [97] and potentially raw potato starch [90] to the diets. Similar effects were not described for any of the phase 2 enzymes [11].

Interesting differences between expression levels, protein concentrations and final enzyme activities were reported [52], since a 10% increase in mRNA for CYP1a2, CYP2a and CYP2E1 was measurable after addition of 10% dried chicory to diets. However, protein concentrations and enzyme activity were only higher for CYP1a2, and CYP2a in that experiment. A similar effect on the metabolism of androstenone after feeding chicory roots via an influence on hepatic 3ß-HSD (3ß-hydroxy steroid dehydrogenase) has been recently reported [11]. However, this feedstuff seems to exhibit a highly variable effect due to varying concentrations of inulin in the dried roots [94].

Linseed within a diet was also shown to reduce skatole concentrations in adipose tissue of pigs [103]. It was suggested by the authors that myristicin (a benzodioxole compound found in linseed) induces hepatic P450 enzymes such as CYP2E1, which in turn leads to an increased degradation of skatole [104].

3.6. Effects of Feeding on Accumulation of Skatole in Adipose Tissue (Step F, Figure 2)

Only a few studies with parallel monitoring of skatole concentrations in blood and adipose tissue biopsies were carried out [10]. Only very little data are available, which allow conclusions to be drawn about the effect of fatty acid composition of feed and adipose tissue on skatole in relation to tissue turnover rate. For example, feeding a linseed diet which provides an elevated amount of PUFA led to a lower skatole accretion. However, this result was not discussed with turnover phenomena [103]. A contribution of dietary manipulations on skatole concentrations via the influence of an increased fat turnover may be discussed and needs further investigation.

4. CONCLUSIONS

The reliable control of skatole accretion in the fat of boars is one of the main prerequisites for pork production with entire males. Skatole formation in the gut is the result of several independent conditions, leading to elevated TRP availability for specialized bacteria in need of this substrate for energy production. Even high skatole formation in the colon does not necessarily elevate skatole in adipose tissue, as absorption varies and the metabolism mainly in the liver may degrade sufficient amounts of skatole to ensure low adipose tissue concentrations. Only if the production and absorption exceed the ability of the liver to degrade skatole via mainly phase 1 enzymes, then skatole may accumulate in adipose tissue. This accumulation may be further modified by adipose tissue turnover; however, this aspect needs further clarification.

Nutritional effects on almost all steps of the process leading to skatole accumulation in adipose tissue have been described. The effectiveness of a feeding intervention to reduce skatole concentrations in adipose tissue is high, if the feed additive affects several steps of skatole synthesis and accumulation. Thus the most efficient feeding interventions are actually the addition of inulin or raw potato starch, which influence the microbial ecosystem, probably via the intestinal pH, the energy availability for bacteria and the TRP availability via possible anti-apoptotic effects. Inulin seems to be effective at lower doses than raw potato starch. The antimicrobial activity of selected plant extracts or essential oil components from herbs and spices may offer an additional strategy to reduce skatole formation.

ACKNOWLEDGMENTS

We wish to thank Christine Baes (SUISAG, Sempach, Switzerland) for proof reading and Carmen Yvonne Ostertag for support.

REFERENCES AND NOTES

1. Bonneau, M. Use of entire males for pig meat in the European Union. Meat Sci. 1998, 49, 257–272.
2. Patterson, R.L.S. 5-alpha-androst-16-en-3-one: Compound responsible for taint in boar fat. J. Sci. Food Agric. 1968, 19, 31–38.
3. Claus, R.; Hoffman, B.; Karg, H. Determination of 5 -androst-16-en-3-one, a boar taint steroid in pigs, with reference to relationships to testosterone. J. Anim. Sci. 1971, 33, 1293–1297
4. Vold, E. Fleischproduktionseigenschaften bei Ebern und Kastraten. IV. Organoleptische und gaschromatographische Untersuchungen wasserdampfflüchtiger Stoffe des Rückenspecks von Ebern; Report No. 238; Institute of Animal Genetics and Breeding, NLH: Vollebekk, Norway, 1970.
5. Jensen, M.T.; Cox, R.P.; Jensen, B.B. 3-Methylindole (skatole) and indole production by mixed populations of pig fecal bacteria. Appl. Environ. Microb. 1995, 61, 3180–3184.
6. Yokoyama, M.T.; Carlson, J.R. Microbial metabolites of tryptophan in the intestinal tract with special reference to skatole. Am. J. Clin. Nutr. 1979, 32, 173–178.
7. Weiler, U.; Font I Furnols, M.; Fischer, K.; Kemmer, H.; Oliver, M.A.; Gispert, M.; Dobrowoski, A.; Claus, R. Influence of differences in sensitivity of Spanish and German consumers to perceive androstenone on the acceptance of boar meat differing in skatole and androstenone concentrations. Meat Sci. 2000, 54, 297–304.
8. Robic, A.; Larzul, C.; Bonneau, M. Genetic and metabolic aspects of androstenone and skatole deposition in pig adipose Tissue: A Review. Genet. Sel. Evol. 2008, 40, 129–143
9. Zamaratskaia, G.; Squires, E.J. Biochemical, nutritional and genetic effects on boar taint in entire male pigs. Animal 2009, 3, 1508–1521.
10. Claus, R.; Weiler, U.; Herzog, A. Physiological aspects of androstenone and skatole formation in the boar—A review with experimental data. Meat Sci. 1994, 38, 289–305.
11. Rasmussen, M.K.; Brsunius, C.; Zamaratskaia, G.; Ekstrand, B. Feeding dried chicory root to pigs decrease androstenone accumulation in fat by increasing hepatic 3β hydroxysteroid dehydrogenase expression. J. Steroid Biochem. Mol. Biol. 2012.
12. Hansen, L.L.; Mejer, H.; Thamsborg, S.M.; Byrne, D.V.; Roepstorff, A.; Karlsson, A.H.; Hansen-Møller, J.; Jensen, M.T.; Tuomola, M. Influence of chicory roots (Cichorium Intybus L.) on boar taint in entire male and female pigs. Anim. Sci. 2006, 82, 359–368.
13. Zammerini, D.; Wood, J.D.; Whittington, F.M.; Nute, G.R.; Hughes, S.I.; Hazzledine, M.; Matthews, K. Effect of dietary chicory on boar taint. Meat Sci. 2012.
14. Jensen, B.B. Prevention of boar taint in pig production. Factors affecting the level of skatole. Acta Vet. Scand. 2006, 48, S6.
15. Prunier, A.; Bonneau, M. Alternatives to piglet castration. Prod. Anim. 2006, 19, 347–356
16. Deslandes, B.; Gariepy, C.; Houde, A. Review of microbiological and biochemical effects of skatole on animal production. Livest. Prod. Sci. 2001, 71, 193–200.

17. Annor-Frempong, I.E.; Nute, G.R.; Whittington, F.W.; Wood, J.D. The problem of taint in pork. III. Odour profile of pork fat and the Interrelationships between androstenone, skatole and indole concentrations. Meat Sci. 1997, 47, 63–76.
18. Jensen, B.B.; Jensen, M.T. Microbial production of skatole in the digestive tract of entire male pigs. In Skatole and Boar Taint; Jensen, W.K., Ed.; Danish Meat Research Institute: Roskilde, Denmark, 1998; pp. 41–76.
19. Cook, K.L.; Rothrock JR., M.J.; Loughrin, J.H.; Doerner, K.C. Characterization of skatole-producing microbial populations in enriched swine lagoon slurry. FEMS Microbiol. Ecol. 2007, 60, 329–340.
20. Whithehead, T.R.; Price, N.P.; Drake, H.L.; Cotta, M.A. Catabolic pathway for the production of skatole and indoleacetic acid by the acetogen clostridium drakei, clostridium scatologenes, and swine manure. Appl. Environ. Microbiol. 2008, 74, 1950–1953.
21. Jensen, M.T; Jensen, B.B. Determination of indole and 3-methylindole (skatole) in bacterial cultures, intestinal content and faeces. J. Chrom. B 1993, 665, 275–280.
22. Yokoyama, M.T.; Johnson, K.A.; Carlson, J.R. Factors influencing the production of p-cresol and skatole by Lactobacillus isolated from the rumen and pig feces. In Proceedings of XVII Conference on Rumen Function, Chicag, IL, USA, 1–2 December 1983; 17, pp. 19–20.
23. Doerner, K.C.; Cook, K.L.; Mason, B.P. 3-Methylindole production is regulated in clostridium scatologenes ATCC 25775. Lett. Appl. Microbiol. 2009, 48, 125–132.
24. Li, C.-Y.; Wu, C.; Liu, J.-X.; Wang, Y.-Z.; Wang, J.-K. Spatial variation of intestinal skatole production and microbial community in Jinhua and Landrace pigs. J. Sci. Food Agr. 2009, 89, 639–644.
25. Li, C.-Y.; Liu, J.-X.; Wang, Y.-Z.; Wu, Y.-M.; Wang, J.-K.; Zhou, Y.-Y. Influence of differing carbohydrate sources on l-tryptophan metabolism by porcine fecal microbiota studied in vitro. Livest. Sci. 2009, 120, 43–50.
26. Jensen, B.B.; Jorgensen, H. Effect of dietary fiber on microbial activity and microbial gas production in various regions of the gastrointestinal tract of pigs. Appl. Environ. Microbiol. 1994, 60, 1897–1904.
27. Claus, R.; Dehnhard, M.; Herzog, A.; Bernal-Barragan, H.; Giménez, T. Parallel measurements of indole and skatole (3-methylindole) in feces and blood plasma of pigs by HPLC. Livest. Prod. Sci. 1993, 34, 115–126
28. Knarreborg, A.; Beck, J.; Jensen, M.T.; Laue, A.; Agergaard, N.; Jensen, B.B. Effect of non-starch polysaccharides on production and absorption of indolic compounds in entire male pigs. Anim. Sci. 2002, 74, 445–453.
29. Lösel, D.; Lacorn, M.; Büttner, D.; Claus, R. Flavor improvement in pork from barrows and gilts via inhibition of intestinal skatole formation with resistant potato starch. J. Agr. Food Chem. 2006, 54, 5990–5995.
30. Lanthier, F.; Lou, Y.; Squires, E.J. Skatole metabolism in the intact pre-pubescent male pig: The relationship between hepatic enzyme activity and skatole concentrations in plasma and fat. Livest. Sci. 2007, 106, 145–153.
31. Hawe, S.M.; Walker, N.; Moss, B.W. Effects of infusing skatole into the terminal ileum of growing male pigs. Livest. Prod. Sci. 1993, 33, 267–276.
32. Agergaard, N.; Laue, A. Absorption of skatole to portal vein blood and liver turnover in entire male pigs using an in vivo animal model. In Skatole and Boar Taint; Jensen, W.K., Ed.; Danish Meat Research Institute: Roskilde, Denmark, 1998; pp. 77–96.
33. Nocerini, M.R.; Honeyfield, D.C.; Carlson, J.R.; Breeze, R.G. Reduction of 3-methylindole production and prevention of acute bovine pulmonary edema and emphysema with lasalocid. J. Anim. Sci. 1985, 60, 232–238.
34. Schreurs, N.M.; Tavendale, M.H.; Lane, G.A.; Barry, T.N.; Marotti, D.M.; McNabb, W.C. Postprandial indole and skatole formation in the rumen when feeding white clover, perennial ryegrass and lotus corniculatus. Proc. New Zealand Soc. Anim. Prod. 2003, 63, 14–17
35. Pedersen, J.K.; Mortensen, A.B.; Madsen, A.; Mortensen, H.P.; Hyldgaard-Jensen, J. Foderets indflydelse pa ornelugt i svinekod. Statens Hysdyrbrugsforsog. Meddelse 1986, 638, 1–7.

36. Mortensen, H.P. Influence of Breed Energy and Protein in the Feed on Skatole Content in Female Pigs, Castrates and Entire Male Pigs; Manuscript No. 837E; Danish Meat Research Institute: Roskilde, Denmark, 1989.
37. Bernal-Barragan, H. Physiologische und nutritive Einflüsse auf die Bildung von Skatol (3-Methylindol) im Dickdarm von Schweinen. Ph.D. Thesis, Universität Hohenheim, Stuttgart, Germany, 1992.
38. Leong, J.; Morel, P.C.H.; Purchas, R.W.; Wilkinson, B.H.P. Effects of dietary components including garlic on concentrations of skatole and indole in subcutaneous fat of female pigs. Meat Sci. 2011, 88, 45–50.
39. Claus, R.; Raab, S. Influences on skatole formation from tryptophan in the pig colon. Adv. Exp. Med. Biol. 2000, 467, 679–684.
40. Raab, S.; Leiser, R.; Kemmer, H.; Claus, R. Effects of energy and purines in the diet on proliferation, differentiation, and apoptosis in the small intestine of the pig. Clin. Exp.Metabolism 1998, 47, 1105–1111.
41. Huang, C.W.; Lee, T.T.; Shih, Y.C.; Yu, B. Effects of dietary supplementation of Chinese medicinal herbs on polymorphonuclear neutrophil immune activity and small intestinal morphology in weanling pigs. J. Anim. Physiol. Anim. Nutr. 2011.
42. Lanthier, F.; Lou, Y.; Terner, M.A.; Squires, E.J. Characterizing developmental changes in plasma and tissue skatole concentrations in the prepubescent intact male pig. J. Anim. Sci. 2006, 84, 1699–1708
43. Zamaratskaia, G. Factors involved in the development of boar taint Influence of breed, age, diet and raising conditions. Ph.D. Thesis, Swedish University of Agricultural Sciences, Uppsala, Swedish, 2004.
44. Agergaard, N.; Laue, A. Absorption from the gastrointestinal tract and liver turnover of skatole. In Measurement and Prevention of Boar Taint in Entire Male Pigs; Bonneau, M., Ed.; Institut National de la Recherche Agronomique: Paris, France, 1993.
45. Babol, J.; Squires, E.J.; Lundström, K. Hepatic metabolism of skatole in pigs by cytochrome P4502E1. J. Anim. Sci. 1998, 76, 822–828.
46. Squires, E.J.; Lundström, K. Relationship between cytochrome P450IIE1 in liver and levels of skatole and its metabolites in intact male pigs. J. Anim. Sci. 1997, 75, 2506–2511
47. Doran, E.; Whittington, F.W.; Wood, J.D.; McGivan, J.D. Cytochrome P450IIE1 (CYP2E1) is induced by skatole and this induction is blocked by androstenone in isolated pig hepatocytes. Chem. Biol. Interact. 2002, 140, 81–92.
48. Sinclair, P.A.; Squires, E.J. Testicular sulfoconjugation of the 16-androstene steroids by hydroxysteroid sulfotransferase: Its effect on the concentrations of 5α-androstenone in plasma and fat of the mature domestic boar. J. Anim. Sci. 2005, 83, 358–365.
49. Wiercinska, P.; Lou, Y.; Squires, E.J. The roles of different porcine cytochrome P450 enzymes and cytochrome b5A in skatole metabolism. Animal 2011
50. Brunius, C.; Rasmussen, M.K.; Lacoutiére, H.; Andersson, K.; Ekstrand, B.; Zamaratskaia, G. Expression and activities of hepatic cytochrome P450 (CYP1A, CYP2A and CYP2E1) in entire and castrated male pigs. Animal 2012, 6, 271–277.
51. Zamaratskaia, G.; Zlabek, V.; Ropstad, E.; Andresen, Ø. In vitro and in vivo association of porcine hepatic cytochrome P450 3A and 2C activities with testicular steroids. Reprod. Dom. Anim. 2012.
52. Rasmussen, M.K.; Zamaratskaia, G.; Ekstrand, B. In vivo effect of dried chicory root (Cichorium intybus L.) on xenobiotica metabolising cytochrome P450 enzymes in porcine liver. Toxicol. Lett. 2011, 200, 88–91.
53. Rasmussen, M.K.; Zamaratskaia, G.; Ekstrand, B. Gender-related differences in cytochrome P450 in porcine liver—Implication for activity, expression and inhibition by testicular steroids. Reprod. Dom. Anim. 2011, 46, 616–623.
54. Terner, M.A.; Gilmore, W.J.; Lou, Y.; Squires, E.J. The role of CYP2A and CYP2E1 in the metabolism of 3-methylindole in primary cultured porcine hepatocytes. Drug Metab. Dispos. 2006, 34, 848–854.
55. Diaz, G.J.; Squires, E.J. Phase II in vitro metabolism of 3-methylindole metabolites in porcine liver. Xenobiotica 2003, 33, 485–498.

56. Diaz, G.J.; Skordos, K.W.; Yost, G.S.; Squires, E.J. Identification of phase I metabolites of 3-methylindole produced by pig liver microsomes. Drug Metab. Dispos. 1999, 27, 1150–1156.
57. Baek, C.; Hansen-Moller, J.; Friis, C.; Cornett, C.; Hansen, S.H. Identification of selected metabolites of skatole in plasma and urine from pigs. J. Agric. Food Chem. 1997, 45, 2332–2340.
58. Sinclair, P.A.; Hancock, S.; Gilmore, W.J.; Squires, E.J. Metabolism of the 16-androstene steroids in primary cultured porcine hepatocytes. J. Steroid Biochem. Mol. Biol. 2005, 96, 79–87.
59. Zamaratskaia, G.; Rasmussen, M.K.; Herbin, I.; Ekstrand, B.; Zlabek, V. In vitro inhibition of porcine cytochrome P450 by 17β-estradiol and 17α-estradiol. Interdis.Toxicol. 2011, 4, 78–84.
60. Zamaratskaia, G.; Zlabek, V.; Chen, G.; Madej, A. Modulation of porcine cytochrome P450 enzyme activities by surgical castration and immunocastration. Animal 2009, 3, 1124–1132.
61. Prusa, K.; Nederveld, H.; Runnels, P.L.; Li, R.; King, V.L.; Crane, J.P. Prevalence and relationships of sensory taint, 5α-androstenone and skatole in fat and lean tissue from the loin (Longissimus dorsi) of barrows, gilts, sows, and boars from selected abattoirs in the United States. Meat Sci. 2011, 88, 96–101.
62. Gillam, E.M.J.; Notley, L.M.; Cai, H.; De Voss, J.J.; Guengerich, F.P. Oxidation of indole by cytochrome P450 enzymes. Biochemistry 2000, 39, 13817–13824.
63. Chilliard, Y. Dietary fat and adipose tissue metabolism in ruminants, pigs, and rodents: A review. J. Dairy Sci. 1993, 76, 3897–3931.
64. Lo Fiego, D.P.; Santoro, P.; Macchioni, P.; De Leonibus, E. Influence of genetic type, live weight at slaughter and carcass fatness on fatty acid composition of subcutaneous adipose tissue of raw ham in the heavy pig. Meat Sci. 2005, 69, 107–114.
65. Metz, S.H.M.; Dekker, R.A. The contribution of fat mobilization to the regulation of fat deposition in growing Large White and Pietrain pigs. Animal Prod. 1981, 33, 149–157.
66. Scott, R.A.; Cornelius, S.G.; Mersmann, H.J. Fatty acid composition of adipose tissue from lean and obese swine. J. Anim. Sci. 1981, 53, 977–981
67. Aluwé, M.; Millet, S.; Bekaert, K.M.; Tuyttens, F.A.M.; Vanhaecke, L.; De Smet, S.; De Brabander, D.L. Influence of breed and slaughter weight on boar taint prevalence in entire male pigs. Animal 2011, 5, 1283–1289.
68. Monziols, M.; Bonneau, M.; Davenel, A.; Kouba, M. Comparison of the lipid content and fatty acid composition of intermuscular and subcutaneous adipose tissues in pig carcasses. Meat Sci. 2007, 76, 54–60.
69. Weiler, U.; Dehnhard, M.; Herbert, E.; Claus, R. Einfluss von Geschlecht, Genotyp und Mastendgewicht auf die Androstenon-und Skatolkonzentrationen im Fett von Mastschweinen. Schriftenreihe BML, Reihe A 1995, 449, 14–32.
70. Lösel, D. Versuche zur Verbesserung der sensorischen Fleischqualität beim Schwein durch nutritive Skatolhemmung. Ph.D. Thesis, Universität Hohenheim, Stuttgart, Germany, 2006.
71. Hansen, L.L.; Mikkelsen, L.L.; Agerhem, H.; Laue, A.; Jensen, M.T.; Jensen, B.B. Effect of fermented liquid food and zinc bacitracin on microbial metabolism in the gut and sensoric profile of m. longissimus dorsi from entire male and female pigs. Anim. Sci. 2000, 71, 65–80.
72. Andersson, K.; Schaub, A.; Andersson, K.; Lundström, K.; Thomke, S.; Hansson, I. The effects of feeding system, lysine level and gilt contact on performance, skatole levels and economy of entire male pigs. Livest. Prod. Sci. 1997, 51, 131–140.
73. Weiler, U.; Götz, M.; Schmidt, A.; Otto, M.; Müller, S. Influences of sex and immunocastration on feed intake behavior, skatole and indole formation in pigs. 2012. Unpublished work.
74. Claus, R.; Lösel, D.; Lacorn, M.; Mentschel, J.; Schenkel, H. Effects of butyrate on apoptosis in the pig colon and its consequences for skatole formation and tissue accumulation. J. Anim. Sci. 2003, 81, 239–248.

75. Lösel, D.; Claus, R. Dose-dependent effects of resistant potato starch in the diet on intestinal skatole formation and adipose tissue accumulation in the pig. J. Vet. Med. A 2005, 52, 209–212.
76. verland, M.; Kjos, N.P.; Borg, M.; Sørum, H. Organic acids in diets for entire male pigs. Livest. Sci. 2007, 109, 170–173.
77. verland, M.; Kjos, N.P.; Borg, M.; Skjerve, E.; Sørum, H. Organic acids in diets for entire male pigs: Effect on skatole level, microbiota in digesta, and growth performance. Livest. Sci. 2008, 115, 169–178.
78. Hansen, L.L.; Larsen, A.E.; Jensen, B.B.; Hansen-Møller, J. Short time effect of zinc bacitracin and heavy fouling with faeces plus urine on boar taint. Anim. Sci. 1997, 64, 351–363.
79. Hansen, L.L.; Larsen, A.E. Effect of antibiotic feed additives on the level of skatole in fat of male pigs. Livest. Prod. Sci. 1994, 39, 269–274.
80. Blank, R.; Müller-Siegwardt, B.; Wolffram, S. Sanguinarine does not influence availability or metabolism of tryptophan in pigs. Livest. Sci. 2010, 134, 24–26.
81. Michiels, J.; Missotten, J.A.M.; Fremaut, D.; De Smet, S.; Dierick, N.A. In vitro characterisation of the antimicrobial activity of selected essential oil components and binary combinations against the pig gut flora. Anim. Feed Sci. Technol. 2009, 151, 111–127.
82. Tavendale, M.H.; Lane, G.A.; Schreurs, N.M.; Fraser, K.; Meagher, L.P. The effects of condensed tannins from Dorycnium rectum on skatole and indole ruminal biogenesis for grazing sheep. Aust. J. Agr. Res. 2005, 56, 1331–1337.
83. Roy, N.C.; Fraser, K.; Lane, G.A.; Sinclair, B.R.; McNabb, W.C. Polyethylene glycol increases intestinal absorption and hepatic uptake of indole and skatole in sheep fed sulla. J. Anim. Feed Sci. 2002, 13, 339–342.
84. Newton, S.M.; Lau, C.; Gurcha, S.S.; Besra, G.S.; Wright, C.W. he evaluation of forty-three plant species for in vitro antimycobacterial activities; Isolation of active constituents from Psoralea corylifolia and Sanguinaria canadensis. J. Ethnopharmacol. 2002, 79, 57–67
85. Ehrlinger, M. Phytogene Zusatzstoffe in der Tierernährung. Ph.D. Thesis, Ludwig-Maximilians-Universität, München, Germany, 2007.
86. Xu, X.; Hu, C.; Wang, M. Effects of fructooligosaccharide on conversion of L-tryptophan to skatole and indole by mixed populations of pig fecal bacteria. J. Gen. Appl. Microbiol. 2002, 48, 83–89.
87. Pauly, C.; Spring, P.; Gahan, D.; Odoherty, J.V. The effect of cereal type and enzyme supplementation on carcass characteristics, volatile fatty acids and intestinal microflora and boar taint in entire male pigs. Animal 2011, 5, 378–386.
88. Øverland, M.; Kjos, N.K.; Fauske, A.K.; Teige, J.; Sørum, H. Easily fermentable carbohydrates reduce skatole formation in the distal intestine of entire male pigs. Livest. Sci. 2011, 40, 206–217.
89. Kim, J.C.; Hansen, C.F.; Mullan, B.P.; Pluske, J.R. Nutrition and pathology of weaner pigs: Nutritional strategies to support barrier function in the gastrointestinal tract. Anim. Feed Sci. Tech. 2011
90. Zamaratskaia, G.; Chen, G.; Lundström, K. Effects of sex, weight, diet and hCG administration on levels of skatole and indole in the liver and hepatic activities of cytochromes P4502E1 and P4502A6 in pigs. Meat Sci. 2006, 72, 331–338.
91. Aluwé, M.; Millet, S.; Nijs, G.; Tuyttens, F.A.M.; Verheyden, K.; De Brabander, H.F.; De Brabander, D.L.; Van Oeckel, M.J. Absence of an effect of dietary fibre or clinoptilolite on boar taint in entire male pigs fed practical diets. Meat Sci. 2009, 82, 346–352.
92. Sun, T.; Lærke, H.N.; Jørgensen, H.; Bach Knudsen, K.E. The effect of extrusion cooking of different starch sources on the in vitro and in vivo digestibility in growing pigs. Anim. Feed Sci. Tech. 2006, 131, 67–86.
93. Kjos, N.P.; Øverland, M.; Fauske, A.K.; Sørum, H. Feeding chicory inulin to entire male pigs during the last period before slaughter reduces skatole in digesta and backfat. Livest. Sci. 2010, 134, 143–145.
94. Hansen, L.L.; Stolzenbach, S.; Jensen, J.A.; Henckel, P.; Hansen-Moller, J.; Syriopoulos, K.; Byrne, D.V. Effect of feeding fermentable fibre-rich feedstuffs on meat quality with

emphasis on chemical and sensory boar taint in entire male and female pigs. Meat Sci. 2008, 80, 1165–1173.
95. Vhile, S.G.; Kjos, N.P.; Sørum, H.; Øverland, M. Feeding Jerusalem artichoke reduced skatole level and changed intestinal microbiota in the gut of entire male pigs. Animal 2011.
96. Van Oeckel, M.J.; Warnants, N.; De Paepe, M.; Casteels, M.; Boucqué, C.V. Effect of fibre-rich diets on the backfat skatole content of entire male pigs. Livest. Prod. Sci. 1998, 56, 173–180.
97. Whittington, F.M.; Nute, G.R.; Hughes, S.I.; McGivan, J.D.; Lean, I.J.; Wood, J.D.; Doran, E. Relationships between skatole and androstenone accumulation, and cytochrome P4502E1 expression in Meishan × Large White Pigs. Meat Sci. 2004, 67, 569–576.
98. Konstantinov, S.R.; Zhu, W.-Y.; Williams, B.A.; Tamminga, S.; De Vos, W.M.; Akkermans, A.D.L. Effect of fermentable carbohydrates on piglet faecal bacterial communities as revealed by denaturing gradient gel electrophoresis analysis of 16S ribosomal DNA. FEMS Microbiol. Ecol. 2003, 43, 225–235.
99. Gibson, P.R.; Newnham, E.; Barrett, J.S.; Shepherd, S.J.; Muir, J.G. Review article: Fructose malabsorption and the bigger picture. Aliment. Pharma. Ther. 2007, 25, 349–363.
100. Baltic, M.; Raicevic, S.; Tadic, I.; Drljacic, A. Influence of zeolite on skatole content of swine fat tissue. In Boar Taint in Entire Male Pigs; Bonneau, M., Lundström, K., Malmors, B., Eds.; Wageningen Academic Publishers: Wageningen, The Netherlands, 1997; pp. 97–99.
101. Jen, K.; Squires, E.J. In vitro assessment of the effectiveness of non-nutritive sorbent materials as binding agents for boar taint compounds. Animal 2011, 5, 1821–1828.
102. Yang, C.S.; Chabra, S.K.; Hong, J.Y.; Smith, T.J. Mechanisms of inhibition of chemical toxicity and carcinogenesis by diallyl sulphide and related compounds from garlic. J. Nutr. 2001, 131, 1041S–1045S.
103. Kouba, M.; Enser, M.; Whittington, F.M.; Nute, G.R.; Wood, J.D. Effect of a high-linolenic acid diet on lipogenic enzyme activities, fatty acid composition, and meat quality in the growing pig. J. Anim. Sci. 2003, 81, 1967–1979.
104. Jeong, H.G.; Yun, C.-H. Induction of rat hepatic cytochrome P450 enzymes by myristicin. Biochem. Biophys. Res. Commun. 1995, 217, 966–971.

CHAPTER 9

Assessment of Blood Concentration of Cardiac Troponin I (cTnI) in Healthy Calves and Stocker Calves Effected with Bovine Respiratory Disease Complex

Anderson DE[*1,4], White B[1], Nickell J[1], Hanzlicek G[2], Blasi D[3] and Fraser B[1,5]

[1]*Department of Clinical Sciences, College of Agriculture, Kansas State University, Manhattan, Kansas, USA*

[2]*Department of Diagnostic Medicine and Pathobiology, College of Agriculture, Kansas State University, Manhattan, Kansas, USA*

[3]*Department of Animal Sciences and Industry, College of Agriculture, Kansas State University, Manhattan, Kansas, USA*

[4]*Department of Large Animal Clinical Sciences, College of Veterinary Medicine, University of Tennessee, River Drive, Knoxville, TN, USA*

[5]*Faculty of Science, School of Veterinary Medicine, University of Queensland, Australia*

ABSTRACT

Bovine respiratory disease complex remains one of the most economically important diseases of stocker cattle. Reports of myocardial injury associated with BRDC have been limited to necropsy findings. An animal-side diagnostic test for cardiac troponin I (cTnI) has been validated for use in cattle. The objective of this study was to determine the association of blood concentration of cTnI and pneumonia in stocker cattle. A proof-of-concept study was conducted using 16 calves (7 healthy; 9 BRDC). Then, a clinical trial was conducted in commercial stocker cattle. Calves were randomly selected at arrival (n = 35) and during treatment for BRDC (n = 66) from a population of 293 stocker calves. In this trial, venous blood samples were obtained and placed in a heparin tube and immediately analyzed using an I-STAT 1 hand held blood analyzer. Statistical analysis included a Student's T-test (proof-of-concept study) and generalized linear models procedures (clinical trial) to evaluate potential associations with cTnl and BRDC. In the first study, cTnI was significantly greater in calves having respiratory disease (0.04 +/- 0.04 ng/ml) when compared with calves not having respiratory disease (0.006 +/- 0.01 ng/ml; P = 0.036). During the clinical trial, blood concentration of cTn-I in calves at arrival (0.01 +/- 0.02 ng/ml) and in calves that remained healthy (0.01 +/- 0.02 ng/ml) were significantly less when compared with calves suffering BRDC at any time point (0.05 +/- 0.01 ng/ml; P = 0.04). Calves that survived BRDC tended to have a lower cTnI (0.03 +/- 0.05 ng/ml) compared with calves that died from BRDC (cTn I 0.10 +/- 0.14 ng/ml; P = 0.055). Based on the results of this study, cTnI may be useful in assessment of the severity of BRDC and may be useful in predicting calves that will suffer prolonged or fatal complications of BRDC.

Keywords: Bovine; Stocker; Pneumonia; Cardiac Troponin I; cTnI

1. INTRODUCTION

Bovine respiratory disease complex (BRDC) remains one of the most detrimental diseases suffered by feedlot calves and is a common clinical problem in all sectors of the cattle industry. BRDC represents economic losses because of diminished body condition, carcass quality, and weight gain [1]. BRDC also creates calf welfare concerns which impact decisions made by the producer regarding treatment, euthanasia, or slaughter of severely affected calves. Although significant advances have been made in research regarding specific pathogens and the interaction among viral and bacterial pathogens, accurate diagnosis and prognosis for individual cattle affected by BRDC in the field remains challenging. Observation of clinical signs such as lethargy, increased respiratory rate, labored breathing, nasal discharge, decreased appetite, and rectal temperature are commonly used to diagnose BRDC in field settings. Research has estimated that these methods of clinical illness scoring have a relatively low sensitivity and specificity for diagnosis of BRDC in cattle [2,3]. A review of feedlot management research identified the lack of accurate diagnostic methods for identifying calves affected with BRDC as a critical need and called for the development of quantitative measures to augment current BRDC diagnostic techniques [4].

Objective animal side measurements may become important tools to differentiate BRDC and non-BRDC cattle, may be helpful to predict response to therapy in BRDC cases, and may be useful in identifying cattle unlikely to recover from BRDC. Although several research trials have compared health outcomes between multiple antimicrobials [5,6], literature evaluating health outcomes compared to objective physiologic measurements is sparse. Recent studies have evaluated changes in available objective physiologic measures (complete blood count, serum biochemistry, arterial blood gas) following induced Mannheimia haemolytica pneumonia [7-9]. Relatively few associations between these parameters and the progression of pneumonia in the calves were found. Work evaluating rectal temperature illustrated that while broad associations between rectal temperature at the time of treatment had some association with outcomes, it did not have highly discriminatory prognostic power [10]. This work highlights the need for additional research to identify diagnostic and prognostic tools for identification of calves that may suffer severe BRDC less likely to be successfully treated. Diagnostic tests facilitating rapid results that may aid disease classification of BRDC and that offer the ability to predict which calves are likely to recover and which calves are more likely to suffer long-term debilitating disease would help to lessen the use of antimicrobial drugs, feed and labor costs, and lessen the number of calves suffering prolonged therapy without benefit. Accurate and timely response to debilitating disease may improve calf welfare by lessening suffering and lessen the futile use of antibiotics. This may also lessen the risk of food contamination by drug residues, meat adulteration by injection site lesions, and improve animal handling efficiency.

Troponin is a contractile protein produced by cardiac and skeletal myocytes. Troponin is part of the tropomyosin complex involved in relaxation and contraction of the muscle. When relaxed, myosin active sites are not available to interact with actin. When muscle contracture is activated and calcium is released into the cell from the sarcoplasmic reticulum, the binding of calcium to troponin results in exposure of myosin to actin causing contraction of the muscle. As such, increases in serum concentration of troponin can be used as an indicator of muscle injury. Cardiac troponin I (cTnI) is a myofibril protein specific to cardiac myocytes and is critical to the regulation of contractility of the myocardium. Troponin I is a myofilament protein, which is vital to myocardial cell function. cTnI is specific to heart muscle and provides a clinical tool to assess myocardial injury. Injury to the heart causes rapid release of troponin into the blood stream from cytosolic stores. cTn I is rapidly eliminated from the blood and continued presence of cTn I indicates ongoing release either associated with damage or myocardial regeneration [11]. Cardiac troponin I (cTn I) has become a common test for myocardial injury in people and is used in veterinary medicine for similar purposes [12-20]. Troponin I assays are increasingly available in table-top or hand-held biochemistry analyzers [21]. This presents an opportunity for field testing of large groups of cattle with rapid results that can aid in disease classification and outcome discrimination. Recently, serum cTn I has been reported for cattle having a variety of cardiogenic and non-cardiogenic disease [21-27].

We hypothesized that cTn I would be useful in assessment of calves having a clinical diagnosis of BRDC and that this serum biomarker would prove useful in predicting the health outcomes (death or re-treatment) of calves having BRDC. Study objectives were accomplished through two independent research projects. The first study (Study I) compared blood cTn I concentration between two independent populations of calves: 1) clinically normal calves and 2) calves confirmed to have BRDC. The second trial (Study II) evaluated Troponin I assays in selected high-risk calves at arrival to a stocker facility and in selected calves at each treatment for BRDC during the first 42-days on-feed after arrival at the stocker facility.

2. MATERIALS AND METHODS

All calves in these studies were handled in accordance with protocols approved by the Kansas State University Animal Care and Use Committee.

2.1 Cardiac Troponin I Analysis

Blood samples were obtained from either the jugular or coccygeal vein of each calf enrolled in the two studies. The blood was placed in a heparin tube and immediately analyzed using an I-STAT 1 hand held blood analyzer. Troponin I assays were performed exactly as described by the manufacturer. The i-STAT cTnI analysis unit uses test cartridges with two-site ELISA technology. The antibodies are specific for cTnI and use reagents derived from bovine and caprine sources. The detection range is 0 to 50 ng/ml (ug/L) with a C.V. of < 8% and an analytical sensitivity of 0.02 ng/ml.

2.2 Study I – cTnI test evaluation

Sixteen calves, ranging in age from 4 to 6 months, were used for proof-of-concept testing of cTnI based on health status. Seven healthy calves (n = 7 serum samples) were selected from a pool of calves that had been acquired for use in an unrelated study. These calves were assigned to the "normal" group. These calves had been observed by veterinarians with cattle expertise for a period ranging from 30 to 60 days before inclusion in the present study and were considered healthy based on prior observation, physical examination, daily observation, and growth performance characteristics. The health status of the calves was assessed further on the day of sampling through physical examination including detailed thoracic auscultation. Although no necropsy examinations were done, all normal calves remained healthy and free of clinical signs of BRDC for a minimum of 30 days after samples had been obtained. Nine calves (n = 9 serum samples) were selected for this study based on a clinical diagnosis of pneumonia. All of these calves had a clinical diagnosis of acute BRDC on the day the blood sample was drawn and tested. Diagnosis of BRDC was made by a veterinarian with cattle expertise based on clinical signs and physical examination including abnormal thoracic auscultation, varying combinations of increased respiratory effort, exercise intolerance, nasal discharge, lethargy, coughing, and increased rectal temperature >40^{0}C (104.0 F). Inclusion criterion for BRDC calves was confirmation of the diagnosis of pneumonia without cardiac disease by necropsy examination by experienced veterinarians with extensive cattle experience including post-mortem examination. An estimate of the reliability of the iSTAT cTnI test for determination of cardiac troponin in calves was done by repeated measures of whole blood samples collected as described above. Serum samples were analyzed in triplicate and the coefficient of variation (CV) determined.

2.3 Study II – High-Risk Stocker Calf study

Troponin assays were conducted in a field trial using calves randomly selected from a group of 293 stocker calves on arrival to the stocker facility. Inclusion criterion for calves in Study II included 1) healthy group (n = 35) - randomly selected calves which were healthy on the day of arrival to the stocker facility; and 2) calves that suffered BRDC (n = 66) at some point during the feeding period after arrival. Serum samples were obtained for cTnI analysis, at the time of arrival, from 35 calves free of any clinical sign of BRDC. Subsequently, serum samples were obtained from 66 calves that suffered BRDC, of which 6 had been selected for sample collection as part of the initial healthy-on-arrival group. Male, mixed breed beef calves were procured from the Southeastern US and shipped by truck to the KSU Beef Stocker Unit in north central Kansas. The cattle arrived in three independent loads and initial processing procedures were the same for all three loads. At arrival, all calves received an individual ear tag and each animal was weighed. Calves in each load were sorted by weight and gonadal status (steers, bulls) and randomly allocated to one of eight pens. The resulting assignments produced 24 pens of 12 or 13 head balanced for weight and gonadal status.

On the first day after arrival, all calves received preventative health products (a bacterin containing Clostridium chauvoei, septicum, novyi, sordelli, tetani, haemolyticum, perfringens Types B, C and D, a modified-live 4-way viral respiratory vaccine, and an injectable antiparasiticide) and intact males were surgically castrated. Based on pen treatment assignment, pens of calves received metaphylaxis using either subcutaneous tilmicosin at 1.5 ml/100 lb (1 mg/kg) of body weight or subcutaneous tulathromycin at 1.1 ml/100 lb (0.5 mg/kg) of body weight. At processing, blood was collected from the coccygeal vein from a convenience sampled population of calves from each load. Convenience sampling refers to the sampling of calves based on the availability of the investigator for collection of blood samples. Blood samples were placed in lithium heparin tubes and immediately placed on ice until transport to the laboratory. Troponin I assays were performed within 12 hours on all samples after calf processing had been completed.

After processing, all calves were monitored twice daily by trained personnel who were blinded to all cTn-I results. All calves showing clinical signs of BRDC (increased respiratory effort, nasal discharge, depression as evidenced by failure to rise and eat, low head and neck posture, or lethargy, rectal temperature > 40 ^{0}C) were treated in accordance with the pre-determined standard protocol (First BRDC treatment: enrofloxacin, second BRDC treatment: florfenicol, third BRDC treatment: oxytetracycline). All antimicrobials were administered according to label instructions. Calves that were treated for BRDC were selected for cTn-I sampling on days investigators were present for treatments and could confirm inclusion criterion. For enrolled calves, a blood sample was obtained from the coccygeal vein prior to administration of antimicrobials. Samples were placed in lithium heparin tubes on ice, and maintained in refrigeration until analyzed later that day.

Necropsy examination was performed on all calves that succumbed to disease or were euthanatized because of progression of BRDC during the course of the study. Necropsy findings of this subset of calves were used as an internal audit for accuracy of clinical case definition.

2.4 Statistics

All hypothesis testing was performed with P-values of < 0.05 considered significant. Serum samples from 4 calves were used to perform quality control on the unit, handling, and testing procedures. The coefficient of variation was calculated and repeatability examined against the expected standards for the analyzer unit. Statistical analysis for the study data where all animals were independent and sampled a single time was done using a Student's T-test to compare mean cTn-I values between clinically normal and diseased animals. Generalized linear models were used to evaluate potential associations among cTnl value with time and BRD status for calves enrolled Study II. The arrival lot and pen were included in the models as random effects to account for lack of independence between test subjects, and calf identification was used as a random variable to account for repeated measures on animals. Associations between serum concentration of cTnI and subsequent health outcomes of re-treatment and death were also evaluated using models specified in the same manner.

3. RESULTS

3.1 Study I – cTnI test evaluation

Results of repeated measurement of serum concentration of bovine cTnI yielded small CV's and small SD (see Table 1). This was followed by measurement of serum concentration of cTnI in 16 calves of which 7 were normal calves and 9 calves had clinical and necropsy signs of BRDC but no evidence of myocardial injury. Calves in the BRD positive group exhibited gross pulmonary and pleural lesions associated with BRD which were in various stages of chronicity including pulmonary consolidation in several calves. Lesions involving the heart were not evident. Troponin-I was significantly (P = 0.036) greater in calves having respiratory disease (0.04 +/- 0.04 ng/ml) when compared with calves not having respiratory disease (0.01 +/- 0.01 ng/ml) (Figure 1).

Table 1: Assessment of repeatability of cTnI measurement in healthy calves using a iSTAT hand-held analyzer; Each analysis was performed in triplicate and mean, SD< and CV determined

Calf ID	cTnI-1 ng/ml	cTnI-2 ng/ml	cTn-3 ng/ml	Mean ng/ml	SD ng/ml	CV (%)
1	0	0	0	0	0	0
2	0.01	0	0	0.003	0.006	17.3
3	0.02	0.02	0.02	0.02	0	0
4	0	0	0	0	0	0

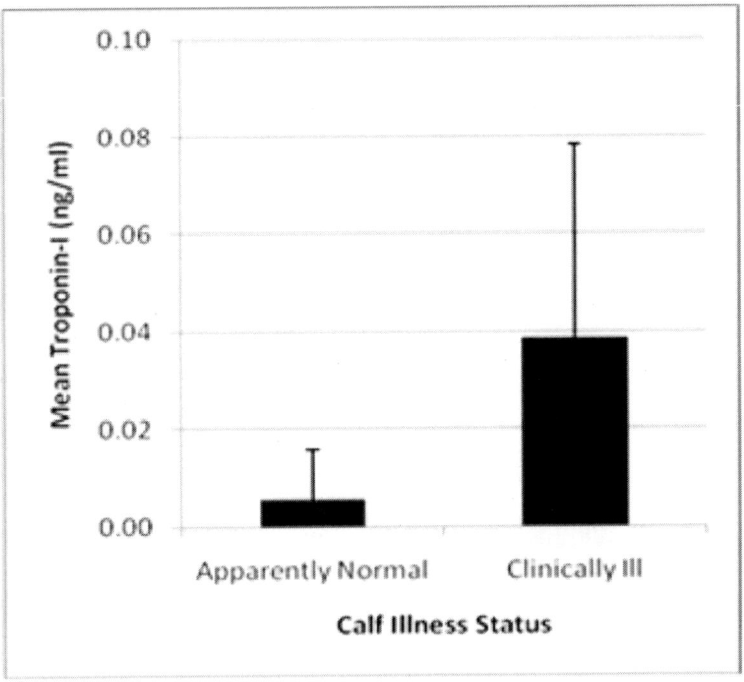

Figure 1: Mean blood concentration of cardiac Troponin I (cTn-I, ng/ml) in mixed breed beef calves (n -16; healthy n = 7; Pneumonia calves n = 9) presented with or without clinical signs of respiratory disease (P = 0.036)

3.2 Study 2 – High-Risk Stocker Calf study

The stocker study included a total of 293 calves that arrived in the three loads over a three day period with a mean arrival weight of 220 kg (485 lbs). Blood was collected for cTn-I analysis at arrival from 35 calves without clinical signs of BRDC selected from each of the three loads (n = 14, 11, and 10). During the 43 day back-grounding phase, 50.8% of the total population was treated for BRDC and 9.6% of the calves died. Necropsies were performed on all animals that died. A total of 28 calves died; 27 calves suffered severe bronchopneumonia and 1 calf died from causes unrelated to BRDC. This calf was excluded from cTn I data analysis.

A total of 66 calves meeting the treatment criterion for BRDC were sampled on 12 occasions during the first 31 days on feed. Of these 66 calves, only 6 had been randomly sampled at arrival to the facility. Samples of treated calves were acquired either at first (n = 47), second (n = 9), or third (n = 10) treatment for BRD. Blood concentration of cTn-I in calves at arrival (0.01

+/- 0.02 ng/ml) and in calves that remained healthy throughout the study period (0.01 +/- 0.02 ng/ml) were significantly less when compared with calves that became ill from BRDC at any time point (0.05 +/- 0.01 ng/ml) (P = 0.04; Figure 2). After accounting for treatment status of calves that developed BRDC, LSM cTn-l values did not differ between samples taken at any of the 3 treatment points (first, second, or third).

When Troponin-I concentrations were examined for samples taken at first treatment for BRDC, no differences in LSM cTn-I were identified based on subsequent retreatment status. However, inclusion of calf death as a factor illustrated an interaction (p = 0.055) between the timing of the sample and subsequent calf death. Calves that developed BRDC and survived had a serum cTn I concentration at first pull of 0.03 +/- 0.05 ng/ml. Calves that developed BRDC and died had a serum cTn I concentration of 0.10 +/- 0.14 ng/ml. cTn-I concentrations were significantly greater for calves that subsequently died when measured on first or third treatment when compared to cohort calves measured at this same time period ($P < 0.05$; Figure 3).

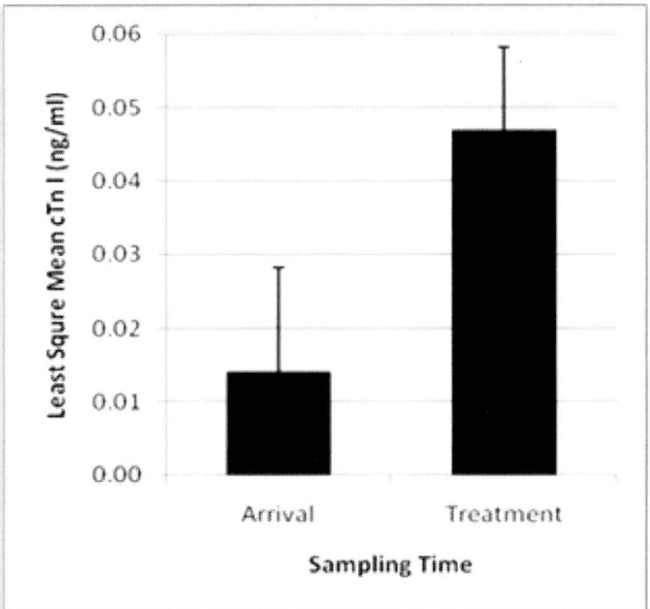

Figure 2: Least square mean blood concentration of cardiac Troponin I (cTn-I, ng/ml) values in stocker calves (total study population, n = 95) at arrival (n = 35) and at treatment for respiratory disease (n = 66, 6 of which were previously sampled as normal at arrival) at any time point during the 42 days on feed after arrival (P = 0.04)

Sample size limited our ability to examine the temporal relationship between blood concentration of cTn I and occurrence of death. However, scatter plot analysis revealed that calves with the largest cTn I died more quickly after treatment than did calves having lower cTn I.

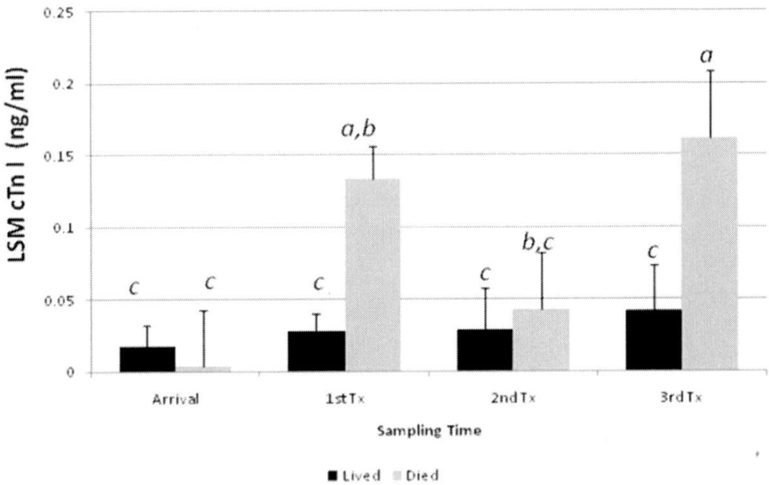

Figure 3: Interactive least square mean blood concentration of cardiac Troponin I (cTn-I, ng/ml) in stocker calves by sampling time period and subsequent health outcome (death). Data represent analysis of serum samples collected at arrival (n = 35); first BRDC treatment (n = 47); second BRDC treatment (n = 9); and third BRDC treatment (n = 10). Columns that differ by superscript are significantly different ($p < 0.05$)

4. DISCUSSION

Assessment of serum concentrations of cTnI may be useful in patients suffering non-cardiac diseases. Troponin assays may have variable accuracy and precision across species. Human cTn T assays have been shown to be unreliable for detection of bovine cTn T [27]. However, the homology of bovine and human cTn I has been reported be > 96% [28]. A commercial immunoassay has been validated for measurement of bovine cTnI [29]. In that study, the authors found that intra-assay precision was 4.48% and inter-assay precision was 13.36%. The authors concluded that this immunoassay, designed for detection of human cTn I was valid for use in cattle. Although cTn I is widely used as a screening test for myocardial infacrtion, increases in serum cTn I has been observed in human patients having a wide variety of cardiogenic (e.g. congestive heart failure, myocarditis, myocardial contusion, atrial fibrillation) and non-cardiogenic (e.g. sepsis, hypovolemia, pulmonary embolism, renal failure) diseases [29]. Despite this wide range of clinical syndromes, cTn I has prognostic value in these patients. Assessment of serum concentrations of Troponin I may be useful in patients suffering non-cardiac diseases. Although cTn I is widely used as a screening test for myocardial infarction, increases in serum cTn I have been observed in human patients having a wide variety of illnesses including cardiogenic (e.g. congestive heart failure, myocarditis, myocardial contusion, atrial fibrillation) and non-cardiogenic (e.g. sepsis, hypovolemia, pulmonary embolism, renal failure) diseases. Despite this wide range of clinical syndromes, cTn I has important prognostic value in these patients. Blood concentration of cTn I was found to be useful in assessment of risk of morbidity and mortality in human patients suffering acute pulmonary embolism [30]. Patients having increased serum concentrations of cTn I were more likely to suffer complications, recurrence of pulmonary embolization, or death. As an example, cTn I has been shown to have utility in assessment of the severity of acute pulmonary embolism in human patients [30]. In that study, patients having increased serum concentrations of cTn I were more likely to suffer complications, recurrence of pulmonary embolization, or death.

These findings suggest that Troponin I may be useful in determining severity of disease in cattle suffering pulmonary disease. Several recently published studies have reported blood cTn I concentration from cattle having cardiac and non-cardiac intra-thoracic diseases [23-27,31]. These reports included pericarditis, endocarditis, congenital heart disease, monensin toxicity, endotoxemia, reticulitis, mediastinal abscess, thymic lymphoma, caudal vena cava thrombosis, and chronic suppurative pneumonia. In a study examining cTn I in 34 normal cattle and 5 cattle having pericarditis, cTn I was markedly increased in 4 of 5 cattle effected with pericarditis [27]. Peek suggested a reference range of cTn I for cattle of < 0.08 ng/ml [26]. This is greater than that suggested by our study in which normal feedlot aged cattle consistently had cTnI of < 0.03 ng/ml. Using this cut-off point, troponin I was increased in 79% of cattle having primary bacterial cardiac disease. In that study, 40% of cattle having thoracic but non-cardiac disease had increased cTn I. However, the authors concluded that cTnI could not differentiate primary cardiac diseases from intrathoracic but non-cardiac disorders. In a study of 20 cattle having traumatic reticuloperitonitis, cTn I was considered normal when < 0.05 ng/ml and cattle having myocardial injury had cTn I of 3.26 +/- 2.1 ng/ml [27]. This cut off point is more consistent with the findings of our study. In these various studies, the authors' noted that cTn I should be further investigated to determine its prognostic value because their study was limited to cattle having post-mortem examination done.

The reported extent of lung involvement in calves necropsied after having died with BRDC varies from < 33% of the lung volume in 30% of cattle, 33 to 66% in 58% of calves, and > 66% in 8.8% of calves [32]. Prolonged respiratory acidosis, hypercapnea, hypoxemia, and pulmonary hypertension may play a role in secondary myocardial injury leading to increasing concentrations of cTn I in the blood [33]. Although BRDC is often isolated to the lung tissues, primary and secondary myocardial injury occurs in some calves [11,34-36]. In a review of necropsy findings from 237 deaths associated with BRDC, findings other than pneumonia included pleuritis in 49% of effected calves, septicemia in 0.9%, and emboli in 0.5% [34]. Interestingly, we showed that cTn I was not statistically different from control calves at the time of the second treatment for BRDC. The significant increase in serum cTn I at the time of first and third treatments may be indicative of the severity and extent of the disease. Lesion severity has been associated with outcome in previous studies in feedlot calves [34]. Significant differences in cTnI between surviving and non-surviving calves treated at the first episode of BRDC may reflect the severity of the pathogen and rapid progression of pneumonia. The lack of difference in cTnI serum concentration among second treatment surviving and non-surviving calves is assumed to have been a result of the stage of progression of the pneumonia. Significant differences in cTnI between surviving and non-surviving calves treated at the third episode of BRDC may reflect chronicity of pneumonia resulting in consolidation, pulmonary fibrosis, and resulting hypertension. The concentration of Troponin I found in blood is expected to be proportional to the damage inflicted. Myocardial injury can be caused by many different insults including ischemia due to myocardial infarction, toxicity caused by ingestion of ionophores, nutritional deficiencies such as selenium deficiency, trauma caused by vehicular accidents, hypoxemia caused by anemia, severe pneumonia, or pulmonary interstitial disease, and increased cardiac work such as occurs with pulmonary hypertension [23-27,32-36]. For the purposes of this study, we hypothesized that many calves that fail to respond to antimicrobial therapy are calves in which BRDC has caused permanent pulmonary debilitation. We speculated that some calves failing to respond to BRDC treatment in feedlot settings are in fact suffering effects of primary, and perhaps permanent, pulmonary tissue injury rather than failure of response to antimicrobials. For calves that lived, the findings in the high-risk study indicated that cTnI concentrations were not elevated at initial treatment compared to that of arrival. Thus, low cTn I was suggestive of BRDC severity expected to respond to therapy or for spontaneous recovery to occur. When the ultimate outcome was death, cTnI levels at the first or third treatment were significantly greater than the baseline established upon arrival (Figure 3). Thus, marked increases in cTn I suggested more severe disease unlikely to respond to therapy or for recovery to occur. Accurate knowledge of animal prognosis based on a test done prior to treatment could result in greater efficiency of health care and reduced calf suffering because of early identification of calves that will become chronic poor performers. An economic benefit may also be realized by the producer and potentially limit the unnecessary use of antimicrobial drugs. In this scenario, cost savings are associated with reduced drug input, reduced feed use, reduced labor use, and improved facilities use.

The similarity in cTnI among groups at arrival suggests that pre-existing BRDC was not prevalent in this group of stocker calves. Some of these calves developed BRDC and subsequently died and this clinical course of disease was concomitant with a significant increase in blood cTnI concentration. These data illustrate that measurement of cTnI may have potential as a prognostic metric for clinical cases of BRD, but appears to be dependent on when the sample is taken relative to disease progression. Further research is warranted regarding the field applications of cTn I in stocker cattle. This test can be performed in field settings with rapid tests results. This makes troponin an attractive alternative to traditional treatment response tests. Further research will be necessary to elucidate the parameters for selection of calves for testing and to develop precise test interpretation cut-points.

5. CONCLUSION

The results of this study suggest that measurement of cTnImay be useful in the discrimination of calves having BRDC and, specifically, those calves at greatest risk of death. Study I demonstrated acceptable repeatability of the hand held device for measurement of bovine serum concentration of cTnI and revealed differences in mean blood cTnI concentration between clinically normal and chronic BRDC calves. These findings were supported by similar findings noted in the high-risk calf trial when comparing samples taken at arrival to blood cTnI concentration at first treatment for BRDC. In study I, calves without clinical signs of disease showed very little variability in cTn I measurements compared to the BRDC calves. This finding may be due to either the stage of disease progression, or individual variability in response subjects. Serum concentration of cTnI is close to the lower detection limit of the hand held device. More research should be performed to evaluate the utility of this test for both diagnostic and prognostic value related to respiratory disease in cattle.

ACKNOWLEDGMENTS

The authors would like to acknowledge the technical support of Dr. Thomas Overbay and the Abaxis provided the cTnI test cartridges.

REFERENCES

1. Cernicchiaro N, White BJ, Renter DG, Babcock AH (2013) Economic and performance consequences associated with the number of treatments for initial individual cases of bovine respiratory disease in commercial feeder cattle. Am J Vet Res 74: 300-9.
2. White BJ, Renter DG (2009) Bayesian estimation of the performance of clinical observations and harvest lung scores for diagnosing bovine respiratory disease in post-weaned beef calves. J Vet Diagn Invest 21: 446-53.
3. White BJ, Goehl DR, Amrine DE, Booker C, Wildman B, et al. (2016) Bayesian evaluation of clinical diagnostic test characteristics of visual observations and remote monitoring to diagnose bovine respiratory disease in beef calves. Prev Vet Med 126: 74-80.
4. Duff GC, Galyean ML (2007) Recent advances in management of highly stressed, newly received feedlot cattle. J Anim Sci 85: 823-40.
5. Wellman NG, O'Connor AM (2007) Meta-analysis of treatment of cattle with bovine respiratory disease with tulathromycin. J Vet Pharmacol Ther 30: 234-41.
6. Schunicht OC, Booker CW, Guichon PT, Jim GK, Wildman BK, et al. (2007) An evaluation of the relative efficacy of tulathromycin for the treatment of undifferentiated fever in feedlot calves in Nebraska. Can Vet J 48: 600-6.
7. Fraser BC, Anderson DE, White BJ, Miesner MD, Lakritz J, et al. (2014) Associations of various physical and blood analysis variables with experimentally induced Mycoplasma bovis pneumonia in calves. Am J Vet Res 75: 200-7.
8. Hanzlicek GA, White BJ, Mosier D, Renter DG, Anderson DE (2010) Serial evaluation of physiological, pathological and behavioral changes related to disease progression in experimentally induced Mannheimia haemolytica pneumonia in postweaned calves. Am J Vet Res 71: 359-69.
9. Skogerboe TL, Rooney KA, Nutsch RG, Weigel DJ, Gajewski K, et al. (2005) Comparative efficacy of tulathromycin versus florfenicol and tilmicosin against undifferentiated bovine respiratory disease in feedlot cattle. Vet Ther 6: 180-96.
10. Theurer ME, White BJ, Larson RL, Holstein K, Amrine DE (2014) Relationship of rectal temperature at first treatment for bovine respiratory disease complex in feedlot calves to probability of not finishing the production cycle normally. J Am Vet Med Assoc 245: 1279-85.
11. Jaffe AS, Landt Y, Parvin CA, Abendschein DR, Geltman EM, et al. (1996) Compartive sensitivity of cardiac troponin I and lactate dehydrogenase isoenzymes for diagnosing acute myocardial infarction. Clin Chem 42: 1770-6.
12. Hasic S, Kiseljaković E, Jadrić R, Radovanović J, Winterhalter-Jadrić M (2003) Cardiac troponin I: the gold standard in acute myocardial infarction diagnosis. Bosn J Basic Med Sci 3: 41-4.
13. Gaze DC, Collinson PO (2005) Cardiac troponins as biomarkers of drug- and toxin-induced cardiac toxicity and cardioprotection. Expert Opin Drug Metab Toxicol 4: 715-25.
14. Schober KE, Cornand C, Kirbach B, Aupperle H, Oechtering G (2002) Serum cardiac troponin I and cardiac troponin T concentrations in dogs with gastric dilatation-volvulus. J AM Vet Med Assoc 221: 381-8.
15. Schober KE, Kirbach B, Oechtering G (1999) Noninvasive assessment of myocardial injury in dogs with suspected cardiac contusion. J Vet Cardiol 1: 17-25.

16. Baumwart RD, Orvalho J, Muers KM (2007) Evaluation of serum cardiac troponin I concentration in Boxers with arrhythmogenic right ventricular cardiomyopathy. Am J Vet Res 68: 524-8.
17. Herndon WE, Kittleson MD, Sanderson K, Drobatz KJ, Clifford CA, et al. (2002) Cardiac troponin I in feline hypertrophic cardiomyopathy. J Vet Intern Med 16: 558-64.
18. Conolly DJ, Cannata J, Boswood A, Archer J, Groves EA, et al. (2003) Cardiac troponin I in cats with hypertrophic cardiomyopathy. J Feline Med Surg 5: 209-16.
19. Schwarzwald CC, Hardy J, Buccellato M (2003) High cardiac troponin I serum concentration in a horse with multiform ventricular tachycardia and myocardial necrosis. J Vet Intern Med 17: 364-8.
20. Slack JA, McGuirk SM, Erb HN, Lien L, Coombs D, et al. (2005) Biocehmical markers of cardiac injury in normal, surviving septic, or non-surviving septic neonatal foals. J Vet inten Med 19: 577-80.
21. Fraser BC, Anderson DE, White BJ, Miesner MD, Wheeler C, et al. (2013) Assessment of a commercially available point-of-care assay for the measurement of bovine cardiac troponin I concentration. Am J Vet Res 74: 870-3.
22. Varga A, Schober KE, Walker WL, Lakritz J, Michael Rings D (2009) Validation of a Commercially Available Immunoassay for the Measurement of Bovine Cardiac Troponin I. J Vet Intern Med 23: 359-65.
23. Varga A, Schober KE, Holloman CH, Stromberg PC, Lakritz J, et al. (2009-b) Correlation of Serum Cardiac Troponin I and Myocardial Damage in Cattle with Monensin Toxicosis. J Vet Intern Med 23: 1108-16.
24. Jeremias A, Gibson CM (2005) Narrative Review: Alternative Causes for Elevated Cardiac Troponin Levels when Acute Coronary Syndromes Are Excluded. Ann Intern Med 142: 786-91.
25. Konstantinides S, Geibel A, Olschewski M, Kasper W, Hruska N, et al. (2002) Importance of Cardiac Troponins I and T in Risk Stratification of Patients With Acute Pulmonary Embolism. Circulation 106: 1263-8.
26. Peek SF, Apple FS, Murakami MA, Crump PM, Semrad SD (2008) Cardiac isoenzymes in healthy Holstein calves and calves with experimentally induced endotoxemia. Can J Vet Res 72: 356-61.
27. Mellanby RJ, Henry JP, Cash R, Ricketts SW, Bexiga JR, et al. (2007) Serum cardiac troponin I concentrations in cattle with pericarditis. Veterinary Record 161: 454-5.
28. Gunes V, Atalan G, Citil M, Erdogan HM (2008) Use of cardiac troponin kits for the qualitative determination of myocardial cell damage due to traumatic reticuloperitonitis in cattle. Veterinary Record 162: 514-7.
29. Willis MS, Snyder JA, Poppenga RH, Grenache DG (2007) Bovine cardiac troponin T is not accurately quantified with a common human clinical immunoassay. J Vet Diagn Invest 19: 106-8.
30. O'Brien JP, Landt Y, Ladenson JH (1997) Differential reactivity of cardiac and skeletal muscle from various species in a cardiac troponin I immunoassay. Clin Chem 43: 2333-8.
31. Mellanby RJ, Henry JP, Cash R, Ricketts SW, Bexiga R, et al. (2009) Serum Cardiac Troponin I Concentrations in Cattle with Cardiac and Noncardiac Disorders. J Vet Intern Med 23: 926-30.
32. Graham WR (1953) The pathology of shipping fever in feedlot cattle. J Am Vet Med Assoc 123: 198-203.
33. Desmecht DJM, Linden AS, Lekeux PM (1996) The relation of ventilator failure to pulmonary, respiratory muscle, and central nervous system disturbances in calves with an experimentally produced pneumonia. J Comp Pathol 115: 203-19.
34. Fulton RW, Blood KS, Panciera RJ, Payton ME, Ridpath JF, et al. (2009) Lung pathology and infectious agents in fatal feedlot pneumonias and relationship with mortality, disease onset, and treatments. J Vet Diagn Invest 21: 464-77.
35. O'Toole D, Allen T, Hunter R, Corbeil LB (2009) Diagnostic exercise: Myocarditis due to Histophilus somni in feedlot and backgrounded cattle. Vet Pathol 46: 1015-7.

36. Haines DM, Moline KM, Sargent RA, Campbell JR, Myers DJ, et al. (2004) Immunohistochemical study of Hemophilus somnus, Mycoplasma bovis, Mannheimia hemolytica, and bovine viral diarrhea virus in death losses due to myocarditis in feedlot cattle. Can Vet J 45: 231-4.

CHAPTER 10

The genetic prehistory of domesticated cattle from their origin to the spread across Europe

Amelie Scheu[1,2*], Adam Powell[1], Ruth Bollongino[1,3], Jean-Denis Vigne[3], Anne Tresset[3], Canan Çakırlar[4], Norbert Benecke[2] and Joachim Burger[1]

[1]*Johannes Gutenberg-University Mainz, Institute of Anthropology, Palaeogenetics Group, 55099 Mainz, Germany.*

[2]*German Archaeological Institute, Scientific Department, Im Dol 2-6, Haus 2, 14195 Berlin, Germany.*

[3]*Muséum National d'Histoire Naturelle, UMR7209, "Archéozoologie, archéobotanique: sociétés, pratiques et environnements", InEE, Département d'Ecologie et Gestion de la Biodiversité, CP 56, 55 rue Buffon, 75005, Paris Cedex 05, France.*

[4]*University of Groningen, Institute of Archaeology, Poststraat 6, NL-9712 ER, Groningen, Netherlands.*

ABSTRACT

Background

Cattle domestication started in the 9th millennium BC in Southwest Asia. Domesticated cattle were then introduced into Europe during the Neolithic transition. However, the scarcity of palaeogenetic data from the first European domesticated cattle still inhibits the accurate reconstruction of their early demography. In this study, mitochondrial DNA from 193 ancient and 597 modern domesticated cattle (*Bos taurus*) from sites across Europe,

Western Anatolia and Iran were analysed to provide insight into the Neolithic dispersal process and the role of the local European aurochs population during cattle domestication.

Results

Using descriptive summary statistics and serial coalescent simulations paired with approximate Bayesian computation we find: (i) decreasing genetic diversity in a southeast to northwest direction, (ii) strong correlation of genetic and geographical distances, iii) an estimated effective size of the Near Eastern female founder population of 81, iv) that the expansion of cattle from the Near East and Anatolia into Europe does not appear to constitute a significant bottleneck, and that v) there is evidence for gene-flow between the Near Eastern/Anatolian and European cattle populations in the early phases of the European Neolithic, but that it is restricted after 5,000 BCE.

Conclusions

The most plausible scenario to explain these results is a single and regionally restricted domestication process of cattle in the Near East with subsequent migration into Europe during the Neolithic transition without significant maternal interbreeding with the endogenous wild stock. Evidence for gene-flow between cattle populations from Southwestern Asia and Europe during the earlier phases of the European Neolithic points towards intercontinental trade connections between Neolithic farmers.

1. BACKGROUND

The transition from foraging to producing economies, also called Neolithisation, was a major turning-point in human prehistory. The process of Neolithisation started in a region spanning from the Zagros Mountains to Central Anatolia and from Palestine to the plains beyond the East Taurus Mountains [1,2]. It was characterized by the successive appearance of sedentism (12th-10th millennia BCE), plant cultivation (mid-10thmillennium), animal husbandry (mid-9th millennium) and pottery (early 7th millennium) [3,4]. Elements of the Neolithic lifestyle expanded into Western Anatolia in the early 7th millennium [5-7], while the earliest signs for Neolithic settlements on the European continent are found in present-day Greece around 6,400 BCE [8]. The subsequent Neolithic spread across the rest of Europe followed at least two main routes: One leading across Southeastern Europe, and the second via the Western Mediterranean [9-

11]. The extent to which this expansion of a new culture and economy was driven by the migration of people has been debated for decades [12-17]. An early study on human ancient DNA emphasized the role of inward migration at the beginning of the Neolithic period in Central Europe [18], a view that is supported by more recent palaeogenomic studies [19,20].

As animal husbandry was an important part of the foundation of the new agricultural lifestyle, remains of domesticated animals can serve as a good proxy for the Neolithic spatial expansion and the presence and activity of farmers in newly populated areas [21]. In recent years, genetic and palaeogenetic studies have increasingly converged on a Southwest Asian origin for the four Neolithic domesticated animals: cattle, sheep, goats, and pigs [22]. For Near Eastern taurine cattle (*Bos taurus*), a recent coalescent-based analysis using ancient Iranian samples suggested a severe Near Eastern domestication bottleneck, with an estimated effective size of just 80 female founders [23]. However, comprehensive data sets of ancient cattle DNA from other areas are so far restricted to Central and Western Europe, for example from Bollongino *et al.* [24]. Thus, detailed and continent-wide evaluation of the early spatiotemporal demography of *Bos taurus* has so far been hindered by the lack of data from the key bridging areas of the Neolithic, namely Anatolia, the Balkans, and the Western Mediterranean.

In this study we greatly extend a previous coalescent-based demographic model, based on 15 ancient Iranian and 27 modern Near Eastern and Anatolian cattle mitochondrial DNA (mtDNA) sequences, in terms of sample size, and geographic and temporal range [23]. To be able to investigate the early population history and migratory patterns of taurine cattle in detail, the present model is now conditioned on a larger ancient ($n = 193$, including the Iranian samples) and modern ($n = 597$) mtDNA dataset that widely covers the area of the Neolithic westwards expansion from the 7th millennium BCE onwards. The focus of the study is on the time period when cattle were first introduced to Europe, thereby allowing us to address the following questions: i) Is the scenario of a single and severe domestication bottleneck in the Near East still supported when adding the much larger dataset from western Anatolia and Europe? ii) Did cattle reach Europe in a single dispersal process or is there evidence for multiple introductions or continuous gene-flow between regions? iii) How much of the genetic diversity from the Near East was introduced to the European continent? iv) Did the spread of cattle coincide with the spread of the Neolithic culture? and v) Are there any signs of admixture with female aurochs along the expansion route of domestic cattle?

2. METHODS

2.1 Material

150 samples of prehistoric domestic cattle from 24 archaeological sites were taken to analyse a 434 bp long mitochondrial d-loop fragment (for detailed information on the archaeological sites and sample age see Additional file 1. The majority of investigated individuals (113) come from Western Anatolia and Southeastern Europe, i.e. a region defined as an "interim zone" [5], pointing to its bridging position between the "Neolithic core zone" and its European fringe.

A further 22 samples from Southern France and Southern Italy represent the first domesticated cattle to reach Europe on the "Mediterranean route" of the Neolithic expansion.

Additionally, new prehistoric samples from Germany (4), Northern and Western France (11) and Syria (1), plus 80 previously published sequences mainly from Central and Western Europe and Iran were used for population genetic analyses ([23-28] and GenBank: KC172647 - KC172649). A total of 597 modern d-loop sequences of 240 bp length were collected from previously published studies [29,30]. They each provide representative sets of sequences that match the European, Anatolian and Near Eastern study area of the present paper, thereby also covering areas which are underrepresented in the aDNA dataset, e.g. Italy and the Iberian Peninsula. For a complete list of GenBank accession numbers of previously published sequences see Additional file 2.

2.2 Ancient DNA work and sequencing

All samples were processed in the ancient DNA facilities at the Institute of Anthropology, Mainz University (Germany), under strict rules for contamination prevention as described in Bramanti *et al.* [18]. Those include strict separation of pre-PCR and post-PCR labs, protective clothes, regular cleaning of surfaces and equipment with detergent and bleach, and UV-irradiation of rooms, laboratory hoods, and equipment. Bone samples were UV-irradiated for 45 min from two sides. The surface was mechanically removed using a sandblaster (P-G 400, Harnisch & Rieth) or rotary saw (Electer Emax IH-300, MAFRA). Bone cubes of about 0.3 cm side length were again UV-irradiated for 45 min from two sides. Samples were pulverized using a mixer mill (MM200, Retsch). Generally, aliquots of 0.5 g bone powder were incubated on a rocking shaker at 37°C in a decalcification and digestion solution containing 2.5 ml EDTA (0.5 M, pH8; Ambion®/Applied Biosystems), 250 µl N-Laurylsarcosine (0.5 %; Merck) and 30 µl Proteinase K (18 U; Roche). DNA extraction was performed using phenol-chloroform-isoamylalcohol (25:24:1; Roth). DNA was washed and concentrated using 50 kDa Centricons or 50 kDa 15 ml Amicons (Millipore). At least two independent extractions per sample were performed. Extraction blank controls were processed during each extraction. Additionally, the cleanness of the grinding jars was tested by extracting hydroxylapatite that was pulverized under the same conditions as the bone samples.

Amplification of 434 bp of the HVSI (positions 15914–9 according to reference sequence V00654) was generally conducted using a PCR primer set consisting of 6 primer pairs as in Bollongino *et al.* [23] (BosU1/L1-BosU6/L6) with slight modifications.

PCR reactions were usually performed with 2.5 U AmpliTaq Gold® (Applied Biosystems), 1x PCR Gold Buffer (Applied Biosystems), 2 mM $MgCl_2$ (Applied Biosystems), 0.2 mM dNTP's (Quiagen), 0.4 µg/µl BSA (Roche), and 0.2 µM primer (Biospring) and HPLC-H_20 (Acros Organics). Initial activation at 90°C for 6 min was followed by 50 cycles of denaturation (40 sec at 94°C), annealing (40 sec at 52-60°C), and elongation (40 sec at 72°C) in a Mastercycler gradient (Eppendorf). Blank controls were processed during each PCR. At least three independent PCRs from two different extracts were performed. Samples were sequenced on an ABI PRISM™ 3130 Genetic Analyzer (Applied Biosystems) using POP-6™ polymer (Applied Biosystems).

Sequences were analysed using the programs SeqMan™ and MegAlign™ (DNASTAR Lasergene® 7.1 and 8). At least three sequences obtained from independent PCRs from two independent DNA extractions per sample per primer pair were usually used to create a majority rule consensus sequence. For further details on ancient DNA work and sequencing including deviations from the general laboratory procedure described see Additional file 3.

2.3 Descriptive and summary statistics

All new and previously published ancient DNA sequences were subdivided into the following geographical groups: Iran/Syria, Western Anatolia, Southeastern Europe, Southeastern Central Europe, Italy, Southern France, Central/Western Europe, and Spain. These groups were further subdivided into chronological subgroups reflecting up to four different Neolithic and post-Neolithic periods per region (see Additional file 4 for detailed information on the groupings). Modern sequences were grouped according to their country of origin (also see Additional file 2).

For statistical analyses, all ancient sequences were cut to a 399 bp fragment to match the fragment sizes of previously published ancient sequences (positions 15,914-16,312 according to reference sequence GenBank V00654). Haplotype diversity, mean number of pairwise differences, Tajima's D, Fu's Fs and population pairwise F_{ST} were calculated using Arlequin 3.5.1.2 [31]. P values are based on 10,000 random permutations. The level of missing data allowed was adjusted in order to include all nucleotide positions even if there were gaps in some ancient sequences. Besides that, default values were used.

The MDS (multidimensional scaling) plot is based on F_{ST} values calculated using Reynolds' genetic distances and running 10,000 permutations in Arlequin 3.5.1.2. The MDS plot was created in R 2.14.2 R [32] using the packages MASS [33], plotrix [34] and shape [35].

Correlation between genetic and geographical distances among defined populations/groups was assessed by a Mantel test [36] under 9,999 random permutations using GENALEX 6.4 [37]. The Mantel test is based on F_{ST}values calculated using Reynolds' genetic distances and running 10,000 permutations in Arlequin 3.5.1.2. Geographical coordinates were determined by eye as the centre of appropriate countries per group for the modern samples and the centre of all archaeological sites per group for the ancient samples.

2.4 Coalescent simulations

Coalescent simulations were performed using Bayes Serial SimCoal [38], by extending the model previously described in Bollongino *et al.* [23]. Similarly, we assume an intergeneration time of 6 years, an ancestral Near Eastern wild aurochs female effective population size of 45,000 [39] and, again, a single domestication process of parameterized size N_D at time 8,500 years BCE (i.e. 1,750 generations BP). Following the domestication bottleneck, this Near Eastern population grows exponentially to a modern Near Eastern and Anatolian effective population size N_{NE} of 1,007,170 (see SI Bollongino *et al.* [23]). At 6,400 years BCE (i.e. 1,400 generations BP) a proportion of the population P is allowed to migrate to form a new and separate European population, which then grows exponentially to a modern European effective population size N_E of 7,942,392 (additional simulations which allow both N_{NE} and N_E to vary by an order of magnitude are described further in Additional file 5). From the split time until 5,000 years BCE migration between the two populations is allowed at per generation rate M_E ('early migration'), after which it is changed to rate M_L ('late migration'). Prior values for N_D are drawn uniformly from the range 1 – 1,000, P uniformly from the range 0 – 1, and both migration parameters uniformly from the range 0 – 0.01. The mutation rate is fixed at 45% per million years, the posterior modal value previously estimated by Bollongino *et al.* [23].

We used the above-mentioned 597 previously published modern sequences of 240 bp length (positions 16,023-16,262 according to reference sequence GenBank V00654) and cut the ancient sequences accordingly. The resulting 790 sequences were grouped into 4 sample groups: ancient Near Eastern and Anatolian ($n = 24$), ancient European ($n = 169$), modern Near Eastern and Anatolian ($n = 100$) and modern European ($n = 497$). We calculated 5 within- and 2 between-sample summary statistics (total = 32, also see Additional file 5 for details), and used approximate Bayesian computation (ABC [40]) to estimate parameter values.

3. RESULTS

Out of 150 newly analysed bones and teeth from prehistoric domesticated cattle, 113 yielded replicable and highly reliable mitochondrial HVR1 sequences, constituting a success rate of 75.3%. The sequences have been deposited in GenBank [GenBank: KF307209 to KF307322]). None of the blank controls contained amplifiable amounts of bovine DNA (for further detailed discussion of the validity of the ancient DNA data see Additional file 6). The successfully analysed samples come from Bosnia-Herzegovina (3 of 5), Bulgaria (52 of 68), France (15 of 19), Germany (4 of 4), Italy (5 of 14), Romania (15 of 16), Syria (1 of 1), and Turkey (18 of 23).

Using the nomenclature of Achilli *et al.* [41], all sequences belong exclusively to lineages from haplogroups that have previously been defined in present-day European domesticated cattle, namely T3 (n = 70), Q (n = 33), T2 (n = 6) and T, T5 or T1'2'3 (n = 4). None of them belongs to a specific mtDNA motif referred to as haplogroup P that is dominating in the indigenous aurochs population of Europe [26,27,42]. It is of note that the high frequency of haplogroup Q in ancient Southeastern Europe (between 50% and 29% in 5,500-5,000 BCE and 2,700-2,200 BCE, respectively) does not match present-day haplogroup distributions of taurine cattle from Europe, and particularly from the same area (combined frequency for T and Q in present-day Balkan and Greece: 1.5 - 2% [43]). It is also markedly higher than in all other ancient European groups (e.g. only 4% in Central/Western Europe (5,400-4,400 BCE)). See Additional file 4 for detail on haplogroup composition and frequency of haplogroup Q across the 13 spatiotemporal groups.

There are 35 different mitochondrial lineages in the 193 prehistoric individuals, eight of which occur more than once in the dataset. Non-unique haplotypes (H) were named according to their haplogroup and numbered consecutively (H1-H8). Additional files 4 and 7 provide a detailed overview on the distribution of haplogroups and shared and unique haplotypes across the 13 spatiotemporal groups. Only haplotypes called T3_H1, Q_H4, and T2_H7 occur more than twice in the dataset (114, 37, and 5 times, respectively), with T3_H1 also being predominant in present-day taurine cattle. Haplotype T3_H1 occurs in all of the ancient 13 spatiotemporal groups, Q_H4 in all except Spain 2,700-1,600 BCE and Southern France 5,500-4,500 BCE. It is of note that Q_H5, T_H6, and T2_H7 are restricted to the geographical groups of Iran/Syria and Southeastern Europe (Q_H5 in Iran/Syria 4,000-1,400 BCE and Southeastern Europe 6,200-5,500 BCE; T_H6 in Iran 7,000-5,000 BCE and Southeastern Europe 2,700-2,200 BCE; T2_H7 in Iran 7,000-5,000 BCE and Southeastern Europe 5,500-5,000 BCE).

3.1 Genetic distances between cattle populations

The MDS plot (Figure 1) reveals a pattern that separates three geographical groups: The four Southeastern European groups cluster with the one from Western Anatolia; both groups from Iran/Syria and from Central/Western Europe are close to each other. However, Southern France and Central/Western Europe are isolated from all other groups and from each other.

Subgroups comprising only the earliest Neolithic cattle of each geographical group were used to further evaluate the influence of sample age and geographical location on genetic distances. Figure 2 maps significant pairwise F_{ST} values between the resulting eight groups. The greatest genetic distances can be observed between Iran 7,000-5,000 BCE and the groups from Central/Western Europe 5,400-4,400 BCE and Southern France 5,500-4,500 BCE with values of 0.47 and 0.40, respectively. These groups also show the greatest geographical distances. The second highest F_{ST} values occur between Southeastern Europe 6,200-5,500 BCE and Central/Western Europe 5,400-4,400 BCE and between Southeastern Europe 6,200-5,500 BCE and Southern France 5,500-4,500 BCE (0.27 and 0.29 respectively). In comparison, the genetic distance between Iran 7,000-5,000 BCE and Southeastern Central Europe 5,100-4,000 BCE is – despite greater geographical distance - slightly lower (0.23). The F_{ST} between Iran 7,000-5,000 BCE and Southeastern Europe 6,200-5,500 BCE is even smaller (0.17). The geographically adjacent groups of Southeastern Europe 6,200-5,500 BCE and 5,500-5,000 BCE and Southeastern Central Europe 5,100-4,000 BCE reveal a distance as high as 0.16 and 0.10, respectively.

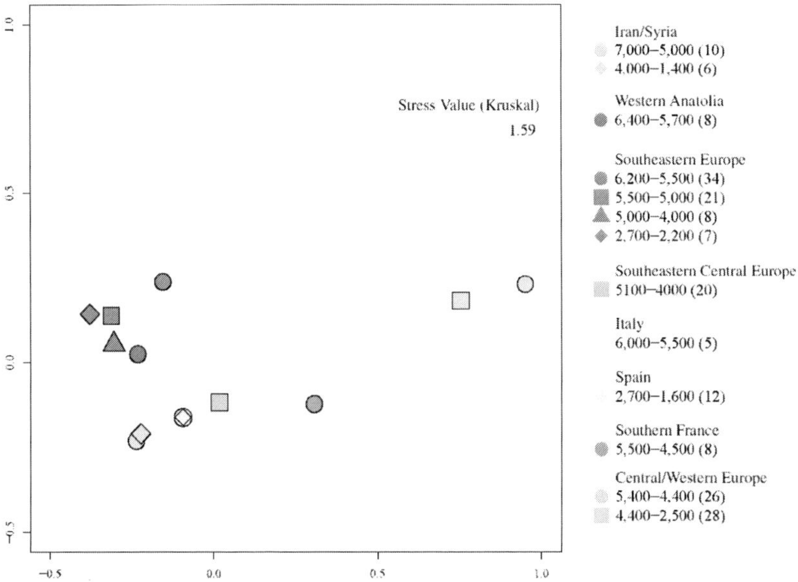

Figure 1 MDS Plot of d-loop sequences from 13 spatiotemporal groups of ancient domesticated cattle. The MDS plot is based on Reynolds' F_{ST}. Numbers represent the age of samples in BCE per group; brackets contain the number of sequences per group.

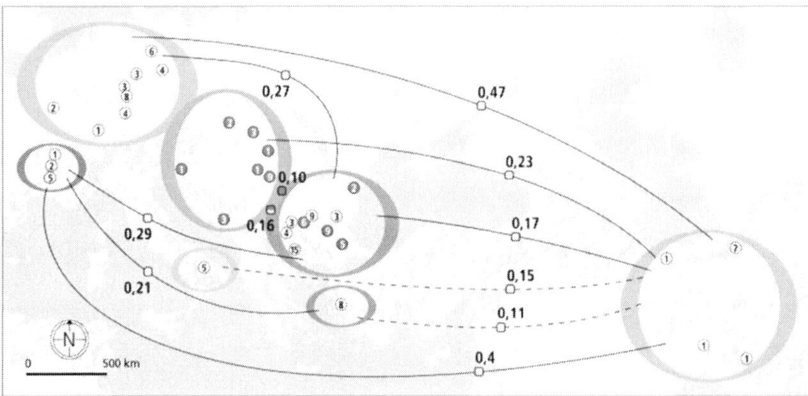

Figure 2 Population pairwise F_{ST}s between d-loop sequences from eight Neolithic groups of ancient domesticated cattle. Coloured rings surround geographical groups. Grey and white dots within the circles represent geographical location of archaeological sites. White dots stand for the oldest Neolithic samples per group, grey dots for Middle/Late Neolithic samples.

Numbers within dots represent the number of d-loop sequences per site. Orange: Central/Western Europe 5,400-4,400 BCE, blue: Southern France 5,500-4,500 BCE, green: Southeastern Central Europe 5,100-4,000 BCE, yellow: Italy 6,000-5,500 BCE, purple: Southeastern Europe 6,200-5,500 BCE and 5,500-5,000 BCE, red: Western Anatolia 6,400-5,700 BCE, and grey: Iran 7,000-5,000 BCE. Numbers on the lines between coloured circles are population pairwise F_{ST}s. Solid lines stand for significant F_{ST}s at the 0.05 level, dashed lines stand for significant F_{ST}s at the 0.1 level. Grey and white colours of squares on the lines encode which chronological groups per geographi groups are being compared.

A Mantel Test on the basis of Reynolds' F_{ST} resulted in a strong positive correlation (Rxy: 0.75, P-value 0.001) between geographical and genetic distance among the eight groups. Approximately 56% of the variation can be explained by geographical distance ($R^2 = 0.56$). There is a weaker correlation of genetic and geographical distances in modern samples (Rxy: 0.54, P-value 0.002). Here, only 29% of the variation can be explained by geographical distance ($R^2 = 0.29$). Complete population pairwise F_{ST} matrices can be found in Additional file 8.

3.2 Measurements of molecular diversity ($\hat{H}$, π), Tajima's D and Fu's Fs

The estimates of haplotype diversity ($\hat{H}$), the mean number of pairwise differences (π), Tajima's D, and Fu's Fs are given in Table 1.

Haplotype diversity clearly decreases in a southeast to northwest direction with Iran 7,000-5,000 BCE (0.96) at the high end, and Southern France 5,500-4,500 BCE (0.00) and Central/Western Europe 5,400-4,400 BCE (0.22) at the low end. The haplotype diversity of the earliest domesticated cattle on the European continent in Southeastern Europe 6,200-5,500 BCE (0.62) is much lower than in Iran, and comparable to Western Anatolia 6,400-5,700 BCE (0.64), but higher than in the geographically close European group of Southeastern Central Europe 5,100-4,000 BCE (0.52). Again, diversity in Central/Western Europe 5,400-4,400 BCE is substantially lower (0.22). Following the northern Mediterranean coast, the values also drop sequentially from Western Anatolia 6,400-5,700 BCE (0.64) to Italy 6,000-5,500 BCE (0.40) to Southern France 5,500-4,500 BCE (0.00).

Haplotype diversity estimates increase with time in the two regions where samples from two Neolithic periods are available: from 0.22 to 0.50 in Central/Western Europe and from 0.62 to 0.78 in Southeastern Europe.

Table 1 Summary statistics of d-loop sequences from 13 spatiotemporal groups of ancient domesticated cattle

Ancient samples	# Sequences	# Haplotypes	$\hat{H}$	π	Tajima's D (P-value)	Fu's Fs (P-value)
IR 7,000-5,000 BCE	10	8	0.96 +/- 0.06	2.82 +/- 1.62	−0.49 (0.33)	**−3.77 (0.01)**
IR/S 4,000-1,400 BCE	6	4	0.80 +/- 0.17	3.80 +/- 2.22	−0.79 (0.27)	0.67 (0.59)
IT 6,000-5,500 BCE	5	2	0.40 +/- 0.24	0.80 +/- 0.68	−0.97 (0.20)	1.04 (0.63)
SECE 5,100-4,000 BCE	20	6	0.52 +/- 0.13	1.30 +/- 0.85	−1.12 (0.13)	−1.48 (0.14)
SEE 6,200-5,500 BCE	34	6	0.62 +/- 0.05	1.26 +/- 0.82	0.08 (0.58)	−0.79 (0.32)
SEE 5,500-5,000 BCE	21	7	0.78 +/- 0.06	1.79 +/- 1.08	0.12 (0.57)	0.37 (0.60)
SEE 5,000-4,000 BCE	8	4	0.79 +/- 0.11	2.11 +/- 1.31	−0.42 (0.37)	0.26 (0.54)
SEE 2,700-2,200 BCE	7	5	0.81 +/- 0.13	1.62 +/- 1.08	−0.04 (0.47)	−0.54 (0.22)
SF 5,500-4,500 BCE	8	1	0.00 +/- 0.00	0.00 +/- 0.00	-	-
SP 2,700-1,600 BCE	12	4	0.45 +/- 0.17	1.30 +/- 0.87	**−1.71 (0.04)**	−0.05 (0.47)
WA 6,400-5,700 BCE	8	4	0.64 +/- 0.18	0.79 +/- 0.63	**−2.14 (0.00)**	**−1.73 (0.02)**
CWE 5,400-4,400 BCE	26	4	0.22 +/- 0.11	0.45 +/- 0.41	**−1.53 (0.03)**	−0.24 (0.32)
CWE 4,400-2,500 BCE	28	9	0.50 +/- 0.12	0.91 +/- 0.65	**−2.08 (0.01)**	**−4.55 (0.00)**

IR/S: Iran/Syria, IT: Italy, SECE: Southeastern Central Europe, SEE: Southeastern Europe, SF: Southern France, SP: Spain, WA: Western Anatolia, and CWE: Central/Western Europe. Numbers in the first column indicate the age of the samples in BCE. $\hat{H}$: Haplotype diversity, π: mean number of pairwise differences. Significant Tajima's D and Fu's Fs value at the 0.05 level are highlighted in bold.

In Southeastern Europe, haplotype diversity remains the same in the subsequent Chalcolithic group 5,000-4,000 BCE (0.78), but increases again during the Bronze Age 2,700-2,200 BCE (0.81). Similar patterns are observed when considering the mean numbers of pairwise differences. The Neolithic subgroups also show a tendency of decreasing values with distance from Iran. Regionally, the values increase with time; in Central/Western Europe from 0.45 to 0.91 and in Southeastern Europe from 1.26 to 1.79 to 2.11. In the youngest Southeastern European group (2,700-2,200 BCE) the value drops again (0.54). In Central/Western Europe, where both diversity indices increase with time, Tajima's D is also significantly negative (Fu's Fs only in the younger group). This is not the case for Southeastern Europe. Diversity estimates of the 597 modern cattle sequences only show a slight tendency towards an east to west gradient for both haplotype diversity, and the mean number of pairwise differences. Tajima's D and Fu's Fs are mostly significantly negative. All diversity estimates and graphical visualisations of chronological and geographical diversity trends can be found in the Additional file 9.

3.3 Coalescent simulations

We performed 5 million coalescent simulations under the demographic model described above, and used a tolerance proportion of 0.1%, meaning that we retained the 5,000 best parameter sets. Figure 3 shows the joint posterior density of parameters N_D and P (marginal to the remaining two), with the joint mode found at $N_D = 81$ and $P = 0.73$. The marginal modal value for N_D was 92 (95% credible interval: 29 – 783). Marginal densities for the two migration parameters M_E and M_L are given in Figure 4. While it is not possible to infer much from the relatively uninformative posterior for M_E (top, mode 0.006 94; 95% CI: 0.00033 – 0.00974), we are able to say that migration between the Near East and Europe (M_L) appears to have been greatly reduced, essentially to zero, in the period after 5,000 years BCE (bottom, mode 0.00022; 95% CI: 0.00001 – 0.00946). Further simulations were performed in order to test the sensitivity of these parameter estimations to our assumed fixed values of N_{NE} and N_E. Increasing or decreasing both N_{NE} and N_E by an order of magnitude produced estimates that did not significantly differ from those given above (see Additional file 5 for details).

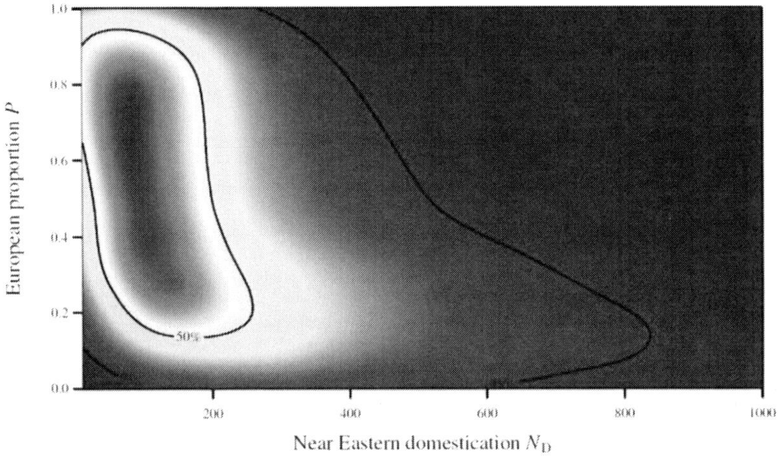

Figure 3 Joint posterior density for the domestication bottleneck (N_D) and the proportion moving into Europe (P). The approximate joint posterior probability density of the proportion of the population P allowed to move into Europe at the time of the split (6,400 BCE) and the effective female population size at the time of the domestication event (N_D). The 50% and 95% credible intervals are overlaid as contours.

4. DISCUSSION

4.1 The domestication process of taurine cattle

Using an ancient ($n = 193$) and modern mtDNA dataset ($n = 597$) of domesticated cattle from the Near East, Anatolia and Europe for coalescent simulation and approximate Bayesian computation, we inferred a (joint) posterior mode of 81 female founder individuals at the beginning of the domestication process. This result is consistent with the previous estimate based on 15 ancient Iranian and 27 modern Near Eastern and Anatolian cattle [23], and demonstrates that this initial finding of a very strong Near Eastern bottleneck is robust even with a greatly expanded continent-wide data set, and is not biased by theoretically possible subsequent introgression from aurochs populations outside the Near East. It can therefore be concluded that domestic cattle indeed have a discrete and rather localised origin, very likely in Southeastern Anatolia and the Near East, a view that is consistent with a huge body of archaeozoological evidence from the 9th millennium BCE [44-46].

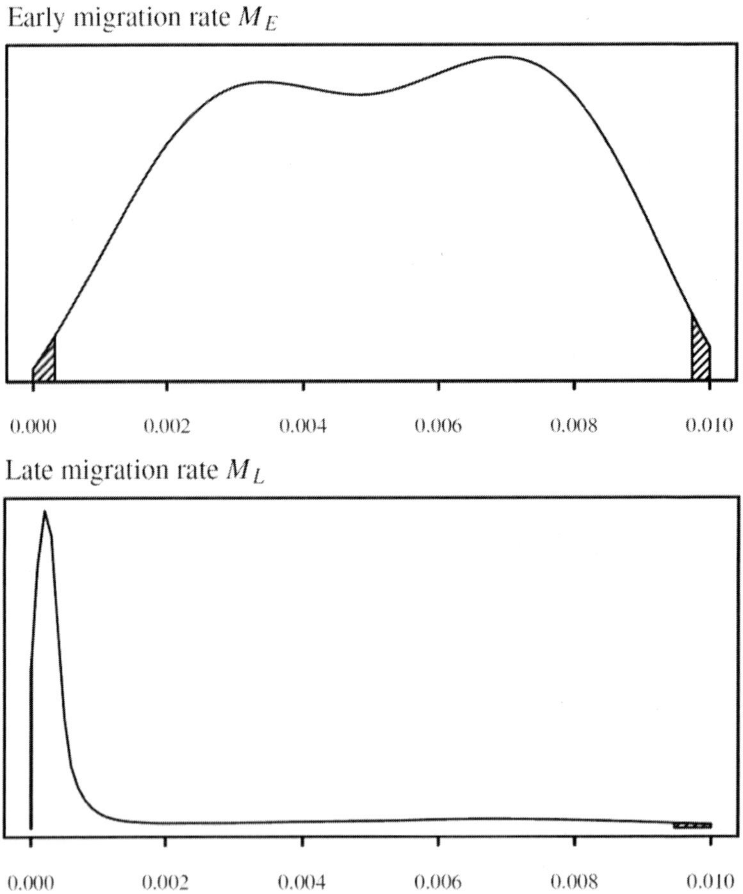

Figure 4 The marginal approximate posterior probability densities of the two migration rate parameters. M_E ('early migration') is the rate from the population split time until 5,000 BCE, and M_L ('late migration') is the rate from 5,000 BCE to present.

Subsequent to the first domestication phase, the ancient DNA data, together with archaeological evidence, point to an intermittent expansion scenario. Expanding from Southeastern Anatolia, cattle reached Western Anatolia and the Aegean not before 7,000 BCE. From here, they spread simultaneously across Southeastern Europe and along the Mediterranean coast into Central, Northern, Southwestern, and Western Europe. In essence, the observed strong correlation between genetic and geographical distances together with decreasing genetic diversity roughly in a southeast to northwest and southwest direction is consistent with the idea of serial dilution of diversity by a series of recurring founder events. The oldest (Neolithic) groups with the greatest geographical distance from each other, namely Iran and Central/Western Europe and Southern France, show the highest F_{ST} values (0.47 and 0.4, respectively). Smaller genetic distances are observed between more adjacent areas, such as between Iran and Western Anatolia and between Iran and Southeastern Europe (0.11 and 0.17, respectively). Other statistics are equally consistent with the serial dilution model: Neolithic cattle from Iran yield the highest value for haplotype diversity in the whole dataset (0.96). Haplotype diversity consistently decreases along the proposed two main Neolithisation routes, with the lowest values in remote areas, i.e. in Neolithic Central/Western Europe and Southern France (0.22 and 0.00, respectively), while intermediate values are observed in between.

Alternative scenarios of secondary domestications or traceable female gene-flow from wild aurochs in Europe have been discussed several times in the literature [29,47-51]. The arguments are mainly based on scarce findings of the mtDNA haplogroup P, pre-dominating in European aurochs, in the domesticated stock [49,50] on the one hand, and the presence of mtDNA lineages in pre-Neolithic Italian aurochs that resemble those of the imported domesticated animals [29,48] on the other, thereby impeding the detection of introgression by mere comparison of haplogroup composition. However, realistic expectations under such models would also include i) a larger inferred founder population due to introgressions of diverse aurochs lineages and ii) significant deviations from the serial dilution of genetic diversity model. None of the two has been observed in or can be inferred from the data presented here. Detection of potential introgression of Italian aurochs through time deserves further attention, e.g. by expanding the existing dataset to encompass finds from diverse archaeological sites and later chronological phases. However, the existing dataset from the rest of Europe suggests that introgressions of local genes into the imported domestic cattle populations are rare and geographically restricted exceptions, or coming from male aurochs. Separate independent domestication(s) of European aurochs can almost with certainty be excluded.

The strict separation of domestic cattle from their wild European relatives is very different to what can be observed in other animals. For example, pigs were imported to Europe in a similar way to cattle, but after a few centuries all their mitochondrial lineages were replaced through admixture with local wild boar [52,53].

4.2 The first domesticated cattle in Europe

The summary statistical patterns described here may be partly biased by the fact that the analysed data come from heterochronous and spatially diverse samples [54]. Therefore, we used coalescent simulations to estimate the key parameters of taurine cattle population history upon their arrival in Europe in a realistic evolutionary demographic framework.

Our model suggests that a high proportion (73%) of domesticated cattle in Anatolia and the Near East may have migrated into Europe. This indicates that the expansion into Europe was a far less severe bottleneck than assumed, and that much of the variation present in the original Anatolian/Near Eastern population survived in initial European cattle populations. Consistent with this, the Western Anatolian and Southeastern European sample groups constitute a cluster in the MDS plot (Figure 1). However, Southern France and Central/Western Europe instead are clearly separated, very likely reflecting genetic diversification along the two main Neolithisation routes. It is noteworthy that the data from Western Anatolia, Southern Italy, and Southern France come from very few sites with less than ten samples each (eight, five, and eight, respectively) and therefore have to be evaluated cautiously. However, the drastic decline in haplotype diversity and mean number of pairwise differences from 0.64/0.80 and 0.40/0.79 in Western Anatolia and Italy to 0.00/0.00 in Southern France is a good fit to a scenario of only few individuals being transported by boat to the Northwestern Mediterranean coast [3,55]. Low diversity estimates are also congruent with the fact that cattle did not play a major role in the domesticated faunal spectrum of Neolithic economies from Mediterranean Europe (Impressa and Cardial), in contrast to Neolithic Cultures in Central Europe, where domesticated cattle were generally well represented [56,57].

Tracing the spread of cattle through the European mainland, there are patterns in the data that point to significant demographic changes connected to the expansion of the Neolithic culture from Southeastern to Central/Western Europe. The genetic distance between Southeastern Europe (6,200-5,500 BCE) and Central/Western Europe (5,400-4,400 BCE) is unexpectedly high (0.27). To put this high value into context: the F_{ST} values between Iran (7,000-5,000 BCE), Southeastern Europe (6,200-5,500 BCE) and Italy (6,000-5,500 BCE) are much lower (0.17 and 0.15, respectively) despite larger geographic distances. A good indicator for this massive demographic change is that the frequency of the mitochondrial Q-lineage drops from 50 % in Southeastern European (6,200-5,500 BCE) to 4% in Central/Western Europe (5,400-4,400 BCE). Haplotype diversity decreases drastically from 0.62 to 0.22. There are several additional lines of evidence that point to the region between Southeastern Europe and Central

Europe as a kind of core area where the Neolithic idea was re-consolidated: i) From archaeology: The Linearbandkeramik culture (LBK, engl. Linear Pottery culture) developed here and spread rapidly over Central Europe starting around 5,600 BCE [58]; ii) From palaeogenetics: A migration of farmers from Southeastern to Central Europe has been inferred using ancient mtDNA [18]; iii) From gene-culture coevolutionary modelling: Spatially-explicit computer simulations of the spread of an allele associated with lactase persistence in humans (i.e. the ability to digest milk sugar as an adult), point to this area as where positive selection started affecting the frequency of this allele in dairying cultures [59]. We therefore suggest that the observed substantial loss of genetic diversity and the increasing genetic distance in prehistoric cattle are the result of a significant founder event along with the spread of the LBK. It probably coincides with a major wave of human migration and is followed by a period of intensified cattle breeding resulting in a rising importance of dairying. This picture becomes even more comprehensive when we look at how patterns change after the early Neolithic.

4.3 After the arrival

Cattle herding becomes more and more important with the onset of the LBK. A few centuries later, cattle bones constitute up to 70% of all domesticated animal bones in faunal assemblages in Central Europe, a value that stays roughly the same for most of the subsequent millennia with some regional fluctuations [56,57]. Accordingly, significantly negative Tajima's D and Fu's Fs values in Neolithic Central/Western Europe and in the majority of the modern sample groups point to extended periods of population growth (see Table 1 and Additional file 9).

Interestingly, there is no indication for population growth in Southeastern Europe. The observed diachronic increase in haplotype diversity in the Southeastern European sample groups appears in tandem with new, previously absent mitochondrial haplotypes (also see Additional file 7). It is of note that two of these new haplotypes (T_H6 and T2_H7) are present here and in the Iranian Neolithic sample but not elsewhere.

According to our demographic modelling, migration between Anatolia/the Near East and Europe was greatly reduced, essentially to zero, in the period after 5,000 BCE. We should expect that accurately estimating the level of early migration between 6,400 and 5,000 BCE to be difficult, as it is somewhat confounded by the proportion of cattle P moving to Europe at the time of the split (indeed the two parameters are very slightly negatively correlated; degree $r = -0.07$, $p = 0.0002$). However, it is clear that there is support at least for *some* level of migration during this early period as the estimated modal migration rate is clearly greater than 0.

We therefore suggest that a probable underlying scenario for our observations is one of continuous gene-flow into Europe following the initial colonization at around 6,400 BCE. This scenario also fits in with archaeological evidence for accelerated westward acculturation occurring in the first half of the 6th millennium BCE [6,60]. This early phase was followed by almost total isolation between the European and Anatolian/Near Eastern cattle populations after 5,000 BCE.

It is of note that this pattern has changed again in later periods. The pattern of decreasing diversity in the direction of the Neolithic expansion and the correlation of genetic and geographical distances is considerably weaker in modern-day cattle breeds than in the Neolithic. It is not clear yet to which extent human migrations from the East as postulated for the Bronze Age [61] influenced the already existing cattle stock in Europe. However, the fading geographical patterns are likely mirroring more recent demographic changes and founder events, such as global trade, exceptional selection pressure on particular high performance breeds and replacement of traditional breeds [62]. Thus, the present study explicitly underlines that ancient demographic and evolutionary processes in selectively bred animals can only be uncovered by using ancient DNA data.

5. CONCLUSIONS

Overall, palaeogenetic together with archaeological and archaeozoological data strongly support the following scenario: taurine cattle were domesticated in a region between Southeastern Anatolia and the Zagros Mountains, Syria and the Lebanon. The domestication process started in the mid-9th millennium BCE, with a small effective number of wild female aurochs (estimated modal value of 81). After 7,000 BCE, domestic cattle populations were transported from the Central Anatolian plateau to Western Anatolia and the Aegean. Much of the original Anatolian and Near Eastern variation (approximately 73%) survived in the first Neolithic cattle that were introduced to Europe around 6,400 BCE. Despite some evidence for subsequent gene-flow with Anatolia and the Near East between 6,400 and 5,000 BCE, most of the initial genetic diversity was lost as cattle spread through Europe along with the Neolithic transition: Via the Mediterranean trajectory, migrating farmers reached i.e. Southern Italy, Northern Africa, the Tyrrhenian Islands, Southern France and the Iberian Peninsula by boat. The low genetic diversity observed in the few genetic data available from these regions points to a significantly low effective population size of cattle arriving in the Western Mediterranean. Along the second trajectory across the European mainland and without major signs of introgression from wild aurochs, cattle finally reached Central, Western (after 5,500 BCE) and

Northern Europe (after 4,100 BCE). Also here, much of the genetic diversity was lost during the move, particularly when cattle were brought to Central Europe by LBK farmers.

Gene-flow between Europe and Anatolia and the Near East appears to have been reduced, essentially to 0, after around 5,000 BCE. In modern breeds however, the genetic effects of the inferred migratory patterns and geographical diversification become far less pronounced, probably due to selective breeding and trade of high performance cows in very recent times.

In summary, the genetic prehistory of domestic cattle seems to consist of a small, localised domestication process, followed by a relatively straightforward series of spasmodic expansion episodes resulting in a serial dilution of genetic diversity from the Near East to Western and Northern Europe. Future genomic multi-locus studies of ancient DNA from prehistoric periods will hopefully add greater detail to this picture, particularly by incorporating the potentially divergent demography of male cattle.

ACKNOWLEDGEMENTS

This work was funded by German Archaeological Institute, Johannes Gutenberg-University Mainz, and CNRS (*Centre national de la recherche scientifique*).

The authors would like to thank Adrian Bălăşescu, Cornelia Becker, Daniel Bradley, Altan Çilingiroğlu, Çiler Çilingiroğlu, Giuliano Cremonesi, Keith Dobney, Ceiridwen Edwards, Ralf Gleser, Angela Graefen, Jean Guilaine, Svend Hansen, Daniel Helmer, Robert Hofmann, Raiko Krauß, Marion Lichardus-Itten, Claire Manen, Ingo Motzenbäcker, Mehmet Özdoğan, Jean Roudil, Silviane Scharl, Nils Müller-Scheeßel, Wolfram Schier, Mona Schreiber, Elisabeth Stephan, and Bernhard Weninger for providing bone material and for helpful discussions.

REFERENCES

1. Özdoğan M. The expansion of the neolithic way of life: What we know and what we do not know. In: *How did farming reach Europe?* Edited by Lichter C, vol. 2. Istanbul: Byzas; 2005. p. 13–27.
2. Özdoğan M. Archaeological Evidence on the Westward Expansion of Farming Communities from Eastern Anatolia to the Aegean and the Balkans. Curr Anthropol. 2011;52(S4):S415–30.
3. Vigne J-D. Zooarchaeological aspects of the Neolithic diet transition in the Near East and Europe, and their putative relationships with the Neolithic Demographic Transition. In:

Bocquet-Appel J-P OB-Y, editor. *The Neolithic Demographic Transition and its Consequences*. New York: Springer Verlag; 2008. p. 179–205.
4. Conolly J, Colledge S, Dobney K, Vigne J-D, Peters J, Stopp B, et al. Meta-analysis of zooarchaeological data from SW Asia and SE Europe provides insight into the origins and spread of animal husbandry. J Archaeol Sci. 2011;38(3):538–45.
5. Özdoğan M. An alternative approach in tracing changes in demographic composition. The westward expansion of the neolithic way of life. In: Bocquet-Appel J-P, Bar-Yosef O, editors. *The neolithic demographic transition and its consequences.* 2008th ed. Berlin: Springer; 2008. p. 139–78.
6. Düring BS. The prehistory of Asia Minor : from complex hunter-gatherers to early urban societies. Cambridge: Cambridge University Press; 2011.
7. Çilingiroğlu A, Cevik O, Çilingiroğlu C. Ulucak Höyük: Towards understanding the early farming communities of Middle West Anatolia: Contribution of Ulucak. In: Özdoğan M, Başgelen N, Kuniholm P, editors. *The Neolithic in Turkey, Western Turkey*. Istanbul: Archaeology & Art Publications; 2012. p. 139–75.
8. Reingruber A. Die deutschen Ausgrabungen auf der Agrissa-Magula in Thessalien II. Die Agrissa Magula. In: Beiträge zur ur- und frühgeschichtlichen Archäologie des Mittelmeer-Kulturraums, Hauptmann H, editors. Das frühe und das beginnende mittlere Neolithikum im Lichte transägäischer Beziehungen. Bonn: Dr. Rudolf Habelt GmbH; 2008.
9. Lüning J. Steinzeitliche Bauern in Deutschland – die Landwirtschaft im Neolithikum. Bonn: Dr. Rudolf Habelt GmbH; 2000.
10. Guilaine J. De la vague à la tombe, La conquête néolithique de la Méditerranée (8000-2000 avant J.-C). In. Paris: Le Seuil; 2003.
11. Tresset A, Vigne J-D. Last hunter-gatherers and first farmers of Europe. C R Biol. 2011;334(3):182–9.
12. Ammerman AJ, Cavalli-Sforza LL. The Neolithic transition and the genetics of populations in Europe. Princeton, Guildford: Princeton University Press; 1984.
13. Zvelebil M, Zvelebil KV. Agricultural transition and Indo-European dispersals. Antiquity. 1988;62(236):574–83.
14. Ammerman AJ. On the Neolithic Transition in Europe - a Comment. Antiquity. 1989;63(238):162–5.
15. Zvelebil M. On the Transition to Farming in Europe, or What Was Spreading with the Neolithic - a Reply. Antiquity. 1989;63(239):379–83.
16. Whittle AWR. Europe in the Neolithic : the creation of new worlds. Cambridge: Cambridge University Press; 1996.
17. Pinhasi R, Thomas MG, Hofreiter M, Currat M, Burger J. The genetic history of Europeans. Trends Genet. 2012;28(10):496–505.
18. Bramanti B, Thomas MG, Haak W, Unterlaender M, Jores P, Tambets K, et al. Genetic discontinuity between local hunter-gatherers and central Europe's first farmers. Science. 2009;326(5949):137–40.
19. Skoglund P, Malmstrom H, Raghavan M, Stora J, Hall P, Willerslev E, et al. Origins and genetic legacy of Neolithic farmers and hunter-gatherers in Europe. Science. 2012;336(6080):466–9.
20. Lazaridis I, Patterson N, Mittnik A, Renaud G, Mallick S, Kirsanow K, et al. Ancient human genomes suggest three ancestral populations for present-day Europeans. Nature. 2014;513(7518):409–13.
21. Tresset A, Bollongino R, Edwards CJ, Hughes S, Vigne J-D. Early diffusion of domestic bovids in Europe: An indicator for human contact, exchanges and migrations? In: Hombert JM, Errico F, editors. *Becoming eloquent, advances in the emergence of language, human cognition, and modern cultures*. Amsterdam: John Benjamins Publ. Comp; 2009. p. 69–90.
22. Larson G, Burger J. A population genetics view of animal domestication. Trends Genet. 2013;29(4):197–205.
23. Bollongino R, Burger J, Powell A, Mashkour M, Vigne J-D, Thomas MG. Modern Taurine Cattle descended from small number of Near-Eastern founders, *Molecular Biology and Evolution*. 2012. doi:10.1093/molbev/mss1092.

24. Bollongino R, Edwards CJ, Alt KW, Burger J, Bradley DG. Early history of European domestic cattle as revealed by ancient DNA. Biol Lett. 2006;2(1):155–9.
25. Anderung C, Bouwman A, Persson P, Carretero JM, Ortega AI, Elburg R, et al. Prehistoric contacts over the Straits of Gibraltar indicated by genetic analysis of Iberian Bronze Age cattle. Proc Natl Acad Sci U S A. 2005;102(24):8431–5.
26. Edwards CJ, Bollongino R, Scheu A, Chamberlain A, Tresset A, Vigne J-D, et al. Mitochondrial DNA analysis shows a Near Eastern Neolithic origin for domestic cattle and no indication of domestication of European aurochs. Proc R Soc B Biol Sci. 2007;274(1616):1377–85.
27. Scheu A, Hartz S, Schmoelcke U, Tresset A, Burger J, Bollongino R. Ancient DNA provides no evidence for independent domestication of cattle in Mesolithic Rosenhof. Northern Germany Journal of Archaeological Science. 2008;35(5):1257–64.
28. Bollongino R, Elsner J, Vigne J-D, Burger J. Y-SNPs do not indicate hybridisation between European aurochs and domestic cattle. PLoS One. 2008;3(10), e3418.
29. Beja-Pereira A, Caramelli D, Lalueza-Fox C, Vernesi C, Ferrand N, Casoli A, et al. The origin of European cattle: evidence from modern and ancient DNA. Proc Natl Acad Sci U S A. 2006;103(21):8113–8.
30. Troy CS, MacHugh DE, Bailey JF, Magee DA, Loftus RT, Cunningham P, et al. Genetic evidence for Near-Eastern origins of European cattle. Nature. 2001;410(6832):1088–91.
31. Excoffier L, Lischer HE. Arlequin suite ver 3.5: a new series of programs to perform population genetics analyses under Linux and Windows. Mol Ecol Resour. 2010;10(3):564–7.
32. R Developement Core Team. R: A Language and Environment for Statistical Computing. R Foundation for Statistical Computing, Vienna, Austria, R Foundation for Statistical Computing. 2012. ISBN: 3-900051-07-0, URL http://www.R-project.org/.
33. Venables WN, Ripley BD. Modern Applied Statistics with S. 4th ed. New York: Springer Verlag; 2002.
34. Lemon J. Plotrix: a package in the red light district of R. R-News. 2006;6(4):8–12.
35. Soetaert K: Shape: Functions for plotting graphical shapes, colors. R package version 1.3.4. http://CRAN.R-project.org/package=shape. 2011.
36. Mantel N. The detection of disease clustering and a generalized regression approach. Cancer Res. 1967;27:209–20.
37. Peakall R, Smouse PE. GENALEX 6: genetic analysis in Excel. Population genetic software for teaching and research Molecular Ecology Notes. 2006;6(1):288–95.
38. Anderson CN, Ramakrishnan U, Chan YL, Hadly EA. Serial SimCoal: a population genetics model for data from multiple populations and points in time. Bioinformatics. 2005;21(8):1733–4.
39. MacEachern S, Hayes B, McEwan J, Goddard M. An examination of positive selection and changing effective population size in Angus and Holstein cattle populations (Bos taurus) using a high density SNP genotyping platform and the contribution of ancient polymorphism to genomic diversity in Domestic cattle. BMC Genomics. 2009;10:181.
40. Beaumont MA, Zhang W, Balding DJ. Approximate Bayesian computation in population genetics. Genetics. 2002;162(4):2025–35.
41. Achilli A, Bonfiglio S, Olivieri A, Malusa A, Pala M, Hooshiar Kashani B, et al. The multifaceted origin of taurine cattle reflected by the mitochondrial genome. PLoS One. 2009;4(6), e5753.
42. Gravlund P, Aaris-Sorensen K, Hofreiter M, Meyer M, Bollback JP, Noe-Nygaard N. Ancient DNA extracted from Danish aurochs (Bos primigenius): genetic diversity and preservation. Ann Anat. 2012;194(1):103–11.
43. Lenstra J, Ajmone-Marsan P, Beja-Pereira A, Bollongino R, Bradley D, Colli L, et al. Meta-Analysis of Mitochondrial DNA Reveals Several Population Bottlenecks during Worldwide Migrations of Cattle. Diversity. 2014;6(1):178–87.
44. Peters J, von den Driesch A, Helmer D. The upper Euphrates-Tigris basin: Cradle of agro-pastoralism? In: Vigne J-D, Helmer D, editors. *The first steps of animal domestication New archaeological approaches Proceedings of the 9th ICAZ Conference, Durham 2002*. Oxford: Oxbow Books; 2005. p. 96–124.

45. Helmer D, Gourichon L, Monchot H, Peters J, Segui MS. Identifying early domestic cattle from Pre-Pottery Neolithic sites on the Middle Euphrates using sexual dimorphism. In: Vigne J-D, Peters J, Helmer D, editors. *The first steps of animal domestication New archaeological approaches Proceedings of the 9th ICAZ Conference, Durham 2002*. Oxford: Oxbow Books; 2005. p. 86–95.
46. Hongo H, Pearson J, Öksük B, Ilgezdi G. The process of ungulate domestication at Çayönü. Southeastern Turkey: A multidisciplinary approach focusing on Bos sp and Cervus elaphus Anthropozoologica. 2009;44:63–73.
47. Achilli A, Olivieri A, Pellecchia M, Uboldi C, Colli L, Al-Zahery N, et al. Mitochondrial genomes of extinct aurochs survive in domestic cattle. Curr Biol. 2008;18(4):R157–8.
48. Mona S, Catalano G, Lari M, Larson G, Boscato P, Casoli A, et al. Population dynamic of the extinct European aurochs: genetic evidence of a north-south differentiation pattern and no evidence of post-glacial expansion. BMC Evol Biol. 2010;10:83.
49. Schibler J, Elsner J, Schlumbaum A. Incorporation of aurochs into a cattle herd in Neolithic Europe: single event or breeding? Sci Rep. 2014;4:5798.
50. Stock F, Edwards CJ, Bollongino R, Finlay EK, Burger J, Bradley DG. Cytochrome b sequences of ancient cattle and wild ox support phylogenetic complexity in the ancient and modern bovine populations. Anim Genet. 2009;40(5):694–700.
51. Bonfiglio S, Achilli A, Olivieri A, Negrini R, Colli L, Liotta L, et al. The enigmatic origin of bovine mtDNA haplogroup R: sporadic interbreeding or an independent event of Bos primigenius domestication in Italy? PLoS One. 2010;5(12), e15760.
52. Ottoni C, Flink LG, Evin A, Geörg C, De Cupere B, Van Neer W, et al. Pig domestication and human-mediated dispersal in western Eurasia revealed through ancient DNA and geometric morphometrics. Mol Biol Evol. 2013;30(4):824–32.
53. Geörg C. Paläopopulationsgenetik von Schwein und Schaf in Südosteuropa und Transkaukasien. ForschungsCluster1. vol. 9. Verlag Marie Leidorf GmbH: Rahden/Westf; 2013.
54. Depaulis F, Orlando L, Hanni C. Using Classical Population Genetics Tools with Heterochroneous Data. Time Matters! PLoS ONE. 2009;4(5):5541.
55. Guilaine J, Manen C. Vigne J-D. Pont de Roque-Haute (Portiragnes, Hérault). Nouveaux regards sur la néolithisation de la France méditerranéenne. Archives d'Ecologie Préhistorique: Toulouse; 2007.
56. Benecke N. Der Mensch und seine Haustiere. Die Geschichte einer jahrtausendealten Beziehung. Stuttgart: Theiss; 1994.
57. Tresset A, Vigne J-D. La chasse, principal élément structurant la diversité des faunes archéologiques du Néolithique ancien, en Europe tempérée comme en Méditerranéenne: tentative d'interprétation fonctionnelle. In: Arbogast RM, Jeunesse C, Schibler J, editors. *Rôle et statut de la chasse dans le Néolithique ancien danubien (5500-4900 av J-C) Actes Premières rencontres danubiennes de Strasbourg, 20-21 nov 96*. Rahden/Westf: Marie Leidorf; 2001. p. 129–51.
58. Pavúk J. Typologische Geschichte der Linearbandkeramik. In: Lüning J, Frirdich C, Zimmermann A, editors. *Die Bandkeramik im 21 Jahrhundert: Symposium in der Abtei Brauweiler bei Köln 2002*. Rahden/Westf: Verlag Marie Leidorf GmbH; 2005. p. 17–39.
59. Itan Y, Powell A, Beaumont MA, Burger J, Thomas MG. The origins of lactase persistence in Europe. PLoS Comput Biol. 2009;5(8), e1000491.
60. Çilingiroğlu C. The appareance of impressed pottery in the Neolithic Aegean and its implications for maritime networks in the Eastern Mediterranean. Tüba-Ar. 2010;13:9–22.
61. Haak W, Lazaridis I, Patterson N, Rohland N, Mallick S, Llamas B, et al. Massive migration from the steppe was a source for Indo-European languages in Europe, *Nature*. 2015.
62. Taberlet P, Coissac E, Pansu J, Pompanon F. Conservation genetics of cattle, sheep, and goats. C R Biol. 2011;334(3):247–54.

CHAPTER 11

Livestock Helminths in a Changing Climate: Approaches and Restrictions to Meaningful Predictions

Naomi J. Fox [1,2,3], Glenn Marion [3], Ross S. Davidson [1], Piran C. L. White [2] and Michael R. Hutchings [1,*]

[1]*SAC, West Mains Road, Edinburgh, EH9 3JG, UK*

[2]*Environment Department, University of York, Heslington, York, YO10 5DD, UK*

[3]*Biomathematics and Statistics Scotland, Kings Buildings, Edinburgh, EH9 3JZ, UK*

SIMPLE SUMMARY

Parasitic helminths represent one of the most pervasive challenges to livestock, and their intensity and distribution will be influenced by climate change. There is a need for long-term predictions to identify potential risks and highlight opportunities for control. We explore the approaches to modelling future helminth risk to livestock under climate change. One of the limitations to model creation is the lack of purpose driven data collection. We also conclude that models need to include a broad view of the livestock system to generate meaningful predictions.

ABSTRACT

Climate change is a driving force for livestock parasite risk. This is especially true for helminths including the nematodes *Haemonchus contortus*, *Teladorsagia circumcincta*, *Nematodirus battus*, and the trematode *Fasciola hepatica,* since survival and development of free-living stages is chiefly affected by temperature and moisture. The paucity of long term predictions of helminth risk under climate change has driven us to explore optimal modelling approaches and identify current bottlenecks to generating meaningful predictions. We classify approaches as correlative or mechanistic, exploring their strengths and limitations. Climate is one aspect of a complex system and, at the farm level, husbandry has a dominant influence on helminth transmission. Continuing environmental change will necessitate the adoption of mitigation and adaptation strategies in husbandry. Long term predictive models need to have the architecture to incorporate these changes. Ultimately, an optimal modelling approach is likely to combine mechanistic processes and physiological thresholds with correlative bioclimatic modelling, incorporating changes in livestock husbandry and disease control. Irrespective of approach, the principal limitation to parasite predictions is the availability of active surveillance data and empirical data on physiological responses to climate variables. By combining improved empirical data and refined models with a broad view of the livestock system, robust projections of helminth risk can be developed.

Keywords: climate change; prediction; risk; livestock; parasites; helminths; disease; modelling

1. INTRODUCTION

Climate change has been implicated as a driving force for recent parasite range expansions, and efforts have been made to model the relationship between pathogen levels and climate. The most economically important parasitic helminths of livestock in temperate climes include the nematodes *Haemonchus contortus*, *Teladorsagia circumcincta* and *Nematodirus battus*, and the trematode *Fasciola hepatica*. The increase in these helminths in recent years [1,2,3,4,5] has been attributed to climate change, since the survival of the free-living stages is chiefly affected by temperature and moisture, and larval development rate is highly temperature dependent [6,7,8,9].

The development of evidence-based risk assessments and targeted surveillance are pivotal when the welfare and economic costs of these pathogens are considered. Subclinical infection is characterised by weight loss, lower milk yield, loss of condition, abortion, infertility and veterinary costs, and heavy infections can cause host mortality. Indirect economic losses from pathogen outbreaks are also incurred via export restrictions and surveillance and mitigation costs [10]. Despite the deleterious impacts of helminths on the livestock industry and their dependence on climatic conditions, predictions of long-term threats to animal health from climate change have so far concentrated on heat stress [11,12,13,14,15,16] and viruses spread by volant vectors, such as blue tongue [17,18,19,20,21,22]. Although there have been a number of studies aiming to link the recent changes in helminthiasis abundance and distribution with environmental change [1,3,4,5,23,24,25], there is a lack of predictions for future helminth risk to livestock.

Here we explore the optimal approaches to generating long-term predictions for helminth risk, stratifying modelling approaches as either correlative or mechanistic. We also highlight the obstacles to generating meaningful predictions and the need for a multidisciplinary approach.

2. MODELLING THE CHANGE

2.1. The Correlative Approach

Predictive models of species distribution, for both epidemiology and conservation, are often based on correlative ecological niche models [26,27,28,29,30,31,32]. These models are based on Hutchinson's ecological niche theory where the current geographic distribution is used to infer the environmental requirements of a species [33]. The current, past or future distribution of that species is then predicted based on these requirements [34]. Insights into the biology of parasite dynamics should be used to systematically build the foundations of these models, and the most proximal environmental predictors should be chosen based on known ecological and physiological theory [35].

A number of programmes are available to determine a species' climate envelope by matching current distribution with climatic parameters, such as CLIMEX [36], HABITAT [37], DOMAIN [38] and SPECIES [39]. In addition to these generic models, species-specific correlative models have also been developed, however these models have primarily been applied to species of conservation importance and invasive alien species. To date, correlative predictive models of helminthiasis have focussed on *F. hepatica* (liver fluke) due to the close relationship between weather and fasciolosis outbreaks [2] and the worldwide importance of fasciolosis as a zoonosis. The UK's National Animal Disease Information Service (NADIS) and the Department of Agriculture and Food in Ireland currently predict *F. hepatica* incidence using the Ollerenshaw index, which is a correlative model focusing on temperature and moisture availability [40,41]. Through applying a modified Ollerenshaw index to UKCP09 climate projections, the first long-term predictions of future *F. hepatica* risk for 2020–2070 across the UK have been developed [42]. The resultant risk maps predict unprecedented levels of fasciolosis outbreaks, and possible changes in the timing of disease outbreaks due to increased risk from overwintering larvae. Despite an overall long term increase in all regions of the UK, spatio-temporal variation in risk levels is expected, with infection risk due to reduce in some areas and fluctuate greatly in others. This forecast is the first approximation of the potential impacts of climate change on helminth risk in the UK, and indicates where active disease surveillance should be targeted.

There are a number of factors that made correlative modelling the preferred approach for determining fascioliasis risk. Firstly, the existence of long term prevalence data facilitated the development of a statistical model [41]. Secondly, despite their complex life cycle, the distribution of liver fluke is driven by simple proximal drivers—temperature and water availability. It is these drivers that form the basis of the correlative model. Thirdly, their dispersal by wild hosts and speed of colonisation of new regions makes them ubiquitous where livestock are present across their fundamental niche.

Despite the information that correlative models can provide, there are a number of disadvantages to applying this approach to climate change predictions. A prominent bottleneck to the development of correlative models is the lack of current or past parasite distribution data. The data used to train these models is usually opportunistic, passive surveillance data. Climate parameters are used according to which measurements and predictions are available, rather than those that are most pertinent to disease transmission. In order to build reliable models, purpose-driven active surveillance data is needed. As climate affects long term, large scale trends, these data must be collected over appropriate scales to observe these trends, rather than the patchy distribution visible at finer scales, which is likely to be a consequence of non-climatic factors. It is equally important to apply predictive models at the right scale. It is often tempting to apply models on as fine a scale as the climate projections allow, as this will lead to detailed maps with an illusion of greater accuracy. However this temptation should be resisted and models based on climatic parameters alone should only be applied at an appropriately coarse scale.

The lack of distribution data also impedes the validation of these predictive models, as ideally they should be validated with data fully independent from that used to build the model [26,28,43]. A further issue with validation is that it often indicates the models ability to predict current distribution; models that are effective at predicting current distribution may not be so reliable when predicting future outbreaks [34]. Hence there is a need to validate these models with data outside the spatial and temporal range of the training dataset. To make use of existing datasets for both model creation and validation there are ongoing developments in statistical methodology to account for biases in reported data. Existing datasets can be utilised for modelling distributions if imperfect detection is considered, with irregular sampling intensity and false absences accounted for using hierarchical Bayesian models [29,30,44,45]. Uncertainty in predictor variables can also be addressed through Bayesian analysis [46].

Given sufficient data for creation and validation of correlative models, fundamental disadvantages remain. As correlative models are generated using distribution data they are based on the realized niche [47] which includes competitive exclusion and biotic interactions (e.g., the presence of hosts and other pathogens), rather than the whole fundamental niche. As distribution data are usually only available for a limited area or time window, correlative models are generally fitted using data reflecting a snapshot of the current climate-pathogen relationship. An assumption is therefore made that a species has reached equilibrium with its environment [35]. This ignores any non-climatic dispersal constraints that may be restricting the current range, or declining sink populations where presence is recorded but climate is not actually suitable for long-term population survival. This failure to account for non-equilibrium ranges is an inherent weakness in correlative models [48,49]. This weakness could be especially pertinent for emerging parasites which are unlikely to be in equilibrium with their current environment, so could be absent from climatically suitable areas due to low propagule pressure. Conversely, these models will not identify areas which are not within bounds necessary for pathogen survival, but where deleterious levels of parasitism could be reached given sufficient propagule pressure. This issue posed by the non-equilibrium nature of emerging pathogens could be addressed through training models for emerging parasites using distribution data from their long established native range [35].

Above all aforementioned limitations, the dangers in extrapolating statistical models are especially pertinent to climate projections. The assumption that a species is in equilibrium with its current environment and the reliance on relationships between climate variables that may not exist in novel climate change scenarios leads to equivocal results when models are extrapolated spatially and temporally [29,44,50,51]. We are likely to experience previously unseen climate combinations and climate parameters exceeding their current observed ranges; parasites could be well adapted to these novel situations, however correlative models would assume these conditions to be unsuitable [34]. Additionally, correlative models cannot identify points where the system behaviour undergoes a qualitative change (tipping points). It is this inability to extrapolate that further emphasises the need for an alternative modelling approach.

As a further testament to correlative models' often unreliable outputs, models created using different statistical approaches have been shown to produce conflicting distribution estimates, even when trained with the same distribution and climate data [52,53]. Such differences between distribution estimates can be exacerbated when extrapolating to novel climatic conditions [32]. To address the divergence in model outputs, a framework for selecting the most robust statistical modelling technique has been developed [32]. This approach would be useful for selecting optimal models for predicting changes in helminth distribution. The differences between distributions predicted by different models could also be informative; discovering why predictions differ could improve understanding of the main drivers, and quantification of the differences could inform decision making and risk analysis [29].

Despite their limitations, correlative models can still provide a first indication of how climate will influence helminth distribution, and identify where limited resources and targeted surveillance should be focused. They provide a useful tool when too few elements of the transmission process have been quantified for the creation and parameterisation of mechanistic models. The limitations of correlative models are being addressed through continued development of statistical methodologies, led by work in both conservation and invasive species control; there is scope for these emerging approaches to be applied to predicting livestock parasite risk. Reliability of correlative models will ultimately be governed by the quality of data used for model training and validation, the statistical methods employed, the ecological and physiological knowledge on which inclusion of proximal variables is based, and discrepancies between the realized and fundamental niche.

2.2. The Mechanistic Approach

The second approach to understanding and predicting parasite risk is the process-based mechanistic approach. Mechanistic models are based on detailed knowledge of the physiology of the species [34] and attempt to replicate the underlying mechanisms that drive the species' response to environmental variables [54]. Given sufficient understanding of the parasites physiology, these models can be employed to predict changes in prevalence [55]. Previous mechanistic models have explored the dynamics of helminth infection in livestock [56,57,58,59,60,61,62,63,64,65]. However these models have not been used to assess the impacts of climate change on helminth transmission.

One element of mechanistic modelling that makes it well adapted in a changing climate is that it is less prone to extrapolation problems than correlative models. The mechanistic approach does not rely on relationships between climate variables that may cease to exist under future climate change [34,47,66,67], making them less prone to breaking down when tested outside current observation limits. Consequently, the mechanistic approach is considered superior in extrapolating beyond current conditions and forecasting the impacts of climate change [68]. When predicting a parasite's response to novel conditions it is vital that models account for the possibility of non-linear responses to predictor variables [43]. Through developing a mechanistic model looking at the effects of increasing temperatures on Schistosomiasis prevalence and abundance, Mangal *et al.* (2008) found that *Schistosoma mansoni* showed a non-linear response to temperature change [68]. Extrapolated predictions for novel climates may prove to be unreliable if the response curves are not fully determined, hence empirical data on survival and development under conditions beyond a parasite's current climatic ranges are required.

The mechanistic approach also avoids some of the inherent problems that correlative models have when applied to emerging and invasive parasites since mechanistic models do not assume the species is in equilibrium with the environment [48] and are capable of predicting tipping points. By incorporating key mechanisms into the model structure more complex questions can also be addressed, such as determining the impact of climate on specific stages of the lifecycle.

The severity of helminth infections is often dependent on intensity, rather than prevalence. The economic and welfare implications of outbreaks would be difficult to evaluate using a correlative approach, as data availability typically constrains these models to only look at prevalence. If data on adult worm burden are available they are often based on indirect measures such as faecal egg count. It is more feasible to include models of adult worm burden within a mechanistic framework. Although mechanistic models can predict changing intensity, the relationship between helminth burden and host production losses is non-linear; a doubling of infection intensity does not lead to a doubling of production loss [69]. A deeper understanding of the relationship between parasite burdens and potential production losses is still required.

The survival of larvae on pasture is largely driven by climate [62], but helminth population dynamics are governed by density dependent process and it is vital that these processes are incorporated (e.g., host acquired immunity), as more larvae on pasture do not necessarily translate to higher parasite burden in the livestock [64]. Previous parasite models have concentrated on infection dynamics in a single host, returning the average host infection level and assuming each host interacts with its own private population of parasites [57,59,62,64]. In reality, aggregation of parasites in host populations is a ubiquitous phenomenon [70]. It has been proposed that heterogeneities in host infection levels arise due to numerous factors including variations in host resistance and susceptibility due to behavioural or physiological differences and spatial heterogeneity in infection risk [71,72,73,74]. Average host models ignore this overdispersion of helminths; a potentially critical feature in a system heavily driven by density-dependent processes.

By modelling the average host, it is assumed that the same level of regulatory constraint is experienced by all parasites in the population. In an overdispersed population, aggregation results in an increase in the effective density experienced by individual parasites, increasing the influence of regulatory processes [74,75,76]. These regulatory processes drive a parasite population towards equilibrium [63], and hence aggregated populations reach equilibrium more quickly and at a lower level. As a consequence, models that do not take the overdispersed parasite distribution into account are likely to overestimate burden [75,76]. Variation in host worm burden within a herd has been generated artificially in infection models through assuming a fixed distribution of worms across hosts, and the system having a pre-defined degree of overdispersion [56,75]. There is scope to develop models where an overdispersed distribution is an emergent property, and which subsume both the mechanisms that generate aggregation and the impacts of the resultant aggregation on infection dynamics. Future changes in grazing season, stocking density, host behaviour and host resistance may affect the aggregation of parasites at the suprapopulation level (defined as all parasites of a given species, at all stages of development, within an ecosystem [77]). Models that incorporate the changing distribution of the suprapopulation would also allow identification of scenarios where individuals within a herd are at risk from heavy infections, thus addressing economic and welfare concerns.

Unlike correlative models, mechanistic model development is not restricted by the paucity of accurate distribution data. Instead, the primary restriction to their development is the extensive physiological data needed for parameterisation [35]. Recent empirical work on how livestock parasites respond to changing climatic variables should facilitate their development and parameterisation. For example, the correlation between the timing, distribution and quantity of rainfall and *H. contortus* development have been investigated [6] and threshold rates of precipitation and evaporation required for survival of *H. contortus* larvae to the infective stage have been determined [78]. The impact of temperature on the emergence of the infective stages of trematodes from the intermediate snail hosts have also been reviewed [79,80]. These studies demonstrate that there are quantitative data on the effects of climatic variables on physiological processes, which can be used to parameterise mechanistic models.

In addition to altered temperature and precipitation, changes in UV reaching the extra-mammalian stages will also vary with changing climate (due to e.g., varying cloud cover, changes in development window and seasonality of infection). Van Dijk *et al.* (2009) investigated how ultraviolet (UV) light affects Trichostrongyloid nematodes, *H. contortus*, *N. battus* and *T. circumcinta*, in their infective L3 stages [81]. Mortality rates increased by up to 100 times due to exposure to UVA light equivalent to the maximum levels expected for a summers day. There were inter-species differences observed: *N. battus*, with its arctic origins, and *T. circumcincta* were less resistant than the tropic-adapted *H. contortus* [81]. Previous mechanistic models have focused on changing temperature, as this was the parameter most studied and available in climate projections. However this study highlights the importance of including other environmental variables in predictive models.

Another oversight in some mechanistic models is that they assume Liebig's Law of the Minimum [34], which presumes that the overall response will be determined by the most limiting factor. This is not necessarily true. For example, larvae on pasture could survive otherwise deleteriously low levels of rainfall if temperatures were not too high. This emphasises the need to look at proximal, rather than distal variables—in this case water availability rather than levels of rainfall. It also emphasises the importance of looking at variables in combination.

In contrast with the correlative approach, mechanistic models are based on measured physiological and behavioural parameters, and reflect the fundamental rather than realized niche [35]. This leads to over estimation of risk showing the whole potential range if dispersal restrictions or biotic interactions are not accounted for [34,48,82,83]. The extent of predictive errors will be partly dependent on the proportion of the fundamental niche that the realized niche occupies [35]. This proportion will depend on the parasites dispersal and competitive abilities; if the discrepancies between the species' fundamental and realized niche could be identified, it could give an indication of how accurate predictive models could be for particular species.

To attain realistic predictions of species distribution, models should ultimately integrate constraints from biotic interactions and dispersal [35]. Projections often assume either complete dispersal, where the parasite can reach all areas with permissive climes, or no dispersal where presence is only predicted in areas of its current range where climate remains within viable thresholds [35]. It is overly simplistic to assume that a species will either fully disperse to fill the fundamental niche, or that they will only survive in areas where the projected fundamental niche and current realized niche overlap [26], however, this approach does at least provide (often very wide) bounds on the possible distribution. The importance of incorporating species interactions in prediction models has been demonstrated [51,84,85], with biotic interactions influencing the predictive power of correlative models even at the macro-ecological scale [85]. The incorporation of dispersal also influences predictive power [30,31,86], with the disparity between models that do incorporate dispersal and those that do not increasing under more extreme climate warming scenarios [31]. The integration between model projections, biotic interactions and simulated host/pathogen distribution and dispersal patterns would improve the accuracy of predictive models (both correlative and mechanistic). This could be achieved through the adoption of hierarchical Bayesian frameworks [86]. However, these should take into account that biotic interactions and dispersal abilities could change with time. Due to the high selective pressures that novel conditions are likely to impose, climate change is likely to select for phenotypes with enhanced dispersal capabilities that can track the shift in their climate envelope. This would affect the parameterisation of models that incorporate dispersal.

Mechanistic models provide a powerful tool in predicting the influence of climate change on helminth risk. They are comparatively robust under spatio-temporal extrapolation and can be developed to answer complex questions. However, a move towards a mechanistic approach should not be seen as a way of alleviating the need for distribution data; to develop and validate models and assess their continued reliability under changing conditions, there is a need for ongoing surveillance.

3. LOOKING BEYOND CLIMATE: THE NEED FOR A PANOPTIC VIEW

Climate is an important driver of helminth distribution, however transmission to livestock at the farm level is also controlled by husbandry. Livestock husbandry is ever changing and the adoption of mitigation and adaptation strategies will be necessary under continuing environmental change. Methods for alleviating heat stress in cattle have been explored including shade provision [14] and different feeding and sprinkling regimes [87]. Changes in the time of highest feed production may alter, resulting in timing of livestock reproduction being managed differently [88]. Housing outdoors instead of indoors in winter will also affect parasite transmission cycles [16]. Additionally, as certain breeds are more susceptible to extreme weather conditions, the breed structure of livestock may have to change to ensure that production levels are maintained and animal welfare needs met.

The influence of these adaptation and mitigation strategies on parasite risk has yet to be incorporated in pathogen transmission models. There is a propensity for predictive models to incorporate ever refined climate prediction, however these will not provide meaningful indications of future risk if they are parameterised for outdated management approaches and redundant host population structures. Predictive models need to have the architecture to allow adaptation and mitigation strategies to be incorporated. For example, as different breeds have varying levels of disease susceptibility [10,13,89], the differing levels of resistance for future breed structures need to be considered in transmission models.

The architecture of mechanistic models should also allow more complex issues to be addressed, for example, determining which of the panoply of potential control strategies resources should be invested into developing. There is currently a strong reliance on anthelmintics for controlling helminths in intensively grazed livestock [90], either through targeting the number of adult parasites in the host, or the number of infectious larvae on pasture. However development of anthelmintic resistance is an ongoing problem [5,23,91,92,93,94] and resistance development could be accelerated in a warming climate [90]. As anthelmintics become decreasingly efficient in combating these parasites, strategies to lower the dependence on chemicals and manage the spread of resistance need adopting [95,96,97,98,99]. There are a variety of alternative control strategies, often linked with organic production. These include grazing management strategies to decrease the number of infective larvae that the host is exposed to, and breeding for resistance in the host [100,101,102,103]. Models should be developed with the architecture to consider interacting effects of changes in climate, management and control strategies.

4. CONCLUSIONS AND RECOMMENDATIONS

Although correlative modelling is a useful tool for establishing baseline predictions, a drive towards mechanistic process-based models will ultimately be needed if we are to foresee the consequences of subtle interactions between various components of a system under climate influence. Although laboratory studies have been done to aid the creation of mechanistic models, the findings have not yet been utilised. To improve the reliability of extrapolating and applying models to novel climates, empirical evidence is also needed on how helminths respond when exposed to conditions beyond their current observed ranges, to account for any non-linearity.

Modelling approaches can be broadly stratified as correlative or mechanistic, but in practice it is difficult to draw an absolute distinction between these modelling techniques. Ultimately, an integration of both modelling approaches may be required. Future improvements in predictions should arise from the continued development of a hybrid approach that combines both mechanistic processes and physiological thresholds with correlative bioclimatic modelling. [28,50,51,104]. Irrespective of the modelling approach, the quality of predictions is critically dependent on the quality of available data [105]. The continued collection of active surveillance data and empirical data on physiological responses to climate variables will ultimately drive the development and validation of meaningful predictions.

The recent expansion of helminth ranges has been mirrored by our increased understanding of the role of climate in their transmission dynamics. However, climate is just one factor affecting disease ecology. The application of mitigation and adaptation strategies, combined with changing control options, need to be considered when determining the overall impacts of climate change. By combining improved empirical data and refined models with a broader view of the livestock system, projections of future disease threats can be improved. This will increase our understanding of these complex systems, identifying potential risks and highlighting opportunities for control.

ACKNOWLEDGEMENTS

SAC and BioSS receive funding from the Scottish Government

REFERENCES

1. Mitchell, G.B.B.; Somerville, D.K. Effects of Climate Change on Helminth Diseases in Scotland. *SAC Publication* **2005**, *1*, 1–11.
2. de Waal, T.; Relf, V.; Good, B.; Gray, J.; Murphy, T.; Forbes, A.; Mulcahy, G. Developing models for the predicion of fasciolosis in Ireland. In *Making Science Work on the Farm: A Workshop on Decision Support Systems for Irish Agriculture*; Holden, N.M., Hochstrasser, T., Schulte, R.P.O., Walsh, S., Eds.; 2007; Joint Working Group in Applied Agricultural Meteorology: Dublin, Ireland, 2007.
3. Kenyon, F.; Sargison, N.D.; Skuce, P.J.; Jackson, F. Sheep helminth parasitic disease in south eastern Scotland arising as a possible consequence of climate change. *Vet. Parasitol.* **2009**, *163*, 293–297.

4. Pritchard, G.C.; Forbes, A.B.; Williams, D.J.L.; Salimi-Bejestani, M.R.; Daniel, R.G. Emergence of fasciolosis in cattle in East Anglia. *Vet. Record* **2005**, *157*, 578–582.
5. van Dijk, J.; David, G.P.; Baird, G.; Morgan, E.R. Back to the future: Developing hypotheses on the effects of climate change on ovine parasitic gastroenteritis from historical data. *Vet. Parasitol.* **2008**, *158*, 73–84.
6. O'Connor, L.J.; Kahn, L.P.; Walkden-Brown, S.W. Moisture requirements for the free-living development of *Haemonchus contortus*: Quantitative and temporal effects under conditions of low evaporation. *Vet. Parasitol.* **2007**, *150*, 128–138.
7. Coyne, M.J.; Smith, G. The development and mortality of the free-living stages of *Haemonchus contortus* in laboratory culture. *Int. J. Parasitol.* **1992**, *22*, 641–650.
8. Barnes, E.H.; Dobson, R.J.; Donald, A.D.; Waller, P.J. Predicting populations of *Trichostrongylus colubriformis* infective larvae on pasture from meteorological data. *Int. J. Parasitol.* **1988**, *18*, 767–774.
9. Armour, J. The epidemiology of helminth disease in farm animals. *Vet. Parasitol.* **1980**, *6*, 7–46.
10. Wilson, A.; Mellor, P. Bluetongue in Europe: Vectors, epidemiology and climate change. *Parasitol. Res.* **2008**, *103*, 69–77.
11. Beatty, D.; Barnes, A.; Taylor, E.; Maloney, S. Do changes in feed intake or ambient temperature cause changes in cattle rumen temperature relative to core temperature? *J. Therm. Biol.* **2008**, *33*, 12–19.
12. García-Ispierto, I.; López-Gatius, F.; Bech-Sabat, G.; Santolaria, P.; Yániz, J.L.; Nogareda, C.; De Rensis, F.; López-Béjar, M. Climate factors affecting conception rate of high producing dairy cows in northeastern Spain. *Theriogenology* **2007**, *67*, 1379–1385.
13. Harle, K.; Howden, S.; Hunt, L.; Dunlop, M. The potential impact of climate change on the Australian wool industry by 2030. *Agr. Syst.* **2007**, *93*, 61–89.
14. Kendall, P.E.; Nielsen, P.P.; Webster, J.R.; Verkerk, G.A.; Littlejohn, R.P.; Matthews, L.R. The effects of providing shade to lactating dairy cows in a temperate climate. *Livestock Sci.* **2006**, *103*, 148–157.
15. Nardone, A.; Ronchi, B.; Lacetera, N.; Bernabucci, U. Climatic Effects on Productive Traits in Livestock. *Vet. Res. Commun.* **2006**, *30*, 75–81.
16. Gregory, N.G. How climatic changes could affect meat quality. *Food Res. Int.* **2010**, *43*, 1866–1873.
17. Caligiuri, V.; Giuliano, G.A.; Vitale, V.; Chiavacci, L.; Travaglio, S.; Salute, V. Bluetongue surveillance in the Campania Region of Italy using a geographic information system to create risk maps. *Epidemiol. Vect.* **2004**, *40*, 385–389.
18. de Koeijer, A.; Hartemink, N.; Boender, G.J.; Elbers, A.; Heesterbeek, H. Epidemiological analysis of the 2006 bluetongue virus serotype 8 epidemic in north-western Europe A risk map for epidemic potential in The Netherlands. *Eur. Food Safety Author.* **2007**, *1*, 1–18.
19. Gubbins, S.; Carpenter, S.; Baylis, M.; Wood, J.L.N.; Mellor, P.S. Assessing the risk of bluetongue to UK livestock: Uncertainty and sensitivity analyses of a temperature-dependent model for the basic reproduction number. *J. Roy. Soc. Interf.* **2008**, *5*, 363–371.
20. Pili, E.; Ciucce, S.; Culurgioni, J.; Figus, V.; Pinna, G.; Marchi, A. Distribution and abundance of Bluetongue vectors in Sardinia: Comparison of field data with prediction maps. *J. Vet. Med.* **2006**, *53*, 312–316.
21. Purse, B.V.; Tatem, A.J.; Caracappa, S.; Rogers, D.J.; Mellow, P.S.; Baylis, M.; Torina, A. Modelling the distributions of *Culicoides* bluetongue virus vectors in Sicily in relation to satellite-derived climate variables. *Med. Vet. Entomol.* **2004**, *18*, 90–101.
22. Racloz, V.; Presi, P.; Vounatsou, P.; Schwermer, H. Use of mapping and statistical modelling for the prediction of bluetongue occurrence in Switzerland based on vector biology. *Vet. Ital.* **2007**, *43*, 513–518.
23. Abbott, K.A.; Taylor, P.M.; Stubbings, L.A. *Sustainable Worm Control Strategies for Sheep*, 2nd ed.; A Technical Manual for Veterinary Surgeons and Advisers; SCOPS (Sustainable Control of Parasites in Sheep),National Sheep Association: Malvern, UK, 2007; pp. 1–48.
24. Mas-Coma, S.; Valero, M.A.A.; Bargues, M.D. Effects of climate change on animal and zoonotic helminthiases. *Rev. Sci. Tech. Off. Int. Epiz* **2008**, *27*, 443–457.

25. O'Connor, L.J.; Walkden-Brown, S.W.; Kahn, L.P. Ecology of the free-living stages of major trichostrongylid parasites of sheep. *Vet. Parasitol.* **2006**, *142*, 1–15.
26. Lawler, J.J.; White, D.; Neilson, R.P.; Blaustein, A.R. Predicting climate-induced range shifts: Model differences and model reliability. *Global Change Biol.* **2006**, *12*, 1568–1584.
27. Leathwick, J.; Elith, J.; Hastie, T. Comparative performance of generalized additive models and multivariate adaptive regression splines for statistical modelling of species distributions. *Ecol. Mode.* **2006**, *199*, 188–196.
28. Heikkinen, R.K.; Luoto, M.; Araújo, M.B.; Virkkala, R.; Thuiller, W.; Sykes, M.T. Methods and uncertainties in bioclimatic envelope modelling under climate change. *Progr. Phys. Geogr.* **2006**, *30*, 1–27.
29. Elith, J.; Leathwick, J.R. Species distribution models: Ecological explanation and prediction across space and time. *Ann. Rev. Ecol. Evol. Syst.* **2009**, *40*, 677–697.
30. Pagel, J.; Schurr, F.M. Forecasting species ranges by statistical estimation of ecological niches and spatial population dynamics. *Global Ecol. Biogeogr.* **2012**, *21*, 293–304.
31. Engler, R.; Guisan, A. MigClim: Predicting plant distribution and dispersal in a changing climate. *Div. Distrib.* **2009**, *15*, 590–601.
32. Thuiller, W. BIOMOD—Optimizing predictions of species distributions and projecting potential future shifts under global change. *Global Change Biol.* **2003**, *9*, 1353–1362.
33. Hutchinson, G.E. Concluding remarks. *Cold Spring Harb. Symp. Quant. Biol.* **1957**, *22*, 415–427.
34. Hijmans, R.J.; Graham, C.H. The ability of climate envelope models to predict the effect of climate change on species distributions. *Global Change Biol.* **2005**, *12*, 2272–2281.
35. Guisan, A.; Thuiller, W. Predicting species distribution: Offering more than simple habitat models. *Ecol. Lett.* **2005**, *8*, 993–1009.
36. Sutherst, R.W.; Maywald, G.F. A computerised system for matching climates in ecology. *Agr. Ecosyst. Environ.* **1985**, *13*, 281–299.
37. Walker, P.A.; Cocks, K.D. HABITAT: A procedure for modelling a disjoint environmental envelope for a plant or animal species. *Global Ecol. Biogeogr. Lett.* **1991**, *1*, 108–118.
38. Carpenter, G.; Gillison, A.N.; Winter, J. DOMAIN: A flexible modelling procedure for mapping potential distributions of plants and animals. *Biodiv. Conserv.* **1993**, *2*, 667–680.
39. Pearson, R.G.; Dawson, T.P.; Berry, P.M.; Harrison, P.A. SPECIES: A spatial evaluation of climate impact on the envelope of species. *Ecol. Model.* **2002**, *154*, 289–300.
40. Ollerenshaw, C.B. The approach to forecasting the incidence of fascioliasis over England and Wales 1958-1962. *Agr. Meteorol.* **1964**, *3*, 35–53.
41. Ollerenshaw, C.B.; Rowlands, W. A method of forecasting the incidence of fascioliasis in Anglesey. *Vet. Record* **1959**, *71*, 591–598.
42. Fox, N.J.; White, P.C.L.; McClean, C.J.; Marion, G.; Evans, A.; Hutchings, M.R. Predicting impacts of climate change on *Fasciola hepatica* risk. *PloS One* **2011**, *6*, e16126.
43. Austin, M. Species distribution models and ecological theory: A critical assessment and some possible new approaches. *Ecol. Model.* **2007**, *200*, 1–19.
44. Elith, J.; Graham, C.H.; Anderson, R.P.; Dudik, M.; Ferrier, S.; Guisan, A.; Hijmans, R.J.; Huettmann, F.; Leathwick, J.R.; Lehmann, A.; Li, J.; Lohmann, L.G.; Loiselle, B.; Manion, G.; Moritz, C.; Nakamura, M.; Nakazawa, Y.; Overton, J.M.; Peterson, A.T.; Phillips, S.J.; Richardson, K.; Scachetti-Pereira, R.; Schapire, R.E.; Soberon, J.; Williams, S.; Wisz, M.S.; Zimmermann, N.E. Novel methods improve prediction of species' distributions from occurrence data. *Ecography* **2006**, *29*, 129–151.
45. Bierman, S.M.; Butler, A.; Marion, G.; Kühn, I. Bayesian image restoration models for combining expert knowledge on recording activity with species distribution data. *Ecography* **2010**, *33*, 451–460.
46. McInerny, G.J.; Purves, D.W. Fine-scale environmental variation in species distribution modelling: Regression dilution, latent variables and neighbourly advice. *Meth. Ecol. Evol.* **2011**, *2*, 248–257.
47. Guisan, A.; Zimmermann, N.E. Predictive habitat distribution models in ecology. *Ecol. Model.* **2000**, *135*, 147–186.

48. Pearson, R.G.; Dawson, T.P. Predicting the impacts of climate change on the distribution of species: Are bioclimate envelope models useful? *Global Ecol. Biogeogr.* **2003**, *12*, 361–371.
49. Sutherst, R.W.; Bourne, A.S. Modelling non-equilibrium distributions of invasive species: A tale of two modelling paradigms. *Biol. Invas.* **2008**, *11*, 1231–1237.
50. Elith, J.; Kearney, M.; Phillips, S. The art of modelling range-shifting species. *Meth. Ecol. Evol.* **2010**, *4*, 330–342.
51. Kearney, M.; Porter, W. Mechanistic niche modelling: Combining physiological and spatial data to predict species' ranges. *Ecol. Lett.* **2009**, *12*, 334–350.
52. Ortega-Huerta, M.A.; Peterson, A.T. Modeling ecological niches and predicting geographic distributions: A test of six presence-only methods. *Revista Mexicana De Biodiversidad* **2008**, *79*, 205–216.
53. Pearson, R.G.; Thuiller, W.; Arau, M.B.; McClean, C.; Miles, L. Model-based uncertainty in species range prediction. *J. Biogeogr.* **2006**, *33*, 1704–1711.
54. Robertson, M.P.; Peter, C.I.; Villet, M.H.; Ripley, B.S. Comparing models for predicting species' potential distributions: A case study using correlative and mechanistic predictive modelling techniques. *Ecol. Model.* **2003**, *164*, 153–167.
55. Anderson, R.M. The role of mathematical models in helminth population biology. *Int. J. Parasitol.* **1987**, *17*, 519–529.
56. Barnes, E.H.; Dobson, R.J. Population dynamics of*Trichostrongylus colubriformis* in sheep: Computer model to simulate grazing systems and the evolution of anthelmintic resistance. *Int. J. Parasitol.* **1990**, *20*, 823–831.
57. Callinan, A.P.L.; Morley, F.H.W.; Arundel, J.H.; White, D.H. A model of the life cycle of sheep nematodes and the epidemiology of nematodiasis in sheep. *Agr. Syst.* **1982**, *9*, 199–225.
58. Cornell, S.J.; Isham, V.S.; Grenfell, B.T. Stochastic and spatial dynamics of nematode parasites in farmed ruminants. *Proc. Roy. Soc. B* **2004**, *271*, 1243–1250.
59. Dobson, R.J.; Donald, A.D.; Barnes, E.H.; Waller, P.J. Population dynamics of*Trichostrongylus colubriformis* in sheep: Model to predict the worm population over time as a function of infection rate and host age. *Int. J. Parasitol.* **1990**, *20*, 365–373.
60. Gordon, G.; O'Callaghan, M.; Tallis, G.M. A Deterministic Model for the Life Cycle of a Class of Internal Parasites of Sheep. *Math. Biosci.* **1970**, *8*, 209–226.
61. Learmount, J.; Taylor, M.A.; Smith, G.; Morgan, C. A computer model to simulate control of parasitic gastroenteritis in sheep on UK farms. *Vet. Parasitol.* **2006**, *142*, 312–329.
62. Leathwick, D.M.; Barlow, N.D.; Vlassoff, A. A model for nematodiasis in New Zealand lambs. *Int. J. Parasitol.* **1992**, *22*, 789–799.
63. Paton, G.; Thomas, R.G.; Waller, P.J. A prediction model for parasitic gastro-enteritis in lambs. *Int. J. Parasitol.* **1984**, *14*, 439–445.
64. Roberts, M.G.; Grenfell, B.T. The Population Dynamics of Nematode Infections of Ruminants: Periodic Perturbations as a Model for Management. *J. Math. Appl. Med. Biol.* **1991**, *8*, 83–93.
65. Tallis, G.M.; Donald, A.D. Further models for the distribution on pasture of infective larvae of the strongylid nematode parasites of sheep. *Math. Biosci.* **1970**, *7*, 179–190.
66. Dormann, C.F. Promising the future? Global change projections of species distributions. *Basic Appl. Ecol.* **2007**, *8*, 387–397.
67. Kearney, M.R.; Wintle, B.A.; Porter, W.P. Correlative and mechanistic models of species distribution provide congruent forecasts under climate change. *Conserv. Lett.* **2010**, *3*, 203–213.
68. Mangal, T.D.; Paterson, S.; Fenton, A. Predicting the impact of long-term temperature changes on the epidemiology and control of Schistosomiasis: A mechanistic model. *PLoS One* **2008**, *1*, e1438.
69. Smith, G. The economics of parasite control: Obstacles to creating reliable models. *Vet. Parasitol.* **1997**, *72*, 437–449.
70. Gregory, R.D.; Woolhouse, M.E. Quantification of parasite aggregation: A simulation study. *Acta tropica* **1993**, *54*, 131–139.
71. Barbour, A.D.; Kafetzaki, M. Modeling the Overdispersion of Parasite Loads. *Math. Biosci.* **1991**, *107*, 249–253.

72. Cornell, S.J.; Isham, V.S.; Smith, G.; Grenfell, B.T. Spatial parasite transmission, drug resistance, and the spread of rare genes. *PNAS* **2003**, *100*, 7401–7405.
73. Louie, K.; Vlassoff, A.; Mackay, A. Nematode parasites of sheep: Extension of a simple model to include host variability. *Parasitology* **2005**, *130*, 437–446.
74. Shaw, D.J.; Dobson, A.P. Patterns of macroparasite abundance and aggregation in wildlife populations: A quantitative review. *Parasitology* **1995**, *111*, 111–133.
75. Churcher, T.S.; Ferguson, N.M.; Basáñez, M.G. Density dependence and overdispersion in the transmission of helminth parasites. *Parasitology* **2005**, *131*, 121–132.
76. Smith, G.; Guerrero, J. Mathematical models for the population biology of *Ostertagia ostertagi* and the significance of aggregated parasite distributions. *Vet. Parasitol.* **1993**, *46*, 243–257.
77. Esch, G.W.; Gibbons, J.W.; Bourque, J.E. An analysis of the relationship between stress and parasitism. *Am. Midland Naturalist* **1975**, *93*, 339–353.
78. O'Connor, L.J.; Kahn, L.P.; Walkden-Brown, S.W. Interaction between the effects of evaporation rate and amount of simulated rainfall on development of the free-living stages of *Haemonchus contortus*. *Vet. Parasitol.* **2008**, *155*, 223–234.
79. Poulin, R. Global warming and temperature-mediated increases in cercarial emergence in trematode parasites. *Parasitology* **2006**, *132*, 143–151.
80. Mas-Coma, S.; Valero, M.A.; Bargues, M.D. Climate change effects on trematodiases, with emphasis on zoonotic fascioliasis and schistosomiasis. *Vet. Parasitol.* **2009**, *163*, 264–280.
81. van Dijk, J.; de Louw, M.D.E.; Kalis, L.P.A.; Morgan, E.R. Ultraviolet light increases mortality of nematode larvae and can explain patterns of larval availability at pasture. *Int. J. Parasitol.* **2009**, *39*, 1151–1156.
82. Aurambout, J.P.; Finlay, K.J.; Luck, J.; Beattie, G.A.C. A concept model to estimate the potential distribution of the Asiatic citrus psyllid (*Diaphorina citri* Kuwayama) in Australia under climate change—A means for assessing biosecurity risk. *Ecol. Model.* **2009**, *220*, 2512–2524.
83. Lafferty, K.D. The ecology of climate change and infectious diseases. *Ecology* **2009**, *90*, 888–900.
84. Davis, A.J.; Jenkinson, L.S.; Lawton, J.H.; Shorrocks, B.; Wood, S. Making mistakes when predicting shifts in species range in response to global warming. *Nature Lett.* **1998**, *391*, 783–786.
85. Araújo, M.B.; Luoto, M. The importance of biotic interactions for modelling species distributions under climate change. *Global Ecol. Biogeogr.* **2007**, *16*, 743–753.
86. Hooten, M.B.; Wikle, C.K.; Dorazio, R.M.; Royle, J.A. Hierarchical spatiotemporal matrix models for characterizing invasions. *Biometrics* **2007**, *63*, 558–567.
87. Mader, T.L.; Davis, M.S. Effect of management strategies on reducing heat stress of feedlot cattle: Feed and water intake. *J. Anim. Sci.* **2004**, *82*, 3077–3087.
88. Rivington, M.; Matthews, K.B.; Bellocchi, G.; Buchan, K.; Stöckle, C.O.; Donatelli, M. An integrated assessment approach to conduct analyses of climate change impacts on whole-farm systems. *Environ. Model. Softw.* **2007**, *22*, 202–210.
89. Eguale, T.; Mekonnen, G.A.; Chaka, H. Evaluation of variation in susceptibility of three Ethiopian sheep breeds to experimental infection with *Fasciola hepatica*. *Small Rumin. Res.* **2009**, *82*, 7–12.
90. Wolstenholme, A.J.; Fairweather, I.; Prichard, R.; Samson-Himmelstjerna, G.V.; Sangster, N.C. Drug resistance in veterinary helminths. *Trends Parasitol.* **2004**, *20*, 469–476.
91. DeVaney, J.A.; Craig, T.A.; Rowe, L.D. Resistance to ivermectin by *Haemonchus contortus* in goats and calves. *Int. J. Parasitol.* **1992**, *22*, 369–376.
92. Le Jambre, L.F.; Southcott, W.H.; Dash, K.M. Resistance of selected lines *of Haemonchus contortus* to Thiabendazole, Morantel, Tartrate and Levamisole. *Int. J. Parasitol.* **1976**, *6*, 217–222.
93. Pandey, V.S.; Sivaraj, S. Anthelmintic resistance in *Haemonchus contortus* from sheep in Malaysia. *Vet. Parasitol.* **1994**, *53*, 67–74.
94. Yadav, C.L.; Kumar, R.; Uppal, R.P.; Verma, S.P. Multiple anthelmintic resistance in *Haemonchus contortus* on a sheep farm in India. *Vet. Parasitol.* **1995**, *60*, 355–360.

95. Ketzis, J.K.; Vercruysse, J.; Stromberg, B.E.; Larsen, M.; Athanasiadou, S.; Houdijk, J.G.M. Evaluation of efficacy expectations for novel and non-chemical helminth control strategies in ruminants. *Vet. Parasitol.* **2006**, *139*, 321–335.
96. Torres-Acosta, J.F.J.; Hoste, H. Alternative or improved methods to limit gastro-intestinal parasitism in grazing sheep and goats. *Small Rumin. Res.* **2008**, *77*, 159–173.
97. Waller, P.J. From discovery to development: Current industry perspectives for the development of novel methods of helminth control in livestock. *Vet. Parasitol.* **2006**, *139*, 1–14.
98. Waller, P.J. Sustainable nematode parasite control strategies for ruminant livestock by grazing management and biological control. *Anim. Feed Sci. Technol.* **2006**, *126*, 277–289.
99. Barger, I.A. Prospects for integration of novel parasite control options into grazing systems. *Int. J. Parasitol.* **1996**, *26*, 1001–1007.
100. Woolaston, R.; Barger, I.A.; Piper, L.R. Response to helminth infection of sheep selected for resistance to *Haemonchus contortus*. *Int. J. Parasitol.* **1990**, *20*, 1015–1018.
101. Gicheha, M.; Kosgey, I.; Bebe, B.; Kahi, A. Efficiency of alternative schemes breeding for resistance to gastrointestinal helminths in meat sheep. *Small Rumin. Res.* **2007**, *69*, 167–179.
102. Windon, R.G. Genetic control of resistance to helminths in sheep. *Vet. Immun. Immunopathol.* **1996**, *54*, 245–54.
103. Woolaston, R.R.; Baker, R.L. Prospects of breeding small ruminants for resistance to internal parasites. *Int. J. Parasitol.***1996**, *26*, 845–55.
104. Monahan, W.B. A mechanistic niche model for measuring species' distributional responses to seasonal temperature gradients. *PloS One* **2009**, *4*, e7921.
105. Rocchini, D.; Hortal, J.; Lengyel, S.; Lobo, J. M.; Jiménez-Valverde, A.; Ricotta, C.; Bacaro, G.; Chiarucci, A. Accounting for uncertainty when mapping species distributions: The need for maps of ignorance. *Progr. Phys. Geogr.***2011**, *35*, 211–226.

CHAPTER 12

Health and Welfare in Dutch Organic Laying Hens

Monique Bestman * and Jan-Paul Wagenaar

Louis Bolk Institute, Hoofdstraat 24, 3972 LA, Driebergen, The Netherlands

SIMPLE SUMMARY

Data on animal health and welfare and farm management during rearing and laying periods were collected from 49 flocks of organic laying hens in the Netherlands to establish how farms performed in terms of animal health and welfare and which factors affected health and welfare.

ABSTRACT

From 2007–2008, data on animal health and welfare and farm management during rearing and laying periods were collected from 49 flocks of organic laying hens in the Netherlands. Our aim was to investigate how organic egg farms performed in terms of animal health and welfare and which farm factors affected this performance. The flocks in our study were kept on farms with 34 to 25,000 hens (average 9,300 hens). Seventy-one percent of the flocks consisted of 'silver hybrids': white hens that lay brown eggs. Fifty-five percent of the flocks were kept in floor-based housing and 45% of the flocks in aviaries. No relation was found between the amount of time spent outdoors during the laying period and mortality at 60 weeks. Flocks that used their outdoor run more intensively had better feather scores. In 40% of the flocks there was mortality caused by predators. The average feed intake was 129 g/day at 30 weeks and 133 g/day at 60 weeks of age. The average percentage of mislaid eggs decreased from three at 30 weeks to two at 60 weeks. The average mortality was 7.8% at 60 weeks. Twenty-five percent of the flocks were not treated for worms in their first 50 weeks. Flubenol© was applied to the flocks that were treated. Ten percent of the flocks followed Flubenol© instructions for use and were wormed five or more times. The other 65% percent were treated irregularly between one and four times. Sixty-eight percent of the flocks showed little or no feather damage, 24% showed moderate damage and 8% showed severe damage. The feather score was better if the hens used the free-range area more intensely, the laying percentage at 60 weeks was higher, and if they were allowed to go outside sooner after arrival on the laying farm. In 69% of the flocks, hens had peck wounds in the vent area: on average this was 18% of the hens. Keel bone deformations were found in all flocks, on average in 21% of the birds. In 78% of the flocks, an average of 13% of the hens had foot-sole wounds, mostly a small crust. Combs were darker in flocks that used the range area more intensively. More fearful flocks had lighter combs. We conclude that organic farms are potentially more animal friendly than other poultry systems based on the animal welfare benefits of the free range areas. However, we also observed mortality rates, internal parasites, keel bone deformities, and foot sole lesions on organic farms that were comparable to or worse than in other husbandry systems. It is unclear whether these 'remaining' problems can be attributed to housing or if they are the result of keeping high productive genotypes in an artificial environment. Organic farms use the same high productive genotypes as other husbandry systems.

Keywords: organic; free range; poultry health; poultry welfare; feather pecking; vent pecking; mortality; keel bone deformations; foot sole lesions; comb color

1. INTRODUCTION

The organic egg sector in the Netherlands increased from 150,000 hens on 40 farms in 2001 [1] to 2.1 million hens on 194 farms in 2011 [2]. The most important features of Dutch organic egg production are [3] a maximum group size of 3,000 birds, six birds per m^2 indoors, 4 m^2 per bird outdoors, 18 cm perch per bird, and one third of indoor floor surface covered with litter. Moreover, the hens are given organically grown feed. In 2008, this was 90% organic. In the Netherlands, there is an additional organic hen rearing regulation. Its most important features are a maximum stocking density of 24 birds per m^2 during the first seven weeks and ten birds per m^2 between 7–18 weeks of age, 7 cm perch per bird from seven weeks onwards, and access to free range starting from eight weeks. The main thing organic egg production has in common with free-range production is the free-range area. The main differences are the number of animals per m^2 (six in organic and nine in free-range), organic ingredients in the feed, and farm size (up to 18,000 in organic and up to 40,000 in free range).

Many characteristics of organic poultry husbandry apply to loose housing in general. In the near future, we expect that the number of egg production systems with loose housing will further increase. In Europe, traditional cages are already forbidden. In the United States, the two States of California and Michigan passed state laws to ban battery cages. Although some farmers who have to stop using battery cages will turn to enriched cages, many farmers will switch to loose housing systems. The resemblance of organic systems to other loose housing systems and the expected growth of the alternative egg market, make our results valuable for a larger group of stakeholders than the organic sector alone.

The most recent overview of how Dutch organic farms perform in terms of health and welfare is based on data from 2001 and 2002 [4]. The results of that study were obtained through manure samples and a questionnaire for farmers. No physical assessment of animals took place. The mean flock size was 1,840 (80–5,400), the mean mortality was 11.4% (0–21) and the main health problems were predators, piling, coli, endoparasites, infectious bronchitis, and brachyspira. In the United Kingdom, a 7% mean mortality was reported in 1997 as well as cannibalism and coccidiosis [5]. A German study in 1999 [6] that involved a physical assessment of slaughterhouse hens reported pododermatitis, keel bone deformations, endoparasites, and fatty livers. In 2001 a Swedish study [7] based upon a questionnaire for farmers with 12–1,700 hens found a 9% (1–60) mean mortality, cannibalism, endoparasites, red mites, and leg parasites as health problems. A 2001 study in Austria [8] found a 7.2% (0–32) mean mortality, endoparasites, infectious bronchitis, Salmonella and cannibalism as health problems in 500–700 bird flocks. A Danish study [9] of 18 flocks with 1,200–5,000 hens found a 22% (9–62) mean mortality that was mainly caused by Pasteurella, predatory attacks, and piling. Thirty-three percent of the Danish flocks had little or no feather damage and 22% had severe feather damage. Of course, the type of health problems reported depend on the research methods applied. Interviews with farmers reveal mortality, and physical assessment may reveal pododermatitis and keel bone deformities, whereas manure samples give information about endoparasites. Though this makes the studies mentioned difficult to compare, they do give an impression of the range of health and welfare issues for organic laying hens during a period when flock sizes were much smaller and there was no common European organic poultry keeping legislation.

Organic hen keeping has been subjected to many changes, both in terms of regulations and trends in poultry husbandry since the most recent publication in 2003 [4] about the health and welfare of Dutch organic laying hens. Since then, beak treatments have been banned, aviaries have been introduced next to ground stables, flock sizes and number of hens per farm have increased, percentages of organic ingredients in poultry feed have increased, a national regulation for organic hen rearing has entered into force, different hybrids and genotypes have been introduced, and farmers have started to pay more attention to an attractive and functional free-range area.

Furthermore, farmers with organic hens experienced higher mortality rates than those with barn systems and were keen to understand how the health and welfare of their animals could be optimized. They wanted to know which factors in the rearing period and laying period influenced animal health and welfare parameters. Research questions were: how are organic egg farms performing in terms of health and welfare? How are farm practices during rearing and laying related to hen health and welfare during the laying period?

2. METHODOLOGY

We sent an invitation to 128 farms that kept organic laying hens according to Skal standards. Skal is the Dutch organic certification body. All farms that expressed an interest were included in the study. Our aim was for 50 flocks to take part. This number was a compromise between costs and statistic demands. In order to achieve a minimum of 50 flocks, farms who did not respond to the initial invitation were actively approached by phone in alphabetical order using the Skal's address list. Once 50 flocks had joined the study, new flocks were no longer able to take part. The 25 rearing farms that were approached reared the 'study flocks'.

Different questionnaires were designed for egg farmers and rearing farmers. The questionnaires were based on questionnaires from earlier studies and were discussed with several experts: veterinarians, farmers, poultry researchers, the Dutch Animal Protection organization, poultry advisors, hatcheries, and the Ministry of Economic Affairs, Agriculture and Innovation. The questionnaires covered housing and management, hybrid type, flock size, number of birds per m^2, feed, feed intake, egg production, mortality, mislaid egg percentage, feather pecking damage, health problems, and use of the free-range area. Both open and multiple-choice questions were included. The egg farmer questionnaires were filled in by the researcher during farm visits which took place when the hens were between 50 and 60 weeks old. Rearing farmers received their questionnaires by letter post and filled them out themselves.

A bird assessment protocol for farm visits was developed based on the following criteria:

- scoring should be objective,
- scoring should not demand expensive laboratory work,
- scoring items should be characteristics that farmers can see or feel themselves. This ensured that the study and its results were easy to communicate with the study's target group, the farmers.

The birds were assessed between 50 and 60 weeks of age. Fifty birds were caught individually at each farm. In tame flocks, this was done by hand while walking through the flock and in shy flocks by cornering the birds using a 'catch crate'. The indicators listed in Table 1 were determined for each individual bird. To avoid inter-observer bias, bird assessments on all the farms were done by the same researcher.

A manure sample was obtained by collecting 20 fresh droppings indoors. These were mixed and sent to the GD Animal Health Service (Deventer, the Netherlands). The sample was analyzed using the McMaster flotation method for worm egg counts. Shell and yolk color measurements were done with a Konica Minolta color reader CR-10 on thirty first-grade eggs that were collected from each flock. Pictures were taken of flocks and housing. After individual birds were assessed, the fear level of the whole flock was assessed during the researcher's visit. This was done on a scale from one (calm) to 10 (showing fearful behavior by flying up several times). The one, five and 10 on this scale were described and the observer made a rough estimate of the rest of the scale. Farmers scored health problem occurrence by indicating whether or not the flock had encountered a prelisted health problem (YES) in the flock or not (NO).

To estimate free-range use, the parameter Free Range Use (FRU) was calculated:

FRU = number of weeks free range is available (popholes open) in the period of 17–50 weeks of age × the maximum percentage of hens seen outside simultaneously under most optimal conditions (estimated by the farmer)

Since the flocks were 50–60 weeks old when we enquired about maximum range use, the percentages covered three seasons and all hours of the day. This prevented a 'snapshot' impression of range use.

Table 1. Methods used for body condition scoring.

Indicator	Assessment method	Relevance to welfare
Comb color	Konica Minolta color reader CR-10; measures color and describes it using three parameters: L, A and B-value [1]. L-value ranges from 0 to 100. A-values range from -86 to + 98. B-value ranges from -108 to + 94 [2].	Several diseases and health problems can cause a paler comb. Farmers regard bright red combs as a sign of health.
Feather condition	1. Visual scoring identifying a featherless spot of at least 5 centimeters diameter on a hen's back (yes/no) 2. Visual scoring of flock pictures taken according to Tauson *et al.* [10]. Classification of scoring of 6 body parts on a scale from 1 (completely featherless) to 4 (no or few feathers missing) [3].	Featherless spots are caused by feather pecking. This is a sign of reduced welfare in both actor [11,12] and victim [13]. Feather pecking can be caused by several factors during the rearing and laying period.
Wounds	Visual scoring of wounds on back, vent and tail area (yes/no)	Skin wounds can be caused by pecking behavior or accidents. Wounds are a sign of reduced welfare and make a bird potentially more vulnerable to infections.

Abnormalities on foot soles	Visual scoring of foot soles by observer (normal/scab/abscess).	Foot soles can have wounds or infections, caused by abnormalities in housing and reduced resistance against diseases (see discussion).
Keel bone deformations	Palpation of keel bone: deformity yes/no. Deformities are defined as deviations from the normally straight line (lateral or dorso-ventral) or thickened sections.	Keel bone deformations can be caused by 'metabolic bone disease', accidents and resting on the keel during perches.
Body weight	Weighing scale in grams.	Body weight could be compared to breeding standards and uniformity within a flock could be calculated (see discussion).

[1] Available online: http://sensing.konicaminolta.asia/learning-center/meter-measurement/meter-spaces (accessed on 2 April 2014); [2] Available online: http://stackoverflow.com/questions/19099063/what-are-the-ranges-of-coordinates-in-cielab-meter-space (accessed on 2 April 2014); 3 For each flock one mean was calculated for all body parts from 50 scored hens. To make the Tauson scores more illustrative, three 'qualitative' categories were defined: no/little damage (Tauson score 3.1–4); moderate damage (Tauson score 2.1–3) and severe damage (Tauson score 1–2).

All data were entered into an Excel spreadsheet. Statistics were calculated using the General Linear Model procedure (regression analysis). The RSEARCH procedure to find meaningful models ('all subset regression') was our starting point. Because R^2 was relatively low in most of these models, we moved on to linear regression for single factor relationships. In the regression analysis, relations were considered as not meaningful if the adjusted R^2 was less than 20. The two-sample *t*-test (unpaired, two-sided) and correlations (the two-sided test of difference from zero) were also used. GenStat for Windows, 13th edition, VSN International Ltd. (2010) was used to calculate all statistics.

3. RESULTS

Results were collected from 49 organic laying flocks on 43 farms. Some of these farms volunteered after receiving our invitation letter. The rest joined when the letter was followed by a phone call. We received information about the rearing period of 35 of these 49 flocks.

3.1. Number and Size of Farms

Five of the 49 flocks were kept on four farms with less than 500 hens. Forty-four flocks were kept on 39 farms with more than 500 hens. The average number of hens on all farms was 9,300 (min 34–max 25,450). Table 2 shows the distribution of the number of hens per farm.

Three egg farms had their own hen-rearing facilities.

Table 2. Distribution of farm sizes.

Size of farms	Number of farms	Percentage of farms
0–100	1	2
101–500	4	8
501–1,000	0	0
1,001–5,000	7	14
5,001–10,000	18	37
10,001–15,000	14	29
15,001–20,000	2	4
20,001–25,450	3	6
Total	49	100

3.2. Hybrids

The most-used hybrid was H&N Silver Nick (51%), followed by Hy-line Silver (20%), Hy-line Brown (10%) and Lohmann Brown Lite (8%). Most of the hens were so-called 'silvers': white- feathered hens that lay brown eggs.

3.3. Housing and Use of the Free Range Area

Twenty-two (45%) of the 49 flocks were kept in aviaries and 27 (55%) in a floor-based housing system (see Table 3). The latter had a grid floor with perches above a manure pit. Eleven of the 22 aviaries had a winter garden, of which 10 were 'regarded as stable area'. 'Regarded as stable area' means that the farmer includes its surface when calculating the number of hens allowed in the stable + winter garden combination. Such winter gardens do not provide additional space, but do provide 'different' space in terms of temperature and daylight. Twelve of the 27 floor-based stables had a winter garden, of which 11 were regarded as stable area.

Table 3. Distribution of housing systems.

Housing system	No of flocks	Winter garden yes/no
Aviary	22	11 yes 11 no
Floor	27	12 yes 15 no
Total	49	23 yes 26 no

Thirty-nine (80%) of the 49 flocks had less than 25% sheltered area in the form of bushes, maize or artificial structures in their outdoor ranging area. Seven flocks (14%) had 26–50% sheltered area, and three flocks (6%) had more than 50% sheltered area (see Table 4). 19 out of 48 flocks (40 %) had mortality by predators: 7 flocks by birds of prey (15%), 6 flocks by foxes (13%) and 6 flocks by both (13%).

During the study, national avian influenza alerts resulted in five periods of decreed indoor confinement. Four of these periods were in autumn or winter and one of them was in the summer. Five of the 49 flocks were vaccinated against avian influenza. These flocks were allowed to range during the confinement periods. The hens that were allowed to range shortly after their arrival on the laying farm tended to range more between 50 and 60 weeks of age (correlation −0.33; p = 0.03). When hens were allowed to range, on average 62% of the birds in a flock were seen outside together.

Table 4. Amount of shelter in outdoor runs.

Percentage of surface covered with bushes, maize or artificial structures	Number of flocks	Percentage of flocks
<25	39	80
26-50	7	14
>50	3	6
Total	49	100

Flocks that used the outdoor run more intensively, with higher FRU, had better feather scores (regression analysis; $R^2 = 25$; $p < 0.001$).

The more intensively hens used the outdoor run, the better farmers assessed their general health and performance (regression analysis; $R^2 = 18$; $p = 0.003$).

3.4. Technical Performance

Table 5 is an overview of production parameters at 30 and 60 weeks of age.

Table 5. Production parameters at 30 and 60 weeks of age.

Item	30 weeks Mean (min-max)	60 weeks Mean (min-max)
Feed intake (gram/hen/day)	129 (109–147)	133 (113–160)
Production (laying %)	91 (76–96)	80 (58–92)
Mislaid eggs (%)	3 (0–12)	2 (0–12)
Mortality (%)	2 (0–11)	7.8 (0–34)

Zero % mislaid eggs means 'close to zero'. Zero % mortality was found in one very small flock, where none of the hens died in their first 60 weeks. The one flock with 34% mortality was a 8,800-hen flock that got infected with the bacteria Erysipelotrix.

3.5. Body Weight

Calculations were done for the H&N Silver Nicks because this was the only genotype with enough available flocks. We found a relation between growth during the period of 7 to 11 weeks and body weight in the laying period at 50–60 weeks (regression analysis; R^2 = 36; p = 0.005): the faster the pullets grew during the middle of the rearing period, the higher their body weight was at a later age.

Table 6 shows mortality distribution.

Table 6. Distribution of mortality at 60 weeks of age.

Mortality	Number of flocks	Percentage
0–5	21	43
6–10	15	31
11–15	4	8
16–20	4	8
>20	1	2
Unknown	4	8
Total	49	100

Table 7 is a list of reported health problems. A health problem was defined as such if the farmer perceived it as a problem. Some problems were diagnosed by a veterinarian and some were not. Health problems were not quantified in terms of mortality or production loss.

Table 7. Presence of health problems.

Health problem	Number of flocks
E. Coli	18 (37%)
Red blood mites	16 (33%)
Infectious Bronchitis	15 (31%)
Piling	15 (31%)
Skin infections	11 (22%)
Multi systemic wasting syndrome	11 (22%)
Intestinal parasites	9 (18%)
Chronic gut infection (enteritis)	6 (12%)
Blackhead	5 (10%)
Fatty livers	2 (4%)
Botulism	1 (2%)
Amyloidosis	1 (2%)
Coccidiosis	0 (0%)

3.6. Intestinal Parasites in Relation to Anthelmintic Use

For one flock it was unclear whether it had been wormed. Therefore, results were available for 48 flocks. Twelve of 48 flocks (25%) were not wormed up to 50 weeks of age (see Table 8). Thirty-one flocks (65%) were wormed irregularly or less frequently than recommended with an anthelmintic producer (e.g., Flubenol© every 6 weeks). Only five flocks (10%) were wormed five or more times between 17 to 50 weeks of age. Table 8 shows egg count results. With regression analysis we did not find a meaningful relation between worming of a flock (yes/no) and % lay at 60 weeks (regression analysis; R^2 = 15.2; p = 0.005), or between worming of a flock (yes/no) and mortality at 60 weeks (regression analysis; R^2 = 6.2; p = 0.731).

Table 8. Number of anthelmintic treatments up until 50 weeks of age (n = 48).

	Number of flocks	Positive for Ascaridia and Heterakis	Positive for Capillaria	Positive for Coccidiosis	Positive for Syngamus
No treatment	12 (25%)	83%	25%	25%	0%
Irregular, less frequent than prescribed, *i.e.*, treated 1-4 times	31 (65%)	61%	19%	42%	0%
Treated 5 or more times	5 (10%)	0 (0%)	0%	3 (60%)	0%

3.7. Parasites and Egg Yolk Color

Yolk color measurements determined L-, A- and B-values. In a color series of slightly yellow to orange, the L-value decreases and both the A- and B-values increase. Flocks that were positive for Ascaridia or Heterakis had higher L-values, meaning a lighter yolk color (p = 0.002; two-sample t-test; 47 d.f.). Flocks that were positive for Capillaria also showed a higher L-value, which meant a lighter yolk color (p = 0.035; two-sample *t*-test; 47 d.f.).

3.8. Feather Pecking Damage

On average, 64% (minimum zero; maximum 100) of the hens in all 49 flocks had a featherless spot of at least five centimeters diameter on their backs. The score results using Tauson *et al.*'s [10] method of scoring six body parts of 50 hens are shown in Table 9. The scores between one and four were obtained following Tauson's protocol. In addition we defined qualitative categories to make the outcomes more illustrative. Tauson's scoring was done on the basis of photographs. Photographs that were of sufficient quality to score were available for only 37 flocks.

Table 9. Severity of feather damage according to Tauson *et al.* (2005).

Category	Number of flocks
No/little feather damage (Tauson score 3.1–4)	25 (68%)
Moderate feather damage (Tauson score 2.1–3)	9 (24%)
Severe feather damage (Tauson score 1–2)	3 (8%)
Total	37

Twenty-five of 37 flocks (68%) had little or no feather damage, nine flocks (24%) had moderate damage and three flocks (8%) had severe damage. The feather score was better (RSEARCH procedure (all possible subset selection; R^2 = 52) if the FRU was higher (p = 0.003), the laying percentage at 60 weeks was higher (p = 0.007), and the sooner the hens went outside after arrival on the laying farm (p = 0.024).

3.9. Vent Pecking

Peck wounds in the vent area were observed in 34 of 49 flocks (69%). On average, 18% (min two–max 50) of the birds in these 34 flocks had wounds in the vent area. It was not possible to make quantitative statements about wound severity as scores were based on wound presence (yes/no). Most of the wounds had an approximate diameter of 0.5 cm, a small amount was bigger (up to 2 cm diameter).

3.10. Keel Bone Deformations

Keel bone deformations were found in all 49 flocks. On average, 21% (min four–max 48) of the birds in a flock had a keel bone deformation: a lateral or dorso-ventral deviation, a thickened section or a combination. In our study farmers used either round metal or rectangular wooden perches. With regression analysis we did not find a relation between keel bone deformation and perch type (regression analysis; R^2 = 11; p =0.012) or between keel bone deformation and aviary or floor-based housing (regression analysis; R^2 =11; p = 0.012).

3.11. Foot Sole Lesions

Wounds on foot soles were observed in 38 of the 49 flocks (78%). On average, 13% (min two–max 48) of the birds in these flocks had such wounds. Since we scored hens on the basis of wound presence (yes/no), it is not possible to make quantitative statements on the severity of these wounds. The majority showed wounds of 0.2–0.5 cm diameter. Swellings that are characteristic for bumble foot were rarely observed. No relation was found between foot sole lesions and FRU (regression analysis; R^2 = 11; $p = 0.013$).

3.12. Comb Color

Comb color measurements entailed defining L-, A- and B-values. In a color series of pink to dark red, the L-value decreases and both A- and B-values increase. The hybrid type (with regression analysis) significantly explained the overall difference in comb color (L-value R^2 = 29, $p = 0.003$; A-value R^2 = 24; $p = 0.009$; B-value R^2 = 46, $p < 0.001$). Therefore, further calculations were only done for the most-used hybrid: H&N Silver Nick. A higher FRU was related to a lower L-value, thus darker combs (regression analysis; R^2 = 29; $p = 0.003$). Flocks that reacted more fearfully during our visits had higher L-values, thus lighter combs (regression analysis; R^2 = 34; $p = 0.003$). Flocks that tested positively for one or more intestinal parasites, tended to have a higher B-values, thus darker combs (t-test; $p = 0.036$).

3.13. Rearing

Data regarding the rearing period was available for 35 of the 49 flocks. Not all questionnaires on the rearing period were complete. Six of the 33 flocks (18%) were kept in cages during the first weeks of their lives. Three of those flocks were kept on 'pullet paper' without litter. Twenty-six (79%) of the 33 flocks had litter from day one. In loose housed rearing, the group size was 13,800 (n = 25 flocks). The average pullet density during the first four weeks was 21 pullets/m^2 (n = 33), with 8.5 as minimum and 33.5 as maximum. After four weeks, many flocks were given additional space, resulting in an average density of 14 pullets/m2 (min 6.6 and max 27.5). For 33 flocks, there was data available regarding whether they had access to an outdoor run. Nine flocks had not been outside and 26 flocks had access to an outdoor run. The average age for first time access was 8.4 weeks. At 17 weeks, on average 28% of the animals from the 'outside flocks' was seen outside.

4. DISCUSSION

The study incorporated data from 49 flocks on 43 farms. Some of the farms volunteered after receiving an invitation letter and the rest joined when the letter was followed by a phone call. It is possible that the farmers who volunteer to participate, are those who perform well. If this is the case, our results may overestimate the welfare situation on farms. Five of these 49 flocks were kept on four farms with less than 500 hens. Forty-four flocks were kept on 39 farms with more than 500 hens. According to Loefs [14] there were 107 farms with organic laying flocks in the Netherlands in 2008 and she only counted farms with more than 500 hens. Our 39 flocks represent slightly more than one third of the farms (39/107 = 36%).

4.1. Relation between Outdoor Run Use and Health and Welfare

We did not find any relation between the number of weeks that hens spent outside during the laying period and mortality at 60 weeks, nor was a relation found between the number of confinement periods and mortality at 60 weeks. However, several studies showed a higher mortality or a higher disease incidence in free-range or organic poultry compared to hens kept inside [15,16]. In 19 of 49 (39%) of our flocks there was mortality caused by predators. Other studies also reported mortality caused by predators. These studies focused on broilers [17] or they presented their results differently [18,19]. Therefore, it is not possible to compare these results to our results. However, all studies illustrate that losses caused by predators are a realistic risk of keeping free-ranging poultry. If mortality is a measure for bird welfare, then one could argue that hens on free-range farms have a reduced welfare. Flocks that used the outdoor run more intensively had better feather scores. If the amount of feather-pecking damage is used as a measure for hen welfare, then hens on farms where more hens use the free-range areas have a better welfare [20,21,22,23].

The more intensely hens used the outdoor run, the better the farmer assessed their general health and performance. Since we found this relation with regression analysis, there may not be a causal relation between the two parameters. Which health and performance aspects satisfy farmers regarding their hens remain unknown, but the relation shows that somehow good use of the outdoor run relates to good health and/or performance perceived by farmers.

4.2. Technical Performance

On average birds had a feed intake of 129 g/day (min 109–max 147) at 30 weeks and 133 g/day (min 113–max 160) at 60 weeks of age. The mean intake levels were within an acceptable range. The feed intake depends on ambient temperature, plumage condition, and housing system [24].

The mean number of mislaid eggs decreased during the laying cycle from three to two percent, which might be caused by adequate management interventions. The mean mortality of 7.8% at 60 weeks of age is difficult to compare with other studies, because these studies describe results of the entire laying cycle, which can be up to 70 weeks or more. Mortality is generally higher in alternative systems [25].

4.3. Body Weight

We found a relation between growth in the rearing period and body weight at 50–60 weeks: the faster the pullets grew during the middle of the rearing period (weeks of life 7–11), the higher was their body weight at later age. Although for several other aspects, a relation has been found between conditions in the rearing period and the laying period, for example feather pecking, we did not find literature specifically on laying hen body weight.

4.4. Intestinal Parasites in Relation to the Use of Anthelmintic

Depending on the number of anthelmintic treatments, a considerable number of flocks tested positively for internal parasites. We did not find a relation with production or mortality. For different reasons it is not clear whether zero tolerance of internal parasites is necessary. Firstly, a low infection level might not be harmful. Secondly, when a higher infection level is combined with other health problems, one could ask what came first: the other health problems, which made the birds vulnerable to parasites or the parasites that made the birds vulnerable to other health problems. In 1999, Thamsborg *et al.* [26] already found that the significance of infections in terms of disease and production losses on organic farms had not been assessed. Moreover, the GD Animal Health Service, a Dutch authority on farm animal health, has the following text on their website [27] 'not all internal parasite species are equally harmful' and 'treatment therefore is not always necessary'. They refer to 'your personal vet' for personal advice. Flubendazole is the only registered anthelmintic medicine against internal parasites in poultry, and parasites may become resistant to it. Therefore, zero tolerance in healthy flocks does not seem logical.

Flocks that were positive for Ascaridia or Heterakis had a lighter yolk color. Flocks that were positive for Capillaria also had a lighter yolk color. Although non-scientific on-line publications recognize this relation, we were unable to find scientific articles to support our findings. A general explanation seems to be that parasites damage the intestines, which in turn influences pigment intake.

4.5. Feather Pecking Damage

Sixty-four percent of the hens in all 49 flocks were found to have feather damage on their back when the first method was used. However, we found that 68% of the flocks had 'no or little' feather damage when Tauson's method was applied in combination with categories we defined to qualify the damage. The difference may be caused by the fact that a five-centimeter featherless spot on a hen's back qualifies as little damage if the rest of the hen's feathers are in good condition, which was mostly the case. As Tauson's method reflects variation better, we think that the method is more suitable than a quick scan of only one body part.

Twenty-five of 37 flocks (68%) showed no or little feather damage, nine flocks (24%) showed moderate damage and three flocks (8%) showed severe damage. The feather score was better if the free-range use was better, the laying percentage at 60 weeks was higher, and the sooner the hens went outside after arrival on the laying farm. A relation between better free-range use and better feather cover is also reported in other studies [20,21,22,23]. A free-range area can be considered as environmental enrichment. Other types of environmental enrichment, such as additional foraging materials [28], scattered grains [29], and elevated perches [30] are also associated with less feather pecking. Another explanation could be that if a flock is distributed over a larger area, the stocking density decreases. Lower stocking density (in combination with a smaller group size) is also associated with less feather pecking [31,32,33]. Concerning the relation between higher laying percentage and better feather cover, it is not clear if this is a causal relationship. There is not much literature to be found on this relation, although Hüber-Eicher and Sebö [34] found the same relation in Swiss commercial flocks.

4.6. Vent Pecking

In 34 of 49 flocks (69%) peck wounds were seen in the vent area. On average, 18% (min two–max 50) of the birds had wounds in the vent area. Most of them were about 0.5 cm in diameter, a small number of them was bigger, up to 2 cm. Pötzsch et al. [35] found that 36.9% of non-cage housed flocks in the UK were affected by vent pecking. However, this was based on the farmer's judgments obtained through voluntary questionnaires. Vent wounds are best observed by manually diagnosing individual hens. Lambton et al. [36], who did behavioral observations in 119 British flocks (barn, free-range and organic), observed vent pecking behavior in 24.8% of the flocks. According to Pötzsch et al. [35], vent pecking and feather pecking damage may be caused by shared common environmental risk factors.

4.7. Keel Bone Deformations

Keel bone deformations were seen in all 49 flocks. On average, 21% (min four–max 48) of the birds in a flock had a keel bone deformation: a lateral or dorso-ventral deviation, a thickened section or a combination. We are not sure how many of the deviations we observed were fractures, since we used the palpation method for our assessment instead of animal dissection. Wilkins *et al.* [37] compared palpation and dissection methods and concluded that in 91% of the cases palpation gave a correct result, assuming that the dissection method is the correct result. We are not sure whether the lateral and dorso-ventral deviations were the result of fractures or a slowly-developed curve. Wilkins *et al.* [37] found keel bone fractures in 36% to more than 80% of the animals in flocks kept in different housing systems. Gregory and Wilkins [38] found that 30% of 72-week old hens had a broken keel. Nicol *et al.* [39] found 52–59% of hens with keel breaks in commercial flocks in aviaries. In those studies, keel bone assessment was done on dissected hens. Compared to those studies, our 21% average of hens with a keelbone deviation (not all of them being a fracture) seems low. Twenty-one percent might be an underestimation of the real number of fractures.

In our study, farmers used either round metal or rectangular wooden perches. We did not find any relation between keel bone deformation and perch type. Nor was a relation found between keel bone deformation and aviary or floor-based housing systems.

4.8. Foot Sole Lesions

In 38 of 49 flocks (78%) wounds on foot soles were seen. On average, 13% (min two–max 48) of the birds in these flocks had foot sole wounds. The majority of them were 0.2–0.5 cm in diameter. Swellings characteristic of bumble foot were rarely observed.

Nicol et al. [39] measured foot damage in laying hens. They used a five-point scoring system with one meaning 'no lesions' and five meaning 'very poor foot condition with inflamed and/or bleeding lesions visible over much of the area'. The foot damage in their study never exceeded an average value of 1.5, thus staying between 'no lesion' and 'lesions which are clearly visible but of minor importance and/or frequency'. Although different assessment methods were applied, which means their results are difficult to compare with our results, we assume the results are about the same.

Tauson and Abrahamsson [40] showed that perch design was an important factor for bumble foot in laying hens. Hens in cages without perches had the best scores: no bumble foot at all. When hens from cages with four different perch designs were compared, those with the narrow perches had the best bumble foot scores and those with broader perches the worst. The narrow perches were narrow rectangular (35 mm wide) or flat/round (38 mm wide). The broader perches were plastic 'mushroom' (48 mm wide) and wooden wide rectangular (53 mm wide). The authors' explanation is that a good anatomical grip is not possible on broader perches and this results in too much pressure on the hen's foot at rest. Lay *et al.* [25] concluded that wet litter conditions and high ammonia litter content in litter could cause footpad dermatitis in hens, which could be followed by bacterial infections, which in turn could lead to bumble foot due to penetration of 'a strange body'. However, we also saw foot wounds in hens on dry litter. Perhaps the wet conditions of the free-range area added to the foot problems or it was caused by perch design in these hens as was found by Tauson and Abrahamsson [40]. According to their results, both the rectangular and the round perches in our study did not meet the hens' needs.

4.9. Comb Color

The hybrid significantly explained the overall difference in comb color. Therefore, further calculations were only done for the most-used hybrid: H&N Silver Nick. FRU was related to darker combs. Flocks that were more fearful during our visits had lighter combs, but we could not find comparable results in the literature. Flocks that tested positively for one or more intestinal parasites tended to have darker combs. It is not clear whether this is a causal relation, because comb color in free-living red grouse is lighter in birds infected with internal parasites [41]. Spending more time in the free-range area is related to darker combs. A comparable finding is reported by Whay *et al.* [42].

5. CONCLUSIONS

With a mean flock size of 9,300 hens and 107 farms in 2008, the organic poultry sector has shown that it is able to become a serious alternative for existing intensive animal production. The presence of a free-range area is the most characteristic aspect of the organic system. Our study shows that a flock's welfare benefits from using this free-range area, as is shows less feather pecking damage if it uses it. Concerning the presence of feather pecking damage and thus animal welfare, the organic system has the potential to perform better than 'indoor' poultry husbandry systems because of the presence of the free range area. The organic flocks perform about the same or worse than other commercial systems for several other factors. Improvements are desirable for mortality, internal parasites, keel bone deformations, and foot sole lesions. However, it can be questioned whether these physical consequences can be attributed to the housing system or if they are the result of the interaction between genotype and environment. The organic system uses the same high productive (and perhaps thus vulnerable) genotypes as other commercial systems in an environment that is still artificial.

ACKNOWLEDGMENTS

The authors gratefully thank and acknowledge the funding by the Dutch Ministry of Economic Affairs for project 'Relation between farm factors and animal health', BO-04-002-004.015.

REFERENCES AND NOTES

1. Bestman, M. Keeping Chickens without Feather Pecking (in Dutch); Louis Bolk Institute: Driebergen, The Netherlands, 2002; p. 17.
2. Agricultural Database Dutch National Statistic Institute (CBS). Available online: http://statline.cbs.nl/StatWeb/publication/?VW=T&DM=SLNL&PA=81517NED&D1=a&D2=a&D3=l&HD=120309-1138&HDR=G2,G1&STB=T (accessed on 27 February 2013).
3. Regulation Organic Animal Husbandry. Available online: http://www.skal.nl/LinkClick.aspx?fileticket=an0aCY%2b7kWE%3d&tabid=108&language=nl-NL (accessed on 27 February 2013).
4. Fiks-van Niekerk, T.; Reuvekamp, B.; Landman, W. Monitoring of Organic Farms. More Infections as on Cage Farms (in Dutch). Pluimveehouderij 2003, 33, 10–11.
5. Lampkin, N. Organic Poultry Production; Report to MAFF; Welsh Institute of Rural Studies, University of Wales: Aberystwyth, UK, 1997.
6. Keutgen, H.; Wurm, S.; Ueberschär, S. Pathologisch-anatomische Untersuchungen bei Legehennen aus Verschiedenen Haltungssystemen. Deutsche Tierärztliche Wochenschrift. 1999, 106, 127–133.
7. Berg, C. Health and Welfare in Organic Poultry Production. Acta Vet. Scand. 2001, S95, 37–45.
8. Sommer, F. A Decade of Experience with Free-Range Poultry Farming in Austria. Is This the Future? In Proceedings of the 50th Western Poultry Disease Conference, Davis, CA, USA, 24–26 March 2001; pp. 95–96.
9. Hegelund, L.; Soerensen, J.T.; Hermansen, J.E. Welfare and Productivity of Laying Hens in Commercial Organic Egg Production Systems in Denmark. NJAS 2006, 54, 147–155.

10. Tauson, R.; Kjaer, J.; Maria, G.A.; Cepero, R.; Holm, K.-E. Applied Scoring of Integument and Health in Laying Hens. Anim. Sci. Pap. Rep. 2005, 23 (Suppl 1), 153–159, and references therein.
11. Vestergaard, K.S.; Kruijt, J.P.; Hogan, J.A. Feather Pecking and Chronic Fear in Groups of Red Jungle Fowl: Their Relations to Dustbathing, Rearing Environment and Social Status. Anim. Behav. 1993, 45, 1127–1140.
12. El-Lethey, H.; Aerni, V.; Jungi, T.W.; Wechsler, B. Stress and Feather Pecking in Laying Hens in Relation to Housing Conditions. Brit. Poultry Sci. 2000, 41, 22–28.
13. Gentle, M.J.; Hunter, L.N. Physiological and Behavioural Responses Associated with Feather Removal in Gallus gallus var domesticus. Res. Vet. Sci. 1990, 27, 149–157.
14. Loefs, R. Organic Laying Hen Sector is Marking Time (in Dutch). Pluimveehouderij 2008, 38, 50–51.
15. Fossum, O.; Jansson, D.S.; Etterlin, P.E.; Vagsholm, I. Causes of Mortality in Laying Hens in Different Housing Systems in 2001 to 2004. Acta Vet. Scand. 2009, 51.
16. Sherwin, C.M.; Richards, G.J.; Nicol, C.J. Comparison of the Welfare of Layer Hens in 4 housing systems in the UK. Brit. Poultry Sci. 2010, 51, 488–499.
17. Stahl, P.; Ruette, S.; Groset, L. Predation on Free-ranging Poultry by Mammalian and Avian Predators: Field Loss Estimates in a French Rural Area. Mammal Rev. 2002, 32, 227–234.
18. Häne, M.; Huber-Eicher, B.; Fröhlich, E. Survey of Laying Hen Husbandry in Switzerland. World's Poultry Sci. J. 2000, 56, 21–31.
19. Moberly, R.I.; White, P.C.L.; Harris, S. Mortality due to Fox Predation in Free-range Poultry Flocks in Britain. Vet. Rec. 2004, 155, 48–52.
20. Bestman, M.W.P.; Wagenaar, J. Farm Level Factors Associated with Feather Pecking Damage in Organic Laying Hens. Livest. Prod. Sci. 2003, 80, 133–140.
21. Green, L.E.; Lewis, K.; Kimpton, A.; Nicol, C.J. A Cross Sectional Study of the Prevalence of Feather Pecking Damage in Laying Hens in Alternative Systems and its Association with Management and Disease. Vet. Rec. 2000, 147, 233–238.
22. Mahboub, H.D.H.; Müller, J.; von Borell, E. Outdoor Use, Tonic Immobility, Heterophil/lymphocyte Ratio and Feather Condition in Free-range Laying Hens of Different Genotype. Brit. Poultry Sci. 2004, 45, 738–744.
23. Nicol, C.J.; Pötzsch, C.; Lewis, K.; Green, L.E. Matched Concurrent Case-control Study of Risk Factors for Feather Pecking Damage in Hens on Free-range Commercial Farms in the UK. Brit. Poultry Sci. 2003, 44, 515–523.
24. van Krimpen, M.; Binnendijk, G.P.; van den Anker, I.; Heetkamp, M.J.W.; Kwakkel, R.P.; van den Brand, H. Effects of Ambient Temperature, Plumage Condition and Housing System on Energy Partitioning and Performance in Laying Hens, thereby Predicting Energy Intake. 2014. in preparation.
25. Lay, D.C.; Fulton, R.M.; Hester, P.Y.; Karcher, D.M.; Kjaer, J.B.; Mench, J.A.; Mullens, B.A.; Newberry, R.C.; Nicol, C.J.; O'Sullivan, N.P.; Porter, R.E. Hen Welfare in Different Housing Systems. Poultry Sci. 2011, 90, 278–294.
26. Thamsborg, S.M.; Roepstorff, A.; Larsen, M. Integrated and Biological Control of Parasites in Organic and Conventional Production Systems. Vet. Parasitol. 1999, 84, 169–186.
27. Poultry parasites. Available online: http://www.gddeventer.com/pluimvee/ zoekresultaat?search=wormen (accessed on 27 February 2013).
28. Huber-Eicher, B. A Survey of Layer-type Pullet Rearing in Switzerland. World's Poultry Sci. J. 1999, 55, 83–91.
29. Blokhuis, H.J.; van der Haar, J.W. Effects of Pecking Incentives during Rearing on Feather Pecking of Laying Hens. Brit. Poultry Sci. 1992, 33, 17–24.
30. Wechsler, B.; Huber-Eicher, B. The Effect of Foraging Material and Perch Height on Feather Pecking and Feather Damage in Laying Hens. Appl. Anim. Behav. Sci. 1998, 58, 131–141.
31. Nicol, C.J.; Gregory, N.G.; Knowles, T.G.; Parkman, I.D.; Wilkins, L.J. Differential Effects of Increased Stocking Density, Mediated by Increased Flock Size, on Feather Pecking and Aggression in Laying Hens. Appl. Anim. Behav. Sci. 1999, 65, 137–152.

32. Savory, C.J.; Mann, J.S.; Macleod, M.G. Incidence of Pecking Damage in Growing Bantams in Relation to Food Form, Group Size, Stocking Density, Dietary Tryptophan Concentrations and Dietary Protein Source. Brit. Poultry Sci. 1999, 40, 579–584.
33. Huber-Eicher, B.; Audigé, L. Analysis of Risk Factors for the Occurrence of Feather Pecking in Laying Hen Growers. Brit. Poultry Sci. 1999, 40, 599–604.
34. Hüber-Eicher, B.; Sebö, F. The Prevalence of Feather Pecking Damage and Development in Commercial Flocks of Laying Hens. Appl. Anim. Behav. Sci. 2001, 74, 223–231.
35. Pötzsch, C.J.; Lewis, K.; Nicol, C.J.; Green, L.E. A Cross-sectional Study of the Prevalence of Vent Pecking in Laying Hens in Alternative Systems and its Associations with Feather Pecking Damage, Management and Disease. Appl. Anim. Behav. Sci. 2001, 74, 259–272.
36. Lambton, S.L.; Knowles, T.G.; Yorke, C.; Nicol, C.J. Risk Factors Affecting the Development of Vent Pecking and Cannibalism in Loose Housed Laying Hen Flocks. In Proceedings of the Poultry Welfare Symposium, Cervia, Italy, 18–22 May 2009; p. 13.
37. Wilkins, L.J.; McKinstry, J.L.; Avery, N.C.; Knowles, T.G.; Brown, S.N.; Tarlton, J.; Nicol, C.J. Influence of Housing System and Design on Bone Strength and Keel Bone Fractures in Laying Hens. Vet. Rec. 2011, 169, 414–420.
38. Gregory, N.G.; Wilkins, L.J. Effect of Age on Bone Strength and the Prevalence of Broken Bones in Perchery Laying Hens. N. Z. Vet. J. 1995, 44, 31–32.
39. Nicol, C.J.; Brown, S.N.; Glen, E.; Pope, S.J.; Short, F.J.; Warriss, P.D.; Zimmerman, P.H.; Wilkins, L.J. Effects of Stocking Density, Flock Size and Management on the Welfare of Laying Hens in Single-tier Aviaries. Brit. Poul. Sci. 2006, 47, 135–146.
40. Tauson, R.; Abrahamsson, P. Foot and Skeletal Disorders in Laying Hens. Effects of Perch Design, Hybrid, Housing System and Stocking Density. Acta Agr. Scand. 1994, 44, 110–119.
41. Mougeot, F.; Martínez-Padilla, J.; Bortolotti, J.; Webster, L.M.I.; Piertney, S.B. Physiological Stress Links Parasites to Carotenoid-based Color Signals. J. Evol. Biol. 2010, 23, 643–650.
42. Whay, H.R.; Main, D.C.J.; Green, L.E.; Heaven, G.; Howell, H.; Morgan, M.; Pearson, A.; Webster, A.J.F. Assessment of the Behaviour and Welfare of Laying Hens on Free-range units. Vet. Rec. 2007, 161, 119–128.

CHAPTER 13

Mutilating Procedures, Management Practices, and Housing Conditions That May Affect the Welfare of Farm Animals: Implications for Welfare Research

Rebecca E. Nordquist [1,2,*], Franz Josef van der Staay [1,2], Frank J. C. M. van Eerdenburg [3], Francisca C. Velkers [4], Lisa Fijn [1,2] and Saskia S. Arndt [5]

[1]*Behavior & Welfare Group (Formerly Emotion & Cognition Group), Department of Farm Animal Health, Faculty of Veterinary Medicine, University Utrecht, Utrecht 3584CL, The Netherlands*

[2]*Brain Center Rudolf Magnus, Universiteitsweg 100, Utrecht 3584CG, The Netherlands*

[3]*Department of Farm Animal Health, Faculty of Veterinary Medicine, University Utrecht, Utrecht 3584CL, The Netherland*

[4]*Epidemiology and Poultry Health Care, Department of Farm Animal Health, Faculty of Veterinary Medicine, University Utrecht, Utrecht 3584CL, The Netherlands*

[5]*Animal Welfare and Laboratory Animal Science, Department Animals in Science and Society, Faculty of Veterinary Medicine, University Utrecht, Utrecht 3508TD, The Netherlands*

SIMPLE SUMMARY

Intensive farming systems are confronted with a number of animal welfare issues such as injuries from horns in cattle and feather pecking in poultry. To solve these problems, mutilating procedures, such as dehorning in cattle and goats and beak trimming in laying hens, are applied routinely. These and other procedures such as early maternal separation, overcrowding, and barren housing conditions impair animal welfare. Scientific underpinning of the efficacy of these interventions and management practices is poor. We advocate that all stakeholders, in particular animal scientists and veterinarians, take the lead in evaluating common, putative mutilating and welfare reducing procedures and management practices to develop better, scientifically supported alternatives, focused on adaptation of the environment to the animals, to ensure uncompromised animal welfare.

ABSTRACT

A number of mutilating procedures, such as dehorning in cattle and goats and beak trimming in laying hens, are common in farm animal husbandry systems in an attempt to prevent or solve problems, such as injuries from horns or feather pecking. These procedures and other practices, such as early maternal separation, overcrowding, and barren housing conditions, raise concerns about animal welfare. Efforts to ensure or improve animal welfare involve adapting the animal to its environment, i.e., by selective breeding (e.g., by selecting "robust" animals) adapting the environment to the animal (e.g., by developing social housing systems in which aggressive encounters are reduced to a minimum), or both. We propose adapting the environment to the animals by improving management practices and housing conditions, and by abandoning mutilating procedures. This approach requires the active involvement of all stakeholders: veterinarians and animal scientists, the industrial farming sector, the food processing and supply chain, and consumers of animal-derived products. Although scientific evidence about the welfare effects of current practices in farming such as mutilating procedures, management practices, and housing conditions is steadily growing, the gain in knowledge needs a boost through more scientific research. Considering the huge number of animals whose welfare is affected, all possible effort must be made to improve their welfare as quickly as possible in order to ban welfare-compromising procedures and practices as soon as possible.

Keywords: animal welfare; mutilating procedures; housing conditions; management practices

1. INTRODUCTION

Mutilating procedures [1,2], certain animal management practices (e.g., [3]), and housing conditions (e.g., [4]) may impair animal welfare. Many of these practices and conditions, which have often developed on the basis of on-farm experience, are intended to prevent or solve problems inherent to industrial farming, such as difficulties with handling animals (e.g., disbudding in cows, sheep, and goats) or production losses (e.g., farrowing crates to prevent crushing of piglets). While a number of these practices are claimed to have a positive effect on animal welfare and may solve particular welfare problems (beak trimming—less feather pecking; tail amputation—less tail biting) [5], systematic scientific appraisal is needed to check whether the proposed solutions serve the intended goal and how they affect animal welfare.

This article describes common mutilating procedures, management practices, and housing conditions that may affect farm animal welfare. Furthermore, the role of the veterinarian, who operates at the front line of animal health and welfare, is discussed with respect to improving welfare and directing future developments in farm animal management. Steps to identify and replace interventions and procedures that compromise welfare are suggested.

Although there is some evidence of the welfare effects of a number of the abovementioned procedures and practices, continued, targeted research to investigate which management and breeding practices minimize impacts on animal welfare, while recognizing the reality of the business task faced by food producers, is important and necessary [6]. It is the duty of veterinarians and animal scientists to provide this evidence [7].

2. CONCEPTS OF ANIMAL WELFARE

Animal welfare relates to more than merely the physical health of an animal. While there are numerous concepts of animal welfare (e.g., [8,9,10,11,12,13,14,15]), the current understanding is perhaps best summarized by Webster ([16], p. 117) as follows: "There is now broad agreement amongst academics and real people that the welfare of a sentient animal is defined by how well it feels; how well it is able to cope with the physical and emotional challenges to which it is exposed". Indeed, current welfare concepts put emphasis on the perception of the animal itself (e.g., [11,14,17,18,19,20]). The idea that animals have evolved in adaptation to their environment, so as to optimize their ability to adapt to changes within that environment through the expression of a variety of physiological and/or behavioral responses, was first applied to animal welfare in the 1990s (see, for example, [21]). In this view, an animal's welfare is not at risk as long as the animal is able to meet environmental challenges [11,18,19], elaborated by, among others in [22,23].

Even though there is broad agreement on how to approach the concept of animal welfare, the definition of markers or measures of welfare and how they should be evaluated remains a major challenge. In recent years, a number of studies have attempted to develop and validate tools and protocols to assess animal welfare (e.g., [5,24,25,26]), such as the Welfare Quality® project [27]. Recently, a new initiative, Animal Welfare Indicators (AWIN), financed by the EU VII Framework Program (FP7-KBBE-2010-4), has been initiated to address animal welfare indicators, with special emphasis on sheep, goats, horses, donkeys, and turkeys. The aim of this initiative, as it was for the Welfare Quality® project [27], is developing protocols for assessing animal welfare, studying the animal welfare implications of diseases and pain, examining the effects of different prenatal social environments, social dynamics, and prenatal handling methods on the development and welfare of the offspring, and, finally, "ensuring that the very best scientific information on animal welfare is easily available to stakeholders and the public" [28].

Mellor recently introduced the concept of "The Five Domains", seeking to assess the impact of the social and physical environment on the affective state of an animal [15]. Webster [29] discussed the relative validity and utility of this approach in comparison to "The Five Freedoms" as described by the Farm Animal Welfare Council [30] and based on the seminal work of the Brambell Committee [31].

The "Five Domains" indeed provide a useful tool to evaluate "particular physical/functional disruptions and imbalances, as well as restrictions on behavioral expression, and then to identify the specific negative affects each disruption, imbalance or restriction would be likely to generate" ([15], p. 8). However, in our opinion, concepts of welfare should also take the dynamics of the individual's interaction with its environment over time into consideration. Any attempt to assess animal welfare thus needs to consider changes in measures over time. Thus, the assessment of welfare should focus on whether or not the animal has the freedom and capacity to react appropriately (i.e., adaptively) to environmental challenges.

This approach forms the basis of the concept of animal welfare as introduced by Ohl & van der Staay:

"An individual is in a positive welfare state when it has the freedom adequately to react to

- hunger, thirst or incorrect food;
- thermal and physical discomfort;
- injuries or diseases;
- fear and chronic stress; and thus,
- the freedom to display normal behavioural patterns that allow the animal to adapt to the demands of the prevailing environmental circumstances and enable it to reach a state that it perceives as positive." ([19], p. 17).

In the light of this concept, we need to ask whether current practices and procedures affect the option and capacity of farm animals to react appropriately, and to what extent. It is of outmost importance to search for adequate read-out parameters regarding adaptive capacity and to avoid primarily economically driven judgements. For example, beak-trimmed hens might eat and produce eggs. However, this does not indicate a positive emotional state and the animals might permanently suffer from impaired sensory perception and pain (see Figure 1). To induce positive welfare, adaptive processes must lead to a state that an animal itself perceives as positive [19].

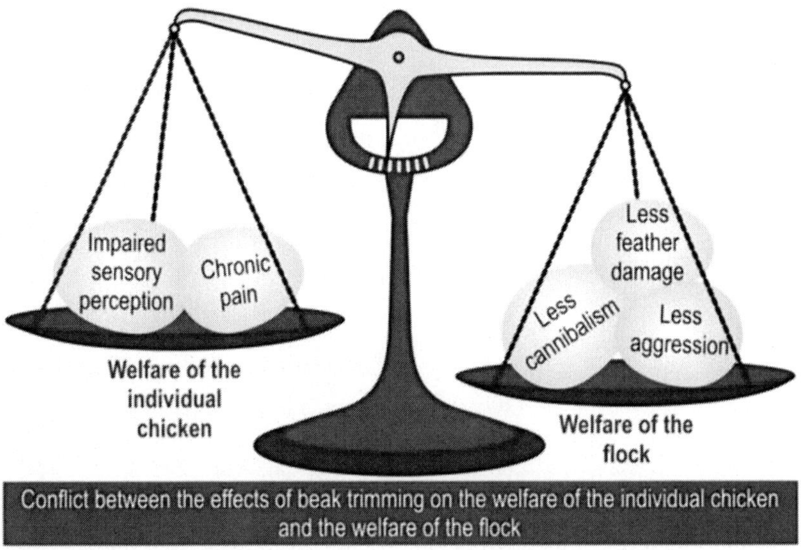

Figure 1. Beak trimming in chickens. This procedure reduces the consequences, but not the incidence, of severe feather pecking. Although it causes discomfort in the individual chicken, the flock may profit from

this intervention because the consequences of feather pecking are generally less severe [33,37]. The general question thus is whether the harm inflicted on the individual outweighs the benefits for the individual and/or the group [2]. This figure shows, as a hypothetical example, the evaluation concluding that the effects of beak trimming are favorable for the welfare of the flock. Depending on a different weighing of the arguments, the conclusion may be that prophylactic beak trimming is unacceptable. For example, in breeds with a very low incidence of feather pecking, trying to reduce the suffering of some of the birds by mutilating all of them seems completely unacceptable from a purely animal welfare perspective, whereas in breeds with a very high incidence of feather pecking, beak trimming might be considered more acceptable. Alternative solutions to beak trimming must be taken into account before even considering the use of mutilating procedures.

3. PROCEDURES, MANAGEMENT PRACTICES, AND HOUSING CONDITIONS THAT AFFECT FARM ANIMAL WELFARE

3.1. Adapting the Animal to Its Environment

Many procedures currently used in the production of animal-derived products (e.g., meat, leather, milk, and eggs) were introduced to counteract problems that emerged when livestock farming became intensified, such as tail docking in fattening pigs in response to tail biting [32], or beak trimming in response to feather pecking in laying hens [33,34]. Other procedures were introduced to improve the handling of animals (e.g., dehorning of cows, gestation crates for sows), or in response to consumer demands (e.g., castration of pigs to avoid boar taint [35]). Thus, in industrial animal farming animals are altered or "adapted" to meet the constraints caused by their housing conditions and the management practices used [36].

3.2. Mutilating Procedures

The aims of mutilating procedures are manifold, one of them being to "help" animals adapt to their environment. When discussing the welfare aspects of these procedures, it is important to distinguish between the welfare of the individual animal and the group as a whole. Consider the examples of beak-trimming in chicken [33,37] (see Figure 1), and the clipping of the canine teeth of pigs. Clipping causes discomfort to the individual pig but is thought to be beneficial to the group—theoretically, it prevents mammary injuries in sows and severe injury during fights with pen mates. However, so far there is no evidence that tooth clipping provides the envisaged benefit for the group [38,39,40]. There is an urgent need for such information, to determine whether benefits for the group outweigh the harm inflicted on the individual animal. Prophylactic mutilating procedures might be performed with the intention to reduce welfare risks. These risks, however, need to be assessed first, regarding their degree of impact (are all animals in the flock affected?), duration, intensity, and probability, which needs to be taken into consideration when determining whether the benefits for the group outweigh the suffering of individuals. From an animal welfare as well as a moral perspective, it would be unacceptable to perform 'unnecessary' mutilations on individual animals.

An overview of mutilating procedures, management practices, and housing conditions is provided in Table 1. The mutilating interventions in Table 1 (A) have already been addressed in a large number of scientific papers. We want to direct attention now to the housing and management practices causing predominantly physical, psychological, and emotional discomfort, and consequently we shall expand on entries in Table 1 (B, C). Not all points raised in these two parts of the table have already gained sufficient scientific attention. Generally speaking, the gaps in scientific knowledge will be closed as soon as there are scientifically valid translations to farming practices that have been proved to improve animal welfare.

Estimated occurrence: ✚In some management systems; ✚✚in many management systems; ✚ ✚✚in virtually all management systems. **a** In laying hens and in breeders of layers and broilers; **b** In a considerable percentage of two beef cattle breeds, Verbeterd Rood Bont ("Improved Red and White") and Belgian Blue, natural calving is nearly impossible, necessitating Caesarean sections. This concerns approximately 15% of the beef cattle stock in the Netherlands; **c** Calves in hutches or igloos; sows in crates; chickens in cages; **d** Absence of natural daylight and/or light intensity or regimens that do not facilitate a diurnal rhythm/resting period; **e** Sows, growing/finishing pigs if fed liquid feed; **f** Mainly in broiler breeders (to prevent high body weights), in layers to induce molting (infrequent), and water restriction in broilers to reduce wet litter (not allowed); **g** Boars (breeding); **h** Housing of dairy goats in deep litter (goats prefer solid and dry surfaces), indoor housing, no pasture. Mutilation is used here for (A) interventions or (B) housing conditions and management methods that produce acute and/or lasting physical discomfort.

Table 1. Mutilating and damaging interventions and housing/management practices, applied sporadically or routinely, which have the potential to impair farm animal welfare (a summary inspired by [2]).

	Interventions	Cattle	Sheep	Goats	Pigs	Chickens
A. Mutilating interventions causing physical discomfort	Castration	+		+	++	
	Docking (tail amputation)				+++	
	Disbudding	+++		+		
	Dehorning		+			
	Ear notching				+	
	Ear tagging—wing band—toe slit	+++	+++	+++	+++	+
	Teeth clipping				+	
	Nose ringing	+				
	Beak trimming					+++ [a]
	Despurring—toe clipping					+
	Dubbing					+
	Caesarean section	+ [b]				
B. Housing and management practices causing predominantly physical discomfort	Confinement [c]	+++			+++	+
	Lighting regimens—artificial lighting				++	+++ [d]
	Feed and water restriction				+ [e]	+++ [f]
C. Housing and management practices causing predominantly psychological and emotional discomfort	Overcrowding/social instability	++		++	++	+++
	(Repeated) mixing	+++	+	+++	+++	++
	Individual housing (of social animals)	+			+ [g]	+
	Early maternal separation	++		+++	+++	+++
	Barren environment			+ [h]	+++	++

In (C), housing and management methods are listed that lead to lasting psychological (emotional) discomfort. Of course, physical discomfort will also affect the emotional state, and thus the overall welfare of the animal, adversely. In turn, emotional discomfort can lead to physical discomfort. Note: The occurrence of these procedures has been estimated on the basis of the expert opinion of five veterinarians and two animal scientists of the Department of Farm Animal Health of the Faculty of Veterinary Medicine at the University of Utrecht. The mutilating procedures are those carried out in the EU. This list is not comprehensive and mutilations/suffering caused by unworkmanlike catching, transportation, slaughtering, bad stockmanship, and bad management, such as infrequent culling of suffering animals, infrequent claw trimming, tardy treatment, and housing conditions such as bad climate conditions, flooring or housing that causes injury or lameness, poor cleanliness, leading to dirty animals covered in manure, etc. are not included. They may occur under all housing conditions and management systems. Also, a number of interventions, such as hot or freeze branding and teeth grinding in sheep that are forbidden in the European Union are not on the list. Such interventions, however, may occasionally still be applied in Europe, and routinely in countries outside Europe.

3.3. Individual Housing of Social Animals and Confinement

Cattle: Calves are usually separated from the cow and housed individually shortly after birth. We will discuss the practice of housing newborn calves in hutches or igloos in detail in the paragraph about maternal separation.

Pigs: In industrial pig farming it is common practice to house farrowing sows in crates. There is some evidence that sows are able to cope with extremely adverse housing conditions, such as long-lasting tethering during gestations, farrowing, and suckling, a practice that is no longer permitted in The Netherlands (e.g., [41,42]). However, even if sows are able to cope with adverse housing conditions, their welfare may be impaired if the required adaptability of the animal is high and approaches the limits of adaptability of the animal [19]. Farrowing crates severely restrict freedom of movement and do not allow normal postural adjustments [43]. Farrowing crates are used in order to prevent piglets being crushed. The challenge is to find alternative systems, giving more freedom to the sow while maintaining a high piglet survival rate and taking the welfare of the sow and her piglets into account (e.g., [44,45]).

3.4. Lighting Regimens—Artificial Lighting

Pigs: Pigs are kept under a large range of light–dark periods and light intensities. Using operant preference tests, pigs are willing to work to switch on a light [46,47], but not to get access to darkness [46]. In both studies, pigs appeared to prefer lower light intensities. Pigs are sometimes kept in semidarkness in order to prevent aggressive behavior. However, this light condition impairs explorative behavior. The welfare of the animals might be improved by an appropriate light regimen, as found, for example, by Martelli et al. [48].

Chickens: A large variety of photoperiods are applied in broiler production, from continuous light to intermittent schedules with a number of successive light on–light off periods [49], based on the assumption that long light exposure will lead to maximal growth. Whereas their effects on performance have been addressed (e.g., [50]), the welfare consequences of these schedules have not yet been well established (a situation nearly unchanged since a review [51] about this topic, written 20 years ago). An optimized light–dark regimen combined with optimal stocking density may improve the health of broilers and positively affect their welfare, e.g., by reducing chronic fear [52]. Very low light intensities (1 lx) may impair welfare [53].

3.5. Feed and Water Restriction

So far, there is no detailed legislation regarding water supply for farm animals. However, scientific evidence for the effects of water restriction is constantly growing:

Pigs: Wet-fed pigs are generally water-restricted in order to reduce the total volume of slurry effluents. This could, however, negatively impact pig welfare. As Vermeer et al. [54] and Nannoni et al. [55] showed, wet-fed pigs are still motivated to obtain additional fresh water from drinkers.

Chickens: Feed restriction is routinely applied in broiler breeders in order to control growth and body mass [56]. These broiler breeders show high motivation to gain access to a foraging area, even if this area does not contain food [56]. They also show deficits in learning and memory: they perform worse in a Y-maze task than mildly food restricted control animals [57]. Indications of hunger are persistent and not reduced by varying feeding schedules (i.e., scattering feed once versus twice per day [58]). Food restriction also affects social interactions, as restricted male broilers show more aggressive behavior than male broilers fed ad libitum [59].

Water restriction, although forbidden, sometimes is applied in broilers to reduce wet litter, a main cause of footpad dermatitis [60]. Chickens may undergo water restriction resulting from adverse social interactions (e.g., chickens lower in the hierarchy being denied access to a limited water supply) or as a consequence of illness or injury preventing a chicken from accessing water. Layer hens deprived of water show a strong motivation to access water as well as remaining in the vicinity of drinking points after 12 h of water deprivation [61]. Increasing the number of drinkers can mitigate this problem.

3.6. Overcrowding/Social Instability

Pigs: Overcrowding in pigs forms a potential risk for the health and therefore the welfare of the animals. Enteric and respiratory disease due to reduced hygiene and air quality when stocking densities are high might occur, as, for example, discussed within the Scientific Opinion of the Panel on Animal Health and Welfare on a request from the Commission on Animal health and welfare in fattening pigs in relation to housing and husbandry [62]. Overcrowding might furthermore lead to problems like leg weakness and a higher prevalence for some claw lesions, as discussed e.g., by Jørgensen [63], and could make aggressive encounters more likely.

Chickens: Large stables, housing thousands of chickens, challenge and exceed the animals' capacity to individually recognize their stable- or pen-mates [64]. This may hamper the establishment of a hierarchy, which safeguards social stability in natural groups. However, unexpectedly, in very large groups aggression decreases [65], i.e., "non-aggressive strategies may be more efficient than fighting for resources when group size is large" ([65], p. 198).

On the other hand, it has been reported that overcrowding in broilers increases serum corticosterone levels [66]. The physical wellbeing of broilers in overcrowded conditions is compromised, as seen in increased injuries, reduced locomotion, and higher mortality rates [67] and in stress parameters such as an increased heterophil/lymphocyte ratio, which is considered to be a reliable stress marker [50].

However, in a large scale study by Dawkins and colleagues, other factors, such as the quality of stockmanship, temperature, and humidity in the stable appeared to affect the welfare of broilers more than stocking density (expect for very high stocking densities [68]). Determining an optimal group size and crowding density may thus contribute to improving welfare in broilers [69] and other farm animal species.

European legislations for adult laying hens dictate that hens in enriched cages must have at least 750 cm^2 of cage area per hen and for hens in alternative systems a maximum stocking density of nine hens per m^2 is allowed. However, there is no legislation in place for stocking densities during the rearing period of laying hens from hatching until approximately 18 weeks of age, which is a substantial part of the life of the laying hen [70]. Evidence-based legislation is urgently needed for the stocking densities during the rearing period that safeguard animal welfare.

3.7. Repeated Mixing

In general, separation from the group, regrouping (mixing), and (re-) introduction to their own or a new group in social animals such as cattle, sheep, goats, pigs, and chickens can negatively affect welfare. In particular, these practices interfere with social recognition, a cornerstone of group cohesiveness and group structure [71]. These species are able to individually recognize a large number of group members (chicken: [72], cattle: [73], goats: [74], sheep: [75], pigs: [76]), although very large groups or flocks may exceed their recognition capacity, which hampers establishing a hierarchy and induces social instability.

Goats: Patt and colleagues showed that separation of goats from and re-introduction in the group [77], and introduction of goats into established groups [78], is stressful. One measure to reduce the effects of separation and re-introduction may be to enable the (separated) goat to still hear and smell the other goats [77].

Pigs: In pigs, repeated mixing is common practice, sometimes driven by the desire to maintain very homogeneous groups per pen. After weaning, sows are returned to group housing. The weaned piglets are transferred to and usually mixed with piglets from other litters, which may lead to aggression [79]. Adverse effects of mixing may increase as a consequence of the intended abandoning of castrating male pigs [80]. Entire male pigs show a higher level of aggression than castrated pigs [81]. However, socializing pigs early [82] and keeping them in intact (sibling) groups until slaughter can significantly reduce aggression and stress [81,82] and, consequently, improve pig welfare.

A study by Fels and colleagues [83] showed that pigs are able to establish a linear (groups of six pigs) or quasi-linear (groups with 12 pigs) social hierarchy within three to four days after mixing, irrespective of whether groups were homo- or heterogeneous. Larger groups may retard or even hamper social hierarchy formation, maintain aggression, and impair welfare.

Chickens: Social mixing occurs in both layer and broiler chickens as they are transferred from one farm type to another: from a hatchery to a rearing farm, and then to a laying farm, in the case of layers; and from a hatchery to a growing farm in the case of broilers. Mixing unfamiliar chickens may provide an unwanted source of stress, as social hierarchies need to be re-established and group stability re-formed. The question of social mixing in chickens after hatching or in transport between farms has yet to be examined in controlled experiments. However, repeated social disruption has been shown to affect serotonergic and dopaminergic systems in the brain in layer hens [84]. Social mixing during the various phases of poultry farming may well be a stressor, though further research is needed to establish whether or not this is the case. Recent innovations in broiler housing may be able to reduce mixing in broiler chickens, such as the 'on-farm hatching concepts' that combine the hatching and brooding phase. This avoids mixing between hatching and rearing, increases hatchability, and reduces mortality in broiler chicks [85]. Furthermore, it provides the chicks with the opportunity to eat and drink directly after hatching.

3.8. Individual Housing (of Social Animals)

Pigs: Boars are housed individually in many sow stables to detect estrous expression in the sows [86]. These boars are usually kept in a small confinement near the group-housed sows. The boars can see and smell the neighboring sows, but cannot contact them. To our knowledge, the welfare consequences of this housing practice on the individually housed boars have not yet been investigated scientifically.

3.9. Early Maternal Separation

Cattle: In most intensive farming systems, dairy calves are separated from their mother in the period immediately after birth to a few days after birth and housed individually indoors in isolated barns, or outdoors, in hutches or igloos [87]. This practice is controversial [3]. The main reason for this practice is to reduce the risk of transferring infectious diseases to the newborns. The consequences of this practice for the welfare of cows and calves have been addressed in a number of publications. It appears that the stress response in both cow and calf is minimal if bonding is forestalled. Investigating the effects of separation starting 6 h, one day, four days, or two weeks after birth, Weary and colleagues [88,89] found that the longer the calf stayed with the cow before separation, the stronger the behavioral responses of both were. On the other hand, socializing the calf may profit from staying with the dam, preferentially in a group. This is one of the pillars of the "family herd" concept [90].

Goats: In goat farming, kids are usually separated from the doe within a few hours after birth, after the kids have ingested colostrum. To our knowledge, the putative welfare implications of early maternal separation for doe and kids have not yet been addressed scientifically.

Chickens, whether layers or broilers, in industrial farming will never see an adult chicken, but are usually hatched, reared, and kept in same-age (peer) groups. Under (semi-)natural conditions, chicks gain information about palatable food from their mother. This information may be relevant at least for chickens that have access to an outdoor run. In industrial farming, chicks are not given the opportunity to learn from an adult. The implications of this practice have not yet been studied systematically.

3.10. Weaning

Pigs: Pigs are usually weaned, i.e., withdrawn from milk supply by the sow and introduced to solid adult feed, at the relatively young age of four weeks. A less abrupt weaning at a higher age may improve the welfare of sow and piglets. During (an extended) pre-weaning period, piglets may learn from the sow to eat novel foods and to increase their intake of solid food. Interaction with the sow may also help reduce the development of damaging behaviors and increase play behavior after weaning [91]. Another measure described to improve welfare in piglets is pre-weaning socialization, in which barriers between farrowing pens are removed, allowing pre-weaning piglets from different litters to interact [92,93].

Cattle/sheep: Due to the common practice of separating the dairy calf from its mother shortly after birth, the weaning process predominantly consists of the transition to solid feed. In beef cattle and sheep, a two-stage weaning procedure has been developed in which the calf/lamb is prevented from drinking milk from its mother for a period of time before separation from the mother. In calves, a nose flap prevents the calf from drinking the cow's milk [94,95]; in sheep, the lamb is prevented from drinking the ewe's milk by covering the ewe's udder with a net [96]. Two-stage weaning may cause less stress and distress than the usual one-stage weaning process, although more research may be needed as the results are not yet unequivocal [94]. More research is also needed regarding alternative methods, e.g., Enriquez et al. [97].

3.11. Barren Environment

Pigs: The behavioral pattern of domesticated pigs is highly conserved and consists of the full behavioral repertoire of the wild boar [44,98,99]. Most pigs are kept in barren pens with a concrete and partially slatted floor. In these environments, pigs are usually provided with enrichment material such as chains and bite sticks [100]. However, pigs are highly motivated to root [101,102], a behavior that cannot be executed if the flooring substrate is absent. It has been found that the lack of rooting material directs pigs' behavior towards pen-mates and increases aggression, ear chewing, licking, and biting pen-mates, It also increased belly nosing, tail biting, and play fighting [103]. Provision of suited rooting material thus may help to control undesirable behavior in pigs.

Domesticated sows engage in complex nest-building behavior starting approximately 1½ to ½ days before farrowing [104,105]. In barren farrowing pens that do not provide substrate suited for nest building, this will translate to restless, abnormal behaviors that are likely redirected nesting behavior, whereas animals provided with substrate will build a nest [106]. Depriving pre-parturient sows of nest building impairs their welfare because it limits expression of their behavioral needs [104]. Moreover, sows in farrowing crates appear to engage in less maternal behavior than loose sows [107]. Considering the biological significance of nest-building behavior in sows, the farrowing systems should facilitate its expression [44]. Post-partum, crushing of piglets is a serious welfare problem. Whereas under (semi-) natural conditions, the sow will perform a behavioral sequence that will guide the piglets away from the area where the sow lies down, the lack of space in most farrowing crates does not allow this behavior, and the loss of piglets increases [108].

Chickens are reared in environments that can vary vastly, ranging from battery cages (currently banned in the EU but still standard practice in many other regions of the world) to furnished cages, aviary systems, and free-ranging. The complexity of the environment does seem to influence chicken welfare, as heterphil:lympocyte ratios, a common physiological measure of stress, are decreased in chickens raised in furnished compared to barren cages [109].

4. IMPROVING ANIMAL WELFARE

Kanis and colleagues [110] distinguished two main strategies for improving animal welfare: (1) selective breeding for desired traits (e.g., breeding for less expression of harmful social behaviors like aggression and savaging in pigs [111]) and (2) improvement of management routines and housing conditions. In general, the aim of breeding and genetic selection may be to decrease an animal's needs and/or to increase its abilities in an attempt to reduce possible negative consequences for welfare. Selective breeding, in combination with improved management practices and stockmanship and more appropriate housing, together with the abandonment of mutilating procedures and other practices that compromise animal welfare, is expected to improve and ensure farm animal welfare (see Figure 2). Animal scientists and veterinarians trained as welfare experts are highly suited to steer and supervise these changes, due to their expert knowledge of animal welfare, in addition to their close contact with farmers.

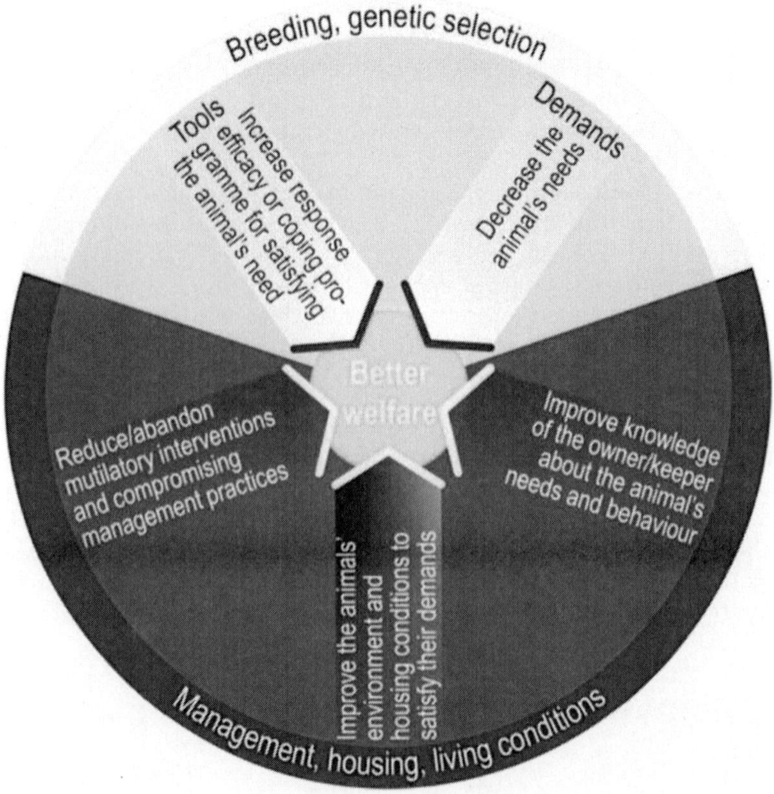

Figure 2. Strategies for improving animal welfare (inspired by [110]). Improving the animal's tools (abilities) to cope efficiently with its

environment or decreasing the animal's demands (needs) via breeding programs and genetic selection are goals of ongoing activities. In particular, the strategy of reducing the animal's demands by using genetic selection procedures is subject to ethical discussions [112,113]. Alternatively, one may reduce or abandon mutilating procedures, improve the animal's housing conditions, and improve the knowledge of owners/keepers/stockmen with respect to the animal's needs and demands. In particular, with respect to management, housing, and living conditions, animal welfare scientists and veterinarians should deploy their expertise, give (un)solicited advice, and assume an active role in the welfare discussion (see also Figure 3).

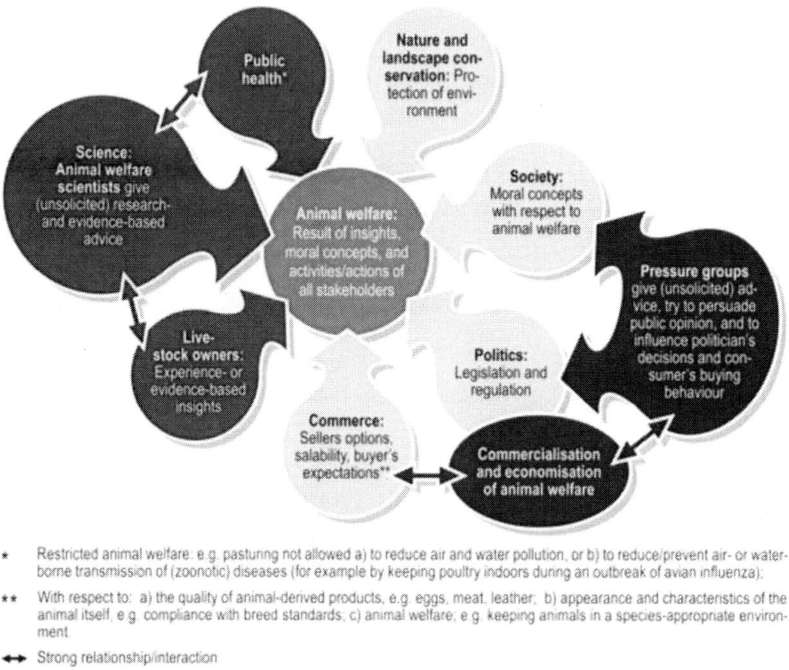

Figure 3. Stakeholders in animal welfare. Animal welfare is the result of (scientific) insights, of moral concepts of scientists and society at large [114,115,116], and of the activities and actions deployed by all stakeholders. Animal welfare scientists have an obligation to perform scientific research on animal health and welfare, to take action if animal welfare is compromised, and to give direction to future developments. Politicians are sensitive to the opinions and sentiments of pressure groups and society at large and may amend legislation and regulations. In parallel, the commercial value of animal welfare is increasingly being

recognized and creates added value for many animal-derived products (e.g., [117]) and services. Note: Veterinarians are involved in ensuring public health, and they are the primary contact persons about animal health/welfare with livestock/animal owners.

5. BREEDING AND SELECTION PROGRAMS

The aim of breeding and selection programs is to adapt animals to their environment [118] (see Figure 2), for example by increasing their adaptability [119]. Many traits that are relevant for biological functioning are under genetic control and are expected to respond to genetic selection [120,121]. Recently, Nicol and colleagues noted that "genetic selection and management strategies derived from a fundamental understanding of the basis of feather pecking behavior compare favorably with current practices of beak-trimming and light reduction as potential control methods" ([34], p. 776) Programs have been initiated to control and diminish harmful social behavior, such as feather pecking in poultry and tail biting in pigs, through selective breeding out of the undesired behaviors [111]. Kjaer and colleagues [122,123] and Rodenburg and colleagues [124,125] initiated selective breeding programs for laying hens to control and reduce feather pecking and cannibalism, and other programs focus on selecting out undesired traits such as aggression [125,126]. Over time, these approaches may indeed lead to a better match between the animal and its environment.

The selection of "robust" farm animals may have a similar effect. Robustness refers to "an increased adaptation to a range of environmental conditions" ([127], p. 343), i.e., an increase and/or restoration of the animal's coping ability or general adaptability, and improvement of its immune system and general health [128]. However, there is no specific "robustness" trait [127] and this strategy is not undisputed, especially for ethical reasons [112,129]. For example, such selection programs may produce unresponsive animals that do not express pain or discomfort even though they are suffering [112].

We suggest that the causes of undesired traits and behaviors should be investigated as the most relevant starting point for research. Then breeding programs could focus on these causes rather than on eliminating undesired traits and behaviors. As mentioned above, the latter approach may generate other, sometimes more urgent, welfare issues.

While selection programs focusing on susceptibility to stress, vitality, aggression, stereotypies [130,131], and sociality [121,132] may be societally desirable, they may not directly translate into economic advantage to the farmer [132]. However, as society is becoming more aware of, and verbal about, the welfare issues involved in industrial animal farming, these programs may provide an economic advantage in the long run. Veterinarians need to explain the long-term advantages to farmers, basing their arguments on scientific evidence.

6. ENABLING THE ANIMAL TO COPE WITH ITS ENVIRONMENT

A different approach from altering the animal that deserves more attention, and scientific substantiation, is to help animals to cope with their environment. A first step might be to create rearing conditions that better "prepare" animals for the environment in which they will be kept in later stages of their life. For example, chickens that are kept in aviaries should be reared in an environment that provides opportunities to learn to use perches [133]. Another option is to train animals to comply with management and husbandry procedures [134,135], such as training pigs to use automatic feeders [136].

7. ADAPTING THE ENVIRONMENT TO THE ANIMAL

7.1. Housing Conditions

Another approach is to adapt the animal's environment to meet its needs and to comply with animal welfare demands. This strategy was suggested more than 3½ decades ago by Faure [137] and has since been discussed by others (e.g., [138]). The artificial housing environment of production animals can be extremely constraining in terms of the animals' freedom to interact with their environment appropriately, for example because of restricted space, overcrowding, lack of retreat possibilities, high levels of noise, ultra- or infrasonic sound, light characteristics (level, spectrum, frequency), aversive routine husbandry, or abnormal social group compositions [139,140] (see Table 1 (B, C)). Consequently, "the farming system must be designed to fit the animal" ([141], p. 1481) and must ensure that animals can function properly.

The European Union took this approach when it made the social housing of gestating sows compulsory in an attempt to improve sow welfare [142]. The ban on gestation crates came into force in January 2013. While this system, with gestation pens or pens with electronic sow feeders, is undoubtedly more appropriate to the animal's needs than individual housing in gestation crates, aggressive encounters and injuries frequently occur in social housing (e.g., [143,144]) and welfare might be at risk. There is thus an urgent need to develop social housing systems in which aggressive encounters are reduced to a minimum (such as, for example, the comfort class concept for pigs: [145]).

It is not sufficient to ban practices, such as specific housing systems, without providing an alternative that constitutes a proper match between the animal's needs and its abilities to interact with an environment and the actual environmental demands and constraints. The improvement of housing conditions and systems to monitor welfare [27,146,147] are steps in the right direction to improving animal welfare by improving the animals' environment.

7.2. Management Practices

Management practices can have a tremendous effect on animal welfare. For example, good stockmanship and non-aversive handling methods can make animals less fearful and more productive [148,149]. Some practices are relatively easy to apply, such as well-maintained bedding in cubicles for dairy cows, which reduces lameness and swollen hocks [150,151,152,153]. The monitoring of lameness, body condition, and flight zone by veterinarians can help improve the welfare of animals on farms (see also [154]).

Veterinarians working in the field need to prevent obvious abuse and appropriate advice might reduce the incidence of lameness on dairy farms, but in many cases it is difficult to convince farmers to implement management changes that improve animal welfare [155]. This requires good interpersonal skills on the part of veterinarians so that they can explain to farmers the benefits of implementing management practices that improve the welfare of their animals. Veterinary education needs to prepare veterinarians for this role.

New concepts for the management and housing of farm animals have been developed, such as the "Rondeel", a round housing system for laying hens that consists of different wedges in which the animals are raised under conditions designed to fulfil the natural needs of the animals [156], and the "Comfort Class" concept for fattening pigs, [145]. "Comfort Class" is defined as a specific minimal level of husbandry conditions of animals, at which the ability of animals to meet their needs is not compromised by husbandry conditions" ([145], p. 172). However, the example of the "Rondeel" shows that years may pass before ideas and concepts are put into practice [156,157]. It is inevitable that adjustments are necessary when concepts are put into practice, but these modifications need to be evaluated scientifically to avoid basing modifications on gut feelings or common practice. Despite the best intentions, changes that are not evidence-based may impair rather than increase animal welfare.

8. WHERE TO GO FROM HERE?

Fundamental research into animal behavior and welfare is poorly funded, probably because it does not provide instant solutions for welfare problems as perceived by society. Yet, basic research is urgently needed to (re-)evaluate the effects of approaches to improve animal welfare, be it through improving the ability of an animal to adapt to its environment, by modifying the environment to better meet the needs of an animal, or by a combination of these approaches. Attempts to resolve serious mismatches between animal and environment are often made on a trial-and-error basis, and rarely shed light on the factors and processes underlying the problem. Unfortunately, many commonly used procedures to improve animal housing and management actually have adverse effects on the animal (see Table 1).

In order to assess an animal's welfare state, it is first necessary to have a thorough knowledge of the animal's normal behavioral repertoire and its behavioral needs and abilities. This will help farmers and stock keepers and veterinarians to identify constraints to an animal's needs and its ability to express natural behavior, constraints that may compromise the animal's welfare [158]. As early as 1978, Kilgour pointed to the need to obtain this basic information: "The ethogram of farm animals should have a high priority for current animal science goals" ([141], p. 1481). Unfortunately, comprehensive ethograms are still not available for farm animals, although ethograms for specific behavioral domains have been published (e.g., [159,160,161]). The behavior of the wild ancestor, or of feral conspecifics, is sometimes used to describe (part of) the behavioral repertoire of domesticated animals [162]. These behaviors are considered to represent the "normal" repertoire of a species. However, this view is based on a number of assumptions [163], which have not yet been sufficiently corroborated scientifically and which can be translated into the following questions:

- Is this "normal" behavior indeed unaltered by domestication?
- What are the effects of (heavy) selection on performance characteristics (e.g., high yield)?
- Does the wild ancestor, or a population that expresses the full pre-domestication genome, exist, and is it available for research?

The current lack of comprehensive ethograms is a serious obstacle to evaluating farm animal welfare, taking steps to improve welfare, and assessing the effects of these actions. In order to fully understand whether behaviors observed are normal or abnormal/harmful, it is imperative to have a full understanding of the behavioral repertoire of a species. This full description of the behavior of a species, in the form of a complete ethogram, is currently lacking for farm animals in general, despite the seminal work of describing, for example, the behavior of pigs in their natural environment [99]. Last but not least, the definition of welfare should integrate scientific insights and moral concepts [16,19], such as those of animal scientists and society at large, e.g., [164,165] (see Figure 3). Quantifiable measures or indicators, for example behavioral measures, are necessary to determine the current welfare state of an animal, to define a target value (based on knowledge about the animal's needs and abilities, i.e., ability to cope), and to monitor progress. Such target values should be updated to reflect new knowledge and insights into animal welfare and the role of the stakeholders in the welfare discussion [166]. The accomplished animal welfare is the result of insights, moral concepts, and activities/actions of all stakeholders (see Figure 3) (see, e.g., [167,168,169,170,171]).

In the final step, the translation of welfare concepts into farm animal practice, the veterinarian is at the forefront, directly advising farmers, steering and supervising changes. This key position makes it necessary to equip veterinarians with fundamental expertise, especially in animal behavior. This consideration has recently been adopted by the faculty of Veterinary Medicine at Utrecht University, The Netherlands, by establishing a curricular focus on animal behavior [172].

9. THE COMMERCIALIZATION OF ANIMAL WELFARE

The commercial potential of animal welfare has been increasingly noticed in recent years. Commercialization and economization may be achieved by awarding quality labels and selling labeled products at a higher price than unlabeled products. For example, the Dutch Animal Welfare Association (Dierenbescherming) awards one, two, or three "stars" of their "Beter leven keurmerk" ("Better Life Quality Mark") to animal-derived products from producers who fulfill the minimum requirements with respect to housing and managing the animals (See for costs of this quality label for the participating companies: [173]). Quality label awarding organizations such as the Dutch Animal Welfare Association generate income with this quality mark (e.g., in 2013:

748,000€ [174], in 2014: 858,000€ [175]). On the other hand, these organizations face considerable costs for controlling and ensuring the welfare level guaranteed by the label.

In a win-win situation—improving animal welfare, gaining money from awarding a welfare quality mark for the awarding organization, and higher prices for these products for the retailers—this may boost the introduction and implementation of measures that improve animal welfare. Selling welfare quality labeled animal-derived products will not only justify higher prices, but may also increase the retailer's reputation.

It can be a time-consuming and difficult process to reach an agreement between different stakeholders to improve animal welfare. However, once an agreement has been reached and implemented, it may be extremely difficult to put negotiations about new steps to further improve animal welfare on the agenda. It remains to be seen whether the economization and commercialization of animal welfare improves animal welfare in the long run.

10. CONCLUSIONS

Farm animals are the largest group of animals kept by humans, and thus the focus of welfare research is likely to stay on farm animals [6]. Procedures that compromise welfare and biological functioning should be identified, abandoned, and replaced by alternatives that do not adversely affect animal welfare. Strategies to meet this goal include selective breeding, improved stockmanship, as well as improving management routines, housing, and rearing conditions. The decision to apply certain management—and especially mutilating procedures—needs to be based on scientific evidence and evaluated against social norms and values. Thus, more research, in particular on the long-term consequences of such procedures, is urgently needed.

The responsibility of humans to safeguard the welfare of domesticated animals is directly associated with the animal's perceived ability to adapt to its environment. This may be influenced by selection/breeding processes, or by the animal husbandry system to which the animal is exposed. The involvement of many stakeholders in the discussion of farm animal welfare (Figure 3) is an added complication, because these stakeholders may have different perceptions of the importance of animal welfare [171,176] and of the measures needed to improve and ensure welfare. Veterinarians and animal scientists are key players who should have fundamental expertise in animal behavior and welfare as well as ethics [177]. They should closely cooperate [178] and should actively participate in directing future developments [7] in farm animal management and animal housing, with a view to improving animal welfare. In this, they should be guided by their professional codes of conduct and ethics (although these codes may require harmonization [179]) in terms of their role in ensuring and improving farm animal welfare. It remains to be seen whether commercialization will benefit animal welfare.

ACKNOWLEDGMENTS

We acknowledge the input of Gerrit Koop about mutilating procedures in small ruminants. We appreciate the useful comments and suggestions from Frauke Ohl, head of the Department of Animals in Science and Society—Animal Welfare & Laboratory Animal Sciences, who passed away much too young on 28 January 2016.

REFERENCES

1. Gregory, N.G. Animal welfare and the meat market. In *Animal Welfare and Meat Production*; Gregory, N.G., Grandin, T., Eds.; CABI: Wallingford, UK, 2007; pp. 1–21.
2. Stafford, K.J.; Mellor, D.J. Painful husbandry procedures in livestock and poultry. In *Improving Animal Welfare: A Practical Approach*; Grandin, T., Ed.; CAB International: Wallingford, UK, 2010; pp. 88–114.
3. Ventura, B.A.; von Keyserlingk, M.A.G.; Schuppli, C.A.; Weary, D.M. Views on contentious practices in dairy farming: The case of early cow-calf separation. *J. Dairy Sci.* **2013**, *96*, 6105–6116.
4. Lidfors, L.; Berg, C.; Algers, B. Integration of natural behavior in housing systems. *AMBIO J. Hum. Environ.* **2005**, *34*, 325–330.
5. Duncan, I.J.H. Science-based assessment of animal welfare: Farm animals. *Rev. Sci. Tech. Int. Off. Epizoot.* **2005**, *24*, 483–492.
6. Walker, M.; Díez-León, M.; Mason, G. Animal welfare science: Recent publication trends and future research priorities. *Int. J. Comp. Psychol.* **2014**, *27*, 80–100.
7. Edwards, J.D. The role of the veterinarian in animal welfare—A global perspective. In Proceedings of the Global Conference on Animal Welfare: An OIE Initiative, Paris, France, 23–25 February 2004; pp. 27–35.
8. Dawkins, M.S. The science of animal suffering. *Ethology* **2008**, *114*, 937–945.
9. Dawkins, M.S. Evolution and animal welfare. *Q. Rev. Biol.* **1998**, *73*, 305–328.
10. Fisher, M.W. Defining animal welfare—Does consistency matter? *N. Z. Vet. J.* **2009**, *57*, 71–73.
11. Fraser, D.; Duncan, I.J.C. "Pleasure", "pains" and animal welfare: Toward a natural history of affect. *Anim. Welf.* **1998**, *7*, 383–396.
12. Nielsen, H.M.; Olesen, I.; Navrud, S.; Kolstad, K.; Amer, P. How to consider the value of farm animals in breeding goals. A review of current status and future challenges. *J. Agric. Environ. Ethics* **2011**, *24*, 309–330.
13. Vanhonacker, F.; Verbeke, W.; Van Poucke, E.; Pieniak, Z.; Nijs, G.; Tuyttens, F. The concept of farm animal welfare: Citizen perceptions and stakeholder opinion in Flanders, Belgium. *J. Agric. Environ. Ethics* **2012**, *25*, 79–101.

14. Webster, J. Zoomorphism and anthropomorphism: Fruitful fallacies? *Anim. Welf.* **2011**, *20*, 29–36.
15. Mellor, D.J. Updating animal welfare thinking: Moving beyond the "five freedoms" towards "a life worth living". *Animals* **2016**, *6*, 21.
16. Webster, J. Critical control points in the delivery of improved animal welfare. *Anim. Welf.* **2012**, *21*, 117–123.
17. Nordenfelt, L. Health and welfare in animals and humans. *Acta Biotheor.* **2011**, *59*, 139–152.
18. Ohl, F.; Putman, R.J. Animal welfare at the group level: More than the sum of individual welfare? *Acta Biotheor.* **2014**, *62*, 35–45.
19. Ohl, F.; van der Staay, F.J. Animal welfare: At the interface between science and society. *Vet. J.* **2012**, *192*, 13–19.
20. Taylor, K.D.; Mills, D.S. Is quality of life a useful concept for companion animals? *Anim. Welf.* **2007**, *16*, 55–66.
21. Barnett, J.L.; Hemsworth, P.H. The validity of physiological and behavioural measures of animal welfare. *Appl. Anim. Behav. Sci.* **1990**, *25*, 177–187.
22. Fraser, D.; Weary, D.M.; Pajor, E.A.; Milligan, B.N. A scientific conception of animal welfare that reflects ethical concerns. *Anim. Welf.* **1997**, *6*, 187–205.
23. Korte, S.M.; Olivier, B.; Koolhaas, J.M. A new animal welfare concept based on allostasis. *Physiol. Behav.* **2007**, *92*, 422–428.
24. Boissy, A.; Manteuffel, G.; Jensen, M.B.; Moe, R.O.; Spruijt, B.; Forkman, B.; Dimitrov, I.; Langbein, J.; Bakken, M.; Veissier, I.; et al. Assessment of positive emotions in animals to improve their welfare. *Physiol. Behav.* **2007**, *92*, 375–397.
25. Van Eerdenburg, F.J.C.M.; Vázquez-Flores, S.; Saltijeral-Oaxaca, J.; Sossidou, E.N. A cow comfort monitoring scheme to increase the milk yield of a dairy farm. In *Livestock Housing*; Aland, A., Banhazi, T., Eds.; Wageningen Academic Publishers: Wageningen, The Netherlands, 2013; pp. 55–74.
26. Yeates, J.W.; Main, D.C.J. Assessment of positive welfare: A review. *Vet. J.* **2008**, *175*, 293–300.
27. Blokhuis, H.J.; Veissier, I.; Miele, M.; Jones, B. The welfare quality® project and beyond: Safeguarding farm animal well-being. *Acta Agric. Scand. Sect. Anim. Sci.* **2010**, *60*, 129–140.
28. Zanella, A. Knowledge is power. *Impact* **2016**, *2016*, 15–17.
29. Webster, J. Animal welfare: Freedoms, dominions and "a life worth living". *Animals* **2016**, *6*, 35.
30. FAWC (Farm Animal Welfare Council). *Report on Priorities for Animal Welfare Research and Development*; Ministry of Agriculture, Fisheries and Food: London, UK, 1993; p. 26.
31. Brambell, F.W.R.; Barbour, D.S.; Lady, B.; Ewer, T.K.; Hobson, A.; Pitchforth, H.; Smith, W.R.; Thorpe, W.H.; Winship, F.J.W. *Report of the Technical Committee to Enquire into the Welfare of Animals Kept under Intensive Livestock Husbandry Systems*; Her Majesty's Stationary Office: London, UK, 1965; p. 85.
32. Zupan, M.; Janczak, A.M.; Framstad, T.; Zanella, A.J. The effect of biting tails and having tails bitten in pigs. *Physiol. Behav.* **2012**, *106*, 638–644.
33. Jendral, M.J.; Robinson, F.E. Beak trimming in chickens: Historical, economical, physiological and welfare implications, and alternatives for preventing feather pecking and cannibalistic activity. *Avian Poult. Biol. Rev.* **2004**, *15*, 9–23.
34. Nicol, C.J.; Bestman, M.; Gilani, A.-M.; de Haas, E.N.; de Jong, I.C.; Lambton, S.; Wagenaar, J.P.; Weeks, C.A.; Rodenburg, T.B. The prevention and control of feather pecking: Application to commercial systems. *Worlds Poult. Sci. J.* **2013**, *69*, 775–788.
35. Prunier, A.; Bonneau, M.; von Borell, E.H.; Cinotti, S.; Gunn, M.; Fredriksen, B.; Giersing, M.; Morton, D.B.; Tuyttens, F.A.M.; Velarde, A. A review of the welfare consequences of surgical castration in piglets and the evaluation of non-surgical methods. *Anim. Welf.* **2006**, *15*, 277–289.
36. Aggrey, S.E. Modification of animals versus modification of the production environment to meet welfare needs. *Poult. Sci.* **2010**, *89*, 852–854.

37. Craig, J.V.; Lee, H.-Y. Beak trimming and genetic stock effects on behavior and mortality from cannibalism in white leghorn-type pullets. *Appl. Anim. Behav. Sci.* **1990**, *25*, 107–123.
38. Bates, R.O.; Hoge, M.D.; Edwards, D.B.; Straw, B.E. The influence of canine teeth clipping on nursing and nursery pig performance. *J. Swine Health Prod.* **2002**, *11*, 75–79.
39. Gallois, M.; Le Cozler, Y.; Prunier, A. Influence of tooth resection in piglets on welfare and performance. *Prev. Vet. Med.* **2005**, *69*, 13–23.
40. Van Beirendonck, S.; Driessen, B.; Verbeke, G.; Permentier, L.; van de Perre, V.; Geers, R. Improving survival, growth rate, and animal welfare in piglets by avoiding teeth shortening and tail docking. *J. Vet. Behav.* **2012**, *7*, 88–93.
41. Barnett, J.L.; Cronin, G.M.; Hemsworth, P.H.; Winfield, C.G. The welfare of confined sows: Physiological, behavioural and production responses to contrasting housing systems and handler attitudes. *Ann. Rech. Vét.* **1984**, *15*, 217–226.
42. Van der Staay, F.J.; Schuurman, T.; Hulst, M.; Smits, M.; Prickaerts, J.; Kenis, G.; Korte, S.M. Effects of recurrent chronic stress: A comparison between tethered and loose sows. *Physiol. Behav.* **2010**, *100*, 154–164.
43. Chidgey, K.L.; Morel, P.C.H.; Stafford, K.J.; Barugh, I.W. Observations of sows and piglets housed in farrowing pens with temporary crating or farrowing crates on a commercial farm. *Appl. Anim. Behav. Sci.* **2016**, *176*, 12–18.
44. Baxter, E.M.; Lawrence, A.B.; Edwards, S.A. Alternative farrowing systems: Design criteria for farrowing systems based on the biological needs of sows and piglets. *Animal* **2011**, *5*, 580–600.
45. Hales, J.; Moustsen, V.A.; Nielsen, M.B.F.; Hansen, C.F. Temporary confinement of loose-housed hyperprolific sows reduces piglet mortality. *J. Anim. Sci.* **2015**, *93*, 4079–4088.
46. Baldwin, B.A.; Start, I.B. Illumination preferences of pigs. *Appl. Anim. Behav. Sci.* **1985**, *14*, 233–243.
47. Taylor, N.; Prescott, N.; Perry, G.; Potter, M.; Le Sueur, C.; Wathes, C. Preference of growing pigs for illuminance. *Appl. Anim. Behav. Sci.* **2006**, *96*, 19–31.
48. Martelli, G.; Nannoni, E.; Grandi, M.; Bonaldo, A.; Zaghini, G.; Vitali, M.; Biagi, G.; Sardi, L. Growth parameters, behavior, and meat and ham quality of heavy pigs subjected to photoperiods of different duration. *J. Anim. Sci.* **2015**, *93*, 758–766.
49. Olanrewaju, H.A.; Thaxton, J.P.; Dozier, W.A., III; Purswell, J.; Roush, W.B.; Branton, S.L. A review of lighting programs for broiler production. *Int. J. Poult. Sci.* **2006**, *5*, 301–308.
50. Das, H.; Lacin, E. The effect of different photoperiods and stocking densities on fattening performance, carcass and some stress parameters in broilers. *Isr. J. Vet. Med.* **2014**, *69*, 211–220.
51. Manser, C.E. Effects of lighting on the welfare of domestic poultry: A review. *Anim. Welf.* **1996**, *5*, 341–360.
52. Sanotra, G.S.; Lund, J.D.; Vestergaard, K.S. In Influence of light-dark schedules and stocking density on behaviour, risk of leg problems and occurrence of chronic fear in broilers. *Br. Poult. Sci.* **2002**, *43*, 344–354.
53. Deep, A.; Schwean-Lardner, K.; Crowe, T.G.; Fancher, B.I.; Classen, H.L. Effect of light intensity on broiler behaviour and diurnal rhythms. *Appl. Anim. Behav. Sci.* **2012**, *136*, 50–56.
54. Vermeer, H.M.; Kuijken, N.; Spoolder, H.A.M. Motivation for additional water use of growing-finishing pigs. *Livest. Sci.* **2009**, *124*, 112–118.
55. Nannoni, E.; Martelli, G.; Cecchini, M.; Vignola, G.; Giammarco, M.; Zaghini, G.; Sardi, L. Water requirements of liquid-fed heavy pigs: Effect of water restriction on growth traits, animal welfare and meat and ham quality. *Livest. Sci.* **2013**, *151*, 21–28.
56. Dixon, L.M.; Brocklehurst, S.; Sandilands, V.; Bateson, M.; Tolkamp, B.J.; D'Eath, R.B. Measuring motivation for appetitive behaviour: Food-restricted broiler breeder chickens cross a water barrier to forage in an area of wood shavings without food. *PLoS ONE* **2014**, *9*, e102322.
57. Buckley, L.A.; McMillan, L.M.; Sandilands, V.; Tolkamp, B.J.; Hocking, P.M.; D'Eath, R.B. Too hungry to learn? Hungry broiler breeders fail to learn a Y-maze food quantity discrimination task. *Anim. Welf.* **2011**, *20*, 469–481.

58. De Jong, I.C.; Fillerup, M.; Blokhuis, H.J. Effect of scattered feeding and feeding twice a day during rearing on indicators of hunger and frustration in broiler breeders. *Appl. Anim. Behav. Sci.* **2005**, *92*, 61–76.
59. Mench, J.A. The development of aggressive behavior in male broiler chicks: A comparison with laying-type males and the effects of feed restriction. *Appl. Anim. Behav. Sci.* **1988**, *21*, 233–242.
60. De Jong, I.C.; Gunnink, H.; van Harn, J. Wet litter not only induces footpad dermatitis but also reduces overall welfare, technical performance, and carcass yield in broiler chickens. *J. Appl. Poult. Res.* **2014**, *23*, 51–58.
61. Rault, J.-L.; Cree, S.; Hemsworth, P. The effects of water deprivation on the behavior of laying hens. *Poult. Sci.* **2016**, *95*, 473–481.
62. Algers, B.; Blokhuis, H.J.; Broom, D.M.; Costa, P.; Domingo, M.; Greiner, M.; Guemene, D.; Hartung, J.; Koenen, F.; Muller-Graf, C.; et al. Animal health and welfare in fattening pigs in relation to housing and husbandry—Scientific opinion of the panel on animal health and welfare. *EFSA J.* **2007**, *564*, 1–14.
63. Jørgensen, B. Influence of floor type and stocking density on leg weakness, osteochondrosis and claw disorders in slaughter pigs. *Anim. Sci.* **2003**, *77*, 439–449.
64. D'Eath, R.B.; Keeling, L.J. Social discrimination and aggression by laying hens in large groups: From peck orders to social tolerance. *Appl. Anim. Behav. Sci.* **2003**, *84*, 197–212.
65. Estevez, I.; Andersen, I.-L.; Nævdal, E. Group size, density and social dynamics in farm animals. *Appl. Anim. Behav. Sci.* **2007**, *103*, 185–204.
66. Najafi, P.; Zulkifli, I.; Jajuli, N.A.; Farjam, A.S.; Ramiah, S.K.; Amir, A.A.; O'Reily, E.; Eckersall, D. Environmental temperature and stocking density effects on acute phase proteins, heat shock protein 70, circulating corticosterone and performance in broiler chickens. *Int. J. Biometeorol.* **2015**, *59*, 1577–1583.
67. Hall, A.L. The effect of stocking density on the welfare and behaviour of broiler chickens reared commercially. *Anim. Welf.* **2001**, *10*, 23–40.
68. Dawkins, M.S.; Donnelly, C.A.; Jones, T.A. Chicken welfare is influenced more by housing conditions than by stocking density. *Nature* **2004**, *427*, 242–244.
69. Škrbić, Z.; Pavlovski, Z.; Lukić, M.; Perić, L.; Milošević, N. The effect of stocking density on certain broiler welfare parameters. *Biotechnol. Anim. Husb.* **2009**, *25*, 11–21.
70. Council of the European Communities. *European Communities Council Directive 1999/74/EC of 19 July 1999 Laying Down Minimum Standards for the Protection of Laying Hens*; Council of the European Communities: Brussels, Belgium, 1999.
71. Miranda-de la Lama, G.C.; Mattiello, S. The importance of social behaviour for goat welfare in livestock farming. *Small Rumin. Res.* **2010**, *90*, 1–10.
72. Bradshaw, R.H. Discrimination of group members by laying hens Gallus domesticus. *Behav. Processes* **1991**, *24*, 143–151.
73. Coulon, M.; Deputte, B.; Heyman, Y.; Bedoin, C. Individual recognition in domestic cattle (*Bos taurus*): Evidence from 2D images of heads from different breeds. *PLoS ONE* **2009**, *4*, e4441.
74. Keil, N.M.; Imfeld-Mueller, S.; Aschwanden, J.; Wechsler, B. Are head cues necessary for goats (*Capra hircus*) in recognising group members? *Anim. Cogn.* **2012**, *15*, 913–921.
75. Kendrick, K.M.; da Costa, A.P.; Leigh, A.E.; Hinton, M.R.; Peirce, J.W. Sheep don't forget a face. *Nature* **2001**, *414*, 165–166.
76. Kristensen, H.H.; Jones, R.B.; Schofield, C.P.; White, R.P.; Wathes, C.M. The use of olfactory and other cues for social recognition by juvenile pigs. *Appl. Anim. Behav. Sci.* **2001**, *72*, 321–333.
77. Patt, A.; Gygax, L.; Wechsler, B.; Hillmann, E.; Palme, R.; Keil, N.M. Factors influencing the welfare of goats in small established groups during the separation and reintegration of individuals. *Appl. Anim. Behav. Sci.* **2013**, *144*, 63–72.
78. Patt, A.; Gygax, L.; Wechsler, B.; Hillmann, E.; Palme, R.; Keil, N.M. The introduction of individual goats into small established groups has serious negative effects on the introduced goat but not on resident goats. *Appl. Anim. Behav. Sci.* **2012**, *138*, 47–59.

79. Rhim, S.-J.; Son, S.-H.; Hwang, H.-S.; Lee, J.-K.; Hong, J.-K. Effects of mixing on the aggressive behavior of commercially housed pigs. *Asian-Australas. J. Anim. Sci.* **2015**, *28*, 1038–1043.
80. Backus, G.B.C.; van den Broek, E.; van der Fels, B.; Heres, L.; Immink, V.M.; Knol, E.F.; Kornelis, M.; Mathur, P.K.; van der Peet-Schwering, C.; van Riel, J.W.; et al. Evaluation of producing and marketing entire male pigs. *NJAS—Wagening. J. Life Sci.* **2016**, *76*, 29–41.
81. Fredriksen, B.; Lium, B.M.; Marka, C.H.; Mosveen, B.; Nafstad, O. Entire male pigs in farrow-to-finish pens—Effects on animal welfare. *Appl. Anim. Behav. Sci.* **2008**, *110*, 258–268.
82. Rydhmer, L.; Hansson, M.; Lundström, K.; Brunius, C.; Andersson, K. Welfare of entire male pigs is improved by socialising piglets and keeping intact groups until slaughter. *Animal* **2013**, *7*, 1532–1541.
83. Fels, M.; Hartung, J.; Hoy, S. Social hierarchy formation in piglets mixed in different group compositions after weaning. *Appl. Anim. Behav. Sci.* **2014**, *152*, 17–22.
84. Cheng, H.-W.; Fahey, A. Effects of group size and repeated social disruption on the serotonergic and dopaminergic system in two genetic lines of White Leghorn hens. *Poult. Sci.* **2009**, *88*, 2018–2025.
85. Van de Ven, L.J.F.; van Wagenberg, A.V.; Groot Koerkamp, P.W.G.; van den Brand, H. Effects of a combined hatching and brooding system on hatchability, chick weight, and mortality in broilers. *Poult. Sci.* **2009**, *88*, 2273–2279.
86. Kemp, B.; Soede, N.M.; Langendijk, P. Effects of boar contact and housing conditions on estrus expression in sows. *Theriogenology* **2005**, *63*, 643–656.
87. Marcé, C.; Guatteo, R.; Bareille, N.; Fourichon, C. Dairy calf housing systems across Europe and risk for calf infectious diseases. *Animal* **2010**, *4*, 1588–1596.
88. Flower, F.C.; Weary, D.M. Effects of early separation on the dairy cow and calf: 2. Separation at 1 day and 2 weeks after birth. *Appl. Anim. Behav. Sci.* **2001**, *70*, 275–284.
89. Weary, D.M.; Chua, B. Effects of early separation on the dairy cow and calf: 1. Separation at 6 h, 1 day and 4 days after birth. *Appl. Anim. Behav. Sci.* **2000**, *69*, 177–188.
90. Van Dixhoorn, I.; Evers, A.; Janssen, A.; Smolders, G.; Spoelstra, S.; Wagenaar, J.P.; Verwer, C. *Familiekudde State of Art*; Wageningen UR Livestock Research (Report 268): Wageningen, The Netherlands, 2010.
91. Oostindjer, M.; Kemp, B.; van den Brand, H.; Bolhuis, J.E. Facilitating "learning from mom how to eat like a pig" to improve welfare of piglets around weaning. *Appl. Anim. Behav. Sci.* **2014**, *160*, 19–30.
92. D'Eath, R.B. Socialising piglets before weaning improves social hierarchy formation when pigs are mixed post-weaning. *Appl. Anim. Behav. Sci.* **2005**, *93*, 199–211.
93. Morgan, T.; Pluske, J.; Miller, D.; Collins, T.; Barnes, A.L.; Wemelsfelder, F. Socialising piglets in lactation positively affects theirpost-weaning behaviour. *Appl. Anim. Behav. Sci.* **2014**, *158*, 23–33.
94. Enríquez, D.H.; Ungerfeld, R.; Quintans, G.; Guidoni, A.L.; Hötzel, M.J. The effects of alternative weaning methods on behaviour in beef calves. *Livest. Sci.* **2010**, *128*, 20–27.
95. Haley, D.B.; Bailey, D.W.; Stookey, J.M. The effects of weaning beef calves in two stages on their behavior and growth rate. *J. Anim. Sci.* **2005**, *83*, 2205–2214.
96. Norouzian, M.A. Effect of weaning method on lamb behaviour and weight gain. *Small Rumin. Res.* **2015**, *133*, 17–20.
97. Enríquez, D.; Hötzel, M.J.; Ungerfeld, R. Minimising the stress of weaning of beef calves: A review. *Acta Vet. Scand.***2011**, *53*, 28.
98. D'Eath, R.B.; Turner, S.P. The natural behaviour of the pig—chapter 2. In *The Welfare of Pigs*; Marchant-Forde, J.N., Ed.; Animal Welfare; Springer: New York, NY, USA, 2009; pp. 13–45.
99. Stolba, A.; Wood-Gush, D.G.M. The behaviour of pigs in a semi-natural environment. *Anim. Prod.* **1989**, *48*, 419–425.
100. Scott, K.; Taylor, L.; Gill, B.P.; Edwards, S.A. Influence of different types of environmental enrichment on the behaviour of finishing pigs in two different housing systems 1. Hanging toy versus rootable substrate. *Appl. Anim. Behav. Sci.* **2006**, *99*, 222–229.

101. Studnitz, M.; Jensen, M.B.; Pedersen, L.J. Why do pigs root and in what will they root? A review on the exploratory behaviour of pigs in relation to environmental enrichment. *Appl. Anim. Behav. Sci.* **2007**, *107*, 183–197.
102. Studnitz, M.; Jensen, K.H. Expression of rooting motivation in gilts following different lengths of deprivation. *Appl. Anim. Behav. Sci.* **2002**, *76*, 203–213.
103. Day, J.E.L.; Burfoot, A.; Docking, C.M.; Whittaker, X.; Spoolder, H.A.M.; Edwards, S.A. The effects of prior experience of straw and the level of straw provision on the behaviour of growing pigs. *Appl. Anim. Behav. Sci.* **2002**, *76*, 189–202.
104. Damm, B.I.; Lisborg, L.; Vestergaard, K.S.; Vanicek, J. Nest-building, behavioural disturbances and heart rate in farrowing sows kept in crates and Schmid pens. *Livest. Prod. Sci.* **2003**, *80*, 175–187.
105. Thodberg, K.; Jensen, K.H.; Herskin, M.S. Nest building and farrowing in sows: Relation to the reaction pattern during stress, farrowing environment and experience. *Appl. Anim. Behav. Sci.* **2002**, *77*, 21–42.
106. Hötzel, M.J.; Machado Filho, L.C.P.; Dalla Costa, O.A. Behaviour of pre-parturient sows housed in intensive outdoor or indoor systems. *Pesqui. Agropecuária Bras.* **2005**, *40*, 169–174.
107. Arey, D.S.; Sancha, E.S. Behaviour and productivity of sows and piglets in a family system and in farrowing crates. *Appl. Anim. Behav. Sci.* **1996**, *50*, 135–145.
108. Blackshaw, J.K.; Hagelsø, A.M. Getting-up and lying-down behaviours of loose-housed sows and social contacts between sows and piglets during day 1 and day 8 after parturition. *Appl. Anim. Behav. Sci.* **1990**, *25*, 61–70.
109. Matur, E.; Eraslan, E.; Akyazi, I.; Ekiz, E.E.; Eseceli, H.; Keten, M.; Metiner, K.; Bala, D.A. The effect of furnished cages on the immune response of laying hens under social stress. *Poult. Sci.* **2015**, *94*, 2853–2862.
110. Kanis, E.; van de Belt, H.; Groen, A.F.; Schakel, J.; de Greef, K.H. Breeding for improved welfare in pigs: A conceptual framework and its use in practice. *Anim. Sci.* **2004**, *78*, 315–329.
111. Turner, S.P. Breeding against harmful social behaviours in pigs and chickens: State of the art and the way forward. *Appl. Anim. Behav. Sci.* **2011**, *134*, 1–9.
112. D'Eath, R.B.; Conington, J.; Lawrence, A.B.; Olsson, I.A.S.; Sandøe, P. Breeding for behavioural change in farm animals: Practical, economic and ethical considerations. *Anim. Welf.* **2010**, *19*, 17–27.
113. Olsson, I.A.; Gamborg, C.; Sandøe, P. Taking ethics into account in farm animal breeding: What can breeding companies achieve? *J. Agric. Environ. Ethics* **2006**, *19*, 37–46.
114. Bayvel, A.C.D.; Cross, N. Animal welfare: A complex domestic and international public-policy issue—Who are the key players? *J. Vet. Med. Educ.* **2010**, *37*, 3–12.
115. Fraser, D. Understanding animal welfare. *Acta Vet. Scand.* **2008**, *50*, A1.
116. Miele, M.; Veissier, I.; Evans, A.; Botreau, R. Animal welfare: Establishing a dialogue between science and society. *Anim. Welf.* **2011**, *20*, 103–117.
117. Buller, H.; Roe, E. Modifying and commodifying farm animal welfare: The economisation of layer chickens. *J. Rural Stud.* **2014**, *33*, 141–149.
118. Olesen, I.; Groen, A.F.; Gjerde, B. Definition of animal breeding goals for sustainable production systems. *J. Anim. Sci.* **2000**, *78*, 570–582.
119. Faure, J.M.; Mills, A.D. Chapter 8: Improving the adaptability of animals by selection. In *Genetics and the Behavior of Domestic Animals*; Grandin, T., Deesing, M.J., Eds.; Elsevier: Amsterdam, The Netherlands, 2014; pp. 291–316.
120. Canario, L.; Mignon-Grasteau, S.; Dupont-Nivet, M.; Phocas, F. Genetics of behavioural adaptation of livestock to farming conditions. *Animal* **2013**, *7*, 357–377.
121. Ellen, E.D.; Rodenburg, T.B.; Albers, G.A.A.; Bolhuis, J.E.; Camerlink, I.; Duijvesteijn, N.; Knol, E.F.; Muir, W.M.; Peeters, K.; Reimert, I.; et al. The prospects of selection for social genetic effects to improve welfare and productivity in livestock. *Front. Genet.* **2014**, *5*, 377.
122. Kjaer, J.B.; Sørensen, P.; Su, G. Divergent selection on feather pecking behaviour in laying hens (*Gallus gallus domesticus*). *Appl. Anim. Behav. Sci.* **2001**, *71*, 229–239.

123. Grams, V.; Wellmann, R.; Preuß, S.; Grashorn, M.A.; Kjaer, J.B.; Bessei, W.; Bennewitz, J. Genetic parameters and signatures of selection in two divergent laying hen lines selected for feather pecking behaviour. *Genet. Sel. Evol.* **2015**, *47*, 77.
124. Nordquist, R.E.; Heerkens, J.L.T.; Rodenburg, T.B.; Boks, S.; Ellen, E.D.; van der Staay, F.J. Laying hens selected for low mortality: Behaviour in tests of fearfulness, anxiety and cognition. *Appl. Anim. Behav. Sci.* **2011**, *131*, 110–122.
125. Rodenburg, T.B.; Bijma, P.; Ellen, E.D.; Bergsma, R.; de Vries, S.; Bolhuis, J.E.; Kemp, B.; van Arendonk, J.A.M. Breeding amiable animals? Improving farm animal welfare by including social effects in breeding programmes. *Anim. Welf.* **2010**, *19*, 77–82.
126. King, T.; Marston, L.C.; Bennett, P.C. Breeding dogs for beauty and behaviour: Why scientists need to do more to develop valid and reliable behaviour assessments for dogs kept as companions. *Appl. Anim. Behav. Sci.* **2012**, *137*, 1–12.
127. Lawrence, A.B.; Conington, J. Sheep welfare: A future perspective. In *The Welfare of Sheep*; Dwyer, C.M., Ed.; Springer Science + Business Media B.V.: Dordrecht, The Netherlands, 2008; pp. 343–360.
128. Star, L.; Ellen, E.D.; Uitdehaaag, K.; Brom, F.W.A. A plea to implement robustness into a breeding goal: Poultry as an example. *J. Agric. Environ. Ethics* **2008**, *21*, 109–125.
129. Thompson, P.B. Why using genetics to address welfare may not be a good idea. *Poult. Sci.* **2010**, *89*, 814–821.
130. Baxter, E.M.; Jarvis, S.; Sherwood, L.; Farisha, M.; Roehea, R.; Lawrence, A.B.; Edwards, S.A. Genetic and environmental effects on piglet survival and maternal behaviour of the farrowing sow. *Appl. Anim. Behav. Sci.* **2011**, *130*, 28–41.
131. Turner, S.P.; D'Eath, R.B.; Roehe, R.; Lawrence, A.B. Selection against aggressiveness in pigs at re-grouping: Practical application and implications for long-term behavioural patterns. *Anim. Welf.* **2010**, *19*, 124–132.
132. Kanis, E.; de Greef, K.H.; Hiemstra, A.; van Arendonk, J.A.M. Breeding for societally important traits in pigs. *J. Anim. Sci.* **2005**, *83*, 948–957.
133. Gunnarsson, S.; Yngvesson, J.; Keeling, L.J.; Forkman, B. Rearing without early access to perches impairs the spatial skills of laying hens. *Appl. Anim. Behav. Sci.* **2000**, *67*, 217–228.
134. Wechsler, B.; Lea, S.E.G. Adaptation by learning: Its significance for farm animal husbandry. *Appl. Anim. Behav. Sci.***2007**, *108*, 197–214.
135. Young, R.J.; Cipreste, C.F. Applying animal learning theory: Training captive animals to comply with veterinary and husbandry procedures. *Anim. Welf.* **2004**, *13*, 225–232.
136. Figueroa, J.; Solà-Oriol, D.; Manteca, X.; Pérez, J.F. Social learning of feeding behaviour in pigs: Effects of neophobia and familiarity with the demonstrator conspecific. *Appl. Anim. Behav. Sci.* **2013**, *148*, 120–127.
137. Faure, J.M. To adapt the environment to the bird or the bird to the environment? In *The Laying Hen and Its Environment*; Moss, R., Ed.; Current Topics in Veterinary Medicine and Animal Science; Martinus Nijhoff Publishers: The Hague, The Netherlands, 1980.
138. Ferguson, D.M. Key features of "environmental fit" that promote good animal welfare in different husbandry systems. *Rev. Sci. Tech. Int. Off. Epizoot.* **2014**, *33*, 161–169.
139. D'Silva, J. Adverse impact of industrial animal agriculture on the health and welfare of farmed animals. *Integr. Zool.***2006**, *1*, 53–58.
140. Morgan, K.N.; Tromborg, C.T. Sources of stress in captivity. *Appl. Anim. Behav. Sci.* **2007**, *102*, 262–302.
141. Kilgour, R. The application of animal behavior and the humane care of farm animals. *J. Anim. Sci.* **1978**, *46*, 1478–1486.
142. Council of the Europen Union. *Council of the Europen Union Council Directive 2001/88/EC of 23 October 2001 Amending Directive 91/630/EEC Laying Down Minimum Standards for the Protection of Pigs*; Council of the Europen Union: Brussels, Belgium, 2001.
143. McGlone, J.J.; von Borell, E.H.; Deen, J.; Johnson, A.K.; Levis, D.G.M.; Meunier-Salaün, M.; Morrow, J.; Reeves, D.; Salak-Johnson, J.L.; Sundberg, P.L. Compilation of the scientific literature comparing housing systems for gestating sows and gilts using measures of physiology, behavior, performance, and health. *Prof. Anim. Sci.* **2004**, *20*, 105–117.

144. Anil, L.; Bhend, K.M.G.; Baidoo, S.K.; Morrison, R.; Deen, J. Comparison of injuries in sows housed in gestation stalls versus group pens with electronic sow feeders. *J. Am. Vet. Med. Assoc.* **2003**, *223*, 1334–1338.
145. De Greef, K.H.; Vermeer, H.M.; Houwers, H.W.J.; Bos, A.P. Proof of principle of the comfort class concept in pigs: Experimenting in the midst of a stakeholder process on pig welfare. *Livest. Sci.* **2011**, *139*, 172–185.
146. Blokhuis, H.J.; Fiks van Niekerk, T.; Bessel, W.; Elson, A.; Guémené, D.; Kjaers, J.B.; Maria Levrino, G.A.; Nicol, C.J.; Tauson, R.; Weeks, C.A.; et al. The LayWel project: Welfare implications of changes in production systems for laying hens. *World's Poult. Sci. J.* **2007**, *63*, 101–104.
147. Tuyttens, F.A.M.; Vanhonacker, F.; Van Poucke, E.; Verbeke, W. Quantitative verification of the correspondence between the Welfare Quality® operational definition of farm animal welfare and the opinion of Flemish farmers, citizens and vegetarians. *Livest. Sci.* **2010**, *131*, 108–114.
148. Hemsworth, P.H. Ethical stockmanship. *Aust. Vet. J.* **2007**, *85*, 194–200.
149. Rushen, J.; de Passillé, A.M. The importance of good stockmanship and its benefits for the animals. In *Improving Animal Welfare: A Practical Approach*; Grandin, T., Ed.; CAB International: Wallingford, UK, 2010; pp. 50–63.
150. Livesey, C.T.; Marsh, C.; Metcalf, J.A.; Laven, R.A. Hock injuries in cattle kept in straw yards or cubicles with rubber mats or mattresses. *Vet. Rec.* **2002**, *150*, 677–679.
151. Mowbray, L.; Vittie, T.; Weary, D.M. Hock lesions and free stall design: Effects of stall surface. In Proceedings of the Fifth International Dairy Housing Conference (ASAE), Fort Worth, TX, USA, 29–31 January 2003; pp. 288–295.
152. Potterton, S.L.; Green, M.J.; Harris, J.; Millar, K.M.; Whay, H.R.; Huxley, J.N. Risk factors associated with hair loss, ulceration, and swelling at the hock in freestall-housed UK dairy herds. *J. Dairy Sci.* **2011**, *94*, 2952–2963.
153. Van Gastelen, S.; Westerlaan, B.; Houwers, D.J.; van Eerdenburg, F.J.C.M. A study on cow comfort and risk for lameness and mastitis in relation to different types of bedding materials. *J. Dairy Sci.* **2011**, *94*, 4878–4888.
154. Grandin, T. (Ed.) *Improving Animal Welfare: A Practical Approach*; CABI: Wallingford, UK, 2010.
155. Whay, H.R.; Barker, Z.E.; Leach, K.A.; Main, D.C.J. Promoting farmer engagement and activity in the control of dairy cattle lameness. *Vet. J.* **2012**, *193*, 617–621.
156. Spoelstra, S.F.; Koerkamp, P.W.G.; Bos, A.P.; Elzen, B.; Feenstra, F.R. Innovation for sustainable egg production: Realigning production with societal demands in The Netherlands. *Worlds Poult. Sci. J.* **2013**, *69*, 279–298.
157. Klerkx, L.; van Bommel, S.; Bos, B.; Holster, H.; Zwartkruis, J.V.; Aarts, N. Design process outputs as boundary objects in agricultural innovation projects: Functions and limitations. *Agric. Syst.* **2012**, *113*, 39–49.
158. Botreau, R.; Bracke, M.B.M.; Perny, P.; Butterworth, A.; Capdeville, J.; van Reenen, C.G.; Veissier, I. Aggregation of measures to produce an overall assessment of animal welfare. Part 2: Analysis of constraints. *Animal* **2007**, *1*, 1188–1197.
159. Chielo, L.I.; Pike, T.; Cooper, J. Ranging behaviour of commercial free-range laying hens. *Animals* **2016**, *6*, 28.
160. Eklund, B.; Jensen, P. Domestication effects on behavioural synchronization and individual distances in chickens (*Gallus gallus*). *Behav. Processes* **2011**, *86*, 250–256.
161. Reimert, I.; Bolhuis, J.E.; Kemp, B.; Rodenburg, T.B. Indicators of positive and negative emotions and emotional contagion in pigs. *Physiol. Behav.* **2013**, *109*, 42–50.
162. Veasey, J.S.; Waran, N.K.; Young, R.J. On comparing the behavior of zoo housed animals with wild conspecifics as a welfare indicator. *Anim. Welf.* **1996**, *5*, 13–24.
163. Price, E.O. Behavioral aspects of animal domestication. *Q. Rev. Biol.* **1984**, *59*, 1–32.
164. Appleby, M.C.; Sandøe, P. Philosophical debate on the nature of well-being: Implications for animal welfare. *Anim. Welf.* **2002**, *11*, 283–294.
165. Haynes, R.P. Competing conceptions of animal welfare and their ethical implications for the treatment of non-human animals. *Acta Biotheor.* **2011**, *59*, 105–120.

166. Bock, B.; Buller, H. Healthy, happy and humane: Evidence in farm animal welfare policy. *Sociol. Rural.* **2013**, *53*, 390–411.
167. Averós, X.; Aparicio, M.A.; Ferrari, P.; Guy, J.H.; Hubbard, C.; Schmid, O.; Ilieski, V.; Spoolder, H.A.M. The effect of steps to promote higher levels of farm animal welfare across the EU. Societal versus animal scientists' perceptions of animal welfare. *Animals* **2013**, *3*, 786–807.
168. Christensen, T.; Lawrence, A.; Lund, M.; Stott, A.; Sandøe, P. How can economists help to improve animal welfare? *Anim. Welf.* **2012**, *21*, 1–10.
169. FAWC (Farm Animal Welfare Committee). *Economics and Farm Animal Welfare*; FAWC Advice to Government, Animal Welfare and Food and Farming Industry; Department for Environment, Food & Rural Affairs: London, UK, 2011.
170. Maciel, C.T.; Bock, B. Modern politics in animal welfare: The changing character of governance of animal welfare and the role of private standards. *Int. J. Sociol. Agric. Food* **2013**, *20*, 219–235.
171. Cornish, A.; Raubenheimer, D.; McGreevy, P. What we know about the public's level of concern for farm animal welfare in food production in developed countries. *Animals* **2016**, *6*, 74.
172. Ohl, F.; Endenburg, N.; Vaarkamp, H.; Pijpers, A.; Rothuizen, J.; Hellebrekers, L.J.; van Sluijs, F.J.; Stegeman, J.A.; van Putten, J.; van Knapen, F.; et al. Animal Welfare—The Veterinary Position. Available online: http://preview.tinyurl.com/zks467k (accessed on 21 February 2017).
173. Costs 2016, Dutch Animal Welfare Association (Dierenbescherming) Costs of Welfare Quality Label. Available online:https://beterleven.dierenbescherming.nl/ fileupload/ Kostendoorberekening_2016_versie_1.2.pdf (accessed on 9 April 2016).
174. Dutch Animal Welfare Association (Dierenbescherming) Report 2013, Annual Report 2013. Available online: http://www.cbf.nl//Uploaded_files/Jaarverslagen/jaarverslag-2013-dierenbescherming.pdf
175. Dutch Animal Welfare Association (Dierenbescherming) Report 2014, Annual Financial Statements 2014—One and a Half Century Standing in the Gap for Animals. Available online: http://www.cbf.nl//Uploaded_files/Jaarverslagen/jaarverlag-2014-dierenbe-scherming. 03e12a.pdf
176. Fraser, D. The globalisation of farm animal welfare. *Rev. Sci. Tech. Int. Off. Epizoot.* **2014**, *33*, 33–38.
177. AWARE FVE & EAEVE Adopted Report on European Veterinary Education in Animal Welfare Science, Ethics and Law—Executive Summary. Available online: http://www.fve.org/uploads/publications/docs/executive_summary_aw_day_one_comp etences_adopted.pdf
178. Christiansen, S.B.; Forkman, B. Assessment of animal welfare in a veterinary context—A call for ethologists. *Appl. Anim. Behav. Sci.* **2007**, *106*, 203–220.
179. Magalhães-Sant'Ana, M.; More, S.J.; Morton, D.B.; Osborne, M.; Hanlon, A. What do European veterinary codes of conduct actually say and mean? A case study approach. *Vet. Rec.* **2015**, *176*, 654.

CHAPTER 14

Animal Rights as a Mainstream Phenomenon

Bernard E. Rollin

Department of Philosophy, Colorado State University, Fort Collins, CO 80523, USA

SIMPLE SUMMARY

The twentieth century has witnessed a bewildering array of ethical revolutions, from civil rights to environmentalism to feminism. Often ignored is the rise of massive societal concern across the world regarding animal treatment. Regulation of animal research exists in virtually all western countries, and reform of "factory farming" is regnant in Europe and rapidly emerging in the United States. Opponents of concern for animals often dismiss the phenomenon as rooted in emotion and extremist lack of appreciation of how unrestricted animal use has improved human life. Such a view totally ignores the rational ethical basis for elevating legal protection for animals, as explained in this essay.

ABSTRACT

Businesses and professions must stay in accord with social ethics, or risk losing their autonomy. A major social ethical issue that has emerged in the past four decades is the treatment of animals in various areas of human use. Society's moral concern has outgrown the traditional ethic of animal cruelty that began in biblical times and is encoded in the laws of all civilized societies. There are five major reasons for this new social concern, most importantly, the replacement of husbandry-based agriculture with industrial agriculture. This loss of husbandry to industry has threatened the traditional fair contract between humans and animals, and resulted in significant amounts of animal suffering arising on four different fronts. Because such suffering is not occasioned by cruelty, a new ethic for animals was required to express social concerns. Since ethics proceed from preexisting ethics rather than ex nihilo, society has looked to its ethic for humans, appropriately modified, to find moral categories applicable to animals. This concept of legally encoded rights for animals has emerged as a plausible vehicle for reform.

Keywords: animal rights; animal husbandry; social ethic; anti-cruelty ethic; recollection

The last 50 years have witnessed a dazzling array of social ethical revolutions in Western society. Such moral movements as feminism, civil rights, environmentalism, affirmative action, consumer advocacy, pro- and anti-abortion activism, homosexual rights, children's rights, the student movement, antiwar activism, public rejection of biotechnology, have forever changed the way governments and public institutions comport themselves. This is equally true for private enterprise: to be successful, businesses must be seen as operating solidly in harmony with changing and emerging social ethics. It is arguable that morally based boycotting of South African business was instrumental in bringing about the end of apartheid, and similar boycotting of some farm products in the U.S. led to significant improvements in the living situations of farm workers. It is de rigeur for major corporations to have reasonable numbers of minorities visibly peopling their ranks, and for liquor companies to advertise on behalf of moderation in alcohol consumption. Cigarette companies now press upon the public a message that cigarettes kill, and extol their involvement in protecting battered women; and forestry and oil companies spend millions (even billions) to persuade the public of their environmental commitments. CNN reported that "green" investment funds grew significantly faster than ordinary funds, and reports of child labor or sweatshop working conditions can literally destroy product markets overnight.

Not only is success tied to accord with social ethics but, even more fundamentally, freedom and autonomy are as well. Every profession—be it medicine, law or agriculture—is given freedom by the social ethic to pursue its aims. In return, society basically says to professions it does not understand well enough to regulate, "you regulate yourselves the way we would regulate you if we understood what you do, which we don't. But we will know if you don't self-regulate properly and then we will regulate you, despite our lack of understanding." For example, some years ago, Congress became concerned about excessive use of antibiotic in animal feeds, and concluded that veterinarians were a major source of the problem. As a result, Congress was about to ban extra-label drug use by veterinarians, a move that would have killed veterinary medicine as we know it. However, through extensive efforts to educate legislators, such legislation did not proceed to law.

One major social ethical concern that has developed over the last four decades is a significant emphasis on the treatment of animals used by society for various purposes. It is easy to demonstrate the degree to which these concerns have seized the public imagination. According to both the U.S. National Cattlemen's Beef Association and the National Institutes of Health (the latter being the source of funding for the majority of biomedical research in the U.S.), both groups not inclined to exaggerate the influence of animal ethics, by the early 1990s Congress had been consistently receiving more letters, phone calls, faxes, e-mails and personal contacts on animal-related issues than on any other topic [1,2].

Whereas 30 years ago one would have found no bills pending in the U.S. Congress relating to animal welfare, recent years have witnessed dozens of such bills annually, with even more proliferating at the state level, as we shall explain in detail. The federal bills have ranged from attempts to prevent duplication in animal research, to saving marine mammals from becoming victims of tuna fishermen, to preventing importation of ivory, to curtailing the parrot trade. State laws passed in large numbers have increasingly prevented the use of live or dead shelter animals for biomedical research and training and have focused on myriad other areas of animal welfare. Eight states have abolished the steel-jawed leghold trap, as have some 90 countries [3]. When Colorado's politically appointed Wildlife Commission failed to act on a recommendation from the Division of Wildlife to abolish the spring bear hunt (because hunters were liable to shoot lactating mothers, leaving their orphaned cubs to die of starvation), the general public ended the hunt through a popular referendum. Seventy percent of Colorado's population voted for this as a constitutional amendment [4]. In Ontario, the environmental minister stopped a similar hunt by executive fiat in response to social ethical concern [5]. California abolished the hunting of mountain lions, and state fishery management agencies have been taking a hard look at catch-and-release programs on humane grounds [6].

In fact, wildlife managers have worried, in academic journals, about "management by referendum" for humane reasons. According to the director of the American Quarter Horse Association, the number of state bills related to horse welfare filled a telephone-book-sized volume in 1998 alone [7]. Public sentiment for equine welfare in California carried a bill through the state legislature making the slaughter of horses or shipping of horses for slaughter a felony in that state and the end of horse slaughter in the U.S. followed. Municipalities have passed ordinances ranging from the abolition of rodeos, circuses, and zoos to the protection of prairie dogs and, in the case of Cambridge, Massachusetts (a biomedical Mecca), the strictest laws in the world regulating research.

Even more dramatic, perhaps, is the worldwide proliferation of laws to protect laboratory animals. In the United States, for example, two major pieces of legislation, which I helped draft and defend before Congress, regulating and constraining the use and treatment of animals in research were passed by the U.S. Congress in 1985, despite vigorous opposition from the powerful biomedical research and medical lobbies. This opposition included well-financed, highly visible advertisements and media promotions indicating that human health and medical progress would be harmed by implementation of such legislation. There was even a less than subtle film titled "Will I Be All Right, Doctor?" the query coming from a sick child, the response coming from a pediatrician who affirmed, in essence, "You will be if 'they' leave us alone to do as we wish with animals." With social concern for laboratory animals unmitigated by such threats, research animal protection laws moved easily through Congress and have been implemented at considerable cost to taxpayers. When I testified before Congress on behalf of this law in 1982, a literature search in the Library of Congress turned up no papers in the scientific literature on laboratory animal analgesia and only two on animal analgesia, one of which said "there ought to be papers." Now there are over 11,000.

In 1986, Britain superseded its pioneering act of 1876 with new laws aimed at strengthening public confidence in the welfare of experimental animals [8]. Many other countries have moved or are moving in a similar direction, despite the fact that some 90% of laboratory animals are rats and mice, not the most cuddly and lovable of animals. Research on Great Apes has been truncated across the world.

Many animal uses seen as frivolous by the public have been abolished without legislation. Toxicological testing of cosmetics on animals has been truncated; companies such as the Body Shop have been wildly successful internationally by totally disavowing such testing, and free-range egg production is a growth industry across the Western world. Greyhound racing in the U.S. has declined, in part for animal welfare reasons, with the Indiana veterinary community spearheading the effort to prevent greyhound racing from coming into the state. Zoos that are little more than prisons for animals (the state of the art during my youth) have all but disappeared, and the very existence of zoos is being increasingly challenged, despite the public's unabashed love of seeing animals. And, as Gaskell and his associates' work has revealed [9], genetic engineering has been rejected in Europe not, as commonly believed, for reasons of risk but for reasons of ethics; in part for reasons of animal ethics. Similar reasons (i.e., fear of harming cattle) have, in part, driven European rejection of bovine somatotropin (BST). Rodeos such as the Houston Livestock Show have, in essence, banned jerking of calves in roping, despite opposition from the Professional Rodeo Cowboys Association, who themselves never show the actual roping of a calf on national television.

Inevitably, agriculture has felt the force of social concern with animal treatment—indeed, it is arguable that contemporary concern in society with the treatment of farm animals in modern production systems blazed the trail leading to a new ethic for animals. As early as 1965, British society took notice of what the public saw as an alarming tendency to industrialize animal agriculture by chartering the Brambell Commission, a group of scientists under the leadership of Sir Rogers Brambell, who affirmed that any agricultural system failing to meet the needs and natures of animals was morally unacceptable [10]. Though the Brambell Commission recommendations enjoyed no regulatory status, they served as a moral lighthouse for European social thought. In 1988, the Swedish Parliament passed, virtually unopposed, what the New York Times call a "Bill of Rights" for farm animals, abolishing in Sweden, in a series of timed steps, the confinement systems currently dominating North American agriculture [11]. Much of northern Europe has followed suit, and the European Union is moving in a similar direction, and sow stalls must be eliminated in by 2011 [12].

Although the U.S. has been a latecomer to agricultural issues, things have moved rapidly, with referenda pressed by HSUS abolishing sow stalls, battery cages, and veal crates across the U.S. My own work attests to this tendency. In 2007, over two days of dialogue, I convinced Smithfield Farms, the world's largest pork producer, to phase out gestation crates. In 2008, the Pew Commission, on which I served as the advocate for farm animal welfare, called for the end of high confinement animal agriculture within ten years, for reasons of animal welfare, environmental despoliation, human and animal health, and social justice. Most dramatically, I was able to broker an agreement between the Humane Society of the United States and the Colorado Livestock Association passing a jointly sponsored farm animal welfare law in Colorado in 2008, abolishing sow stalls and veal crates.

The agriculture community in the U.S. has been far behind societal concern. There is one monumental conceptual error that is omnipresent in the agricultural industry's discussions of animal welfare—an error of such magnitude that it trivializes the industry's responses to ever-increasing societal concerns about the treatment of agricultural animals. When one discusses farm animal welfare with industry groups or with the American Veterinary Medical Association, one finds the same response—animal welfare is solely a matter of "sound science".

Those of us serving on the Pew Commission, better known as the National Commission on Industrial Farm Animal Production, encountered this response regularly during our dealings with industry representatives. This commission studied intensive animal agriculture in the U.S. [13]. For example, one representative of the Pork Producers, testifying before the Commission, answered that while people in her industry were quite "nervous" about the Commission, their anxiety would be allayed were we to base all of our conclusions and recommendations on "sound science". Hoping to rectify the error in that comment, as well as educate the numerous industry representatives present, I responded to her as follows: "Madame, if we on the Commission were asking the question of how to raise swine in confinement, science could certainly answer that question for us. But that is not the question the Commission, or society, is asking. What we are asking is, ought we raise swine in confinement? And to this question, science is not relevant". Judging by her "huh", I assume I did not make my point.

Questions of animal welfare are at least partly "ought" questions, questions of ethical obligation. The concept of animal welfare is an ethical concept to which, once understood, science brings relevant data. When we ask about an animal's welfare, or about a person's welfare, we are asking about what we owe the animal, and to what extent. A document called the CAST report, first published by U.S. Agricultural scientists in the early 1980's, discussed animal welfare, it affirmed that the necessary and sufficient conditions for attributing positive welfare to an animal were represented by the animals' productivity. A productive animal enjoyed positive welfare; a non-productive animal enjoyed poor welfare [14].

This notion was fraught with many difficulties. First of all, productivity is an economic notion predicated of a whole operation; welfare is predicated of individual animals. An operation, such as caged laying hens may be quite profitable if the cages are severely over-crowded, yet the individual hens do not enjoy good welfare. Second, as we shall see, equating productivity and welfare is, to some significant extent, legitimate under husbandry conditions, where the producer does well if and only if the animals do well, and square pegs, as it were, are fitted into square holes with as little friction as possible (as when pigs live outside). Under industrial conditions, however, animals do not naturally fit in the niche or environment in which they are kept, and are subjected to "technological sanders" that allow for producers to force square pegs into round holes—antibiotics, feed additives, hormones, air handling systems—so the animals do not die and produce more and more kilograms of meat or milk. Without these technologies, the animals could not be productive. We will return to the contrast between husbandry and industrial approaches to animal agriculture.

The key point to recall here is that even if the CAST Report definition of animal welfare did not suffer from the difficulties we outlined, it is still an ethical concept. It essentially says "what we owe animals and to what extent is simply what it takes to get them to create profit". This in turn would imply that the animals are well-off if they have only food, water, and shelter, something the industry has sometimes asserted. Even in the early 80's, however, there were animal advocates and others who would take a very different ethical stance on what we owe farm animals. Indeed, the famous five freedoms articulated in Britain by the Farm Animal Welfare Council during the 1970's (even before the CAST Report) represents quite a different ethical view of what we owe animals, when it affirms that:

The welfare of an animal includes its physical and mental state and we consider that good animal welfare implies both fitness and a sense of well-being. Any animal kept by man, must at least, be protected from unnecessary suffering.

We believe that an animal's welfare, whether on farm, in transit, at market or at a place of slaughter should be considered in terms of '**five freedoms**':

1) **Freedom from Hunger and Thirst**—by ready access to fresh water and a diet to maintain full health and vigor.
2) **Freedom from Discomfort**—by providing an appropriate environment including shelter and a comfortable resting area.
3) **Freedom from Pain, Injury or Disease**—by prevention or rapid diagnosis and treatment.
4) **Freedom to Express Normal Behavior**—by providing sufficient space, proper facilities and company of the animal's own kind.
5) **Freedom from Fear and Distress**—by ensuring conditions and treatment which avoid mental suffering.

Clearly, the two definitions contain very different notions of our moral obligation to animals (and there is an indefinite number of other definitions). Which is correct, of course, cannot be decided by gathering facts or doing experiments—indeed which ethical framework one adopts will in fact determine the shape of science studying animal welfare.

To clarify: suppose you hold the view that an animal is well-off when it is productive, as per the CAST Report. The role of your welfare science in this case will be to study what feed, bedding, temperature, etc. are most efficient at producing the most meat, milk, or eggs for the least money—much what animal and veterinary science does today. On the other hand, if you take the FAWC view of welfare, your efficiency will be constrained by the need to acknowledge the animal's natural behavior and mental state, and to assure that there is minimal pain, fear, distress and discomfort—not factors in the CAST view of welfare unless they have a negative impact on economic productivity. Thus, in a real sense, sound science does not determine your concept of welfare; rather, your concept of welfare determines what counts as sound science!

The failure to recognize the inescapable ethical component in the concept of animal welfare leads inexorably to those holding different ethical views talking past each other. Thus, producers ignore questions of animal pain, fear, distress, confinement, truncated mobility, bad air quality, social isolation, and impoverished environment unless any of these factors impact negatively on the "bottom line". Animal advocates, on the other hand, give such factors primacy, and are totally unimpressed with how efficient or productive the system may be.

A major question obviously arises here. If the notion of animal welfare is inseparable from ethical components, and people's ethical stance on obligations to farm animals differ markedly across a highly diverse spectrum, whose ethic is to predominate and define, in law or regulation, what counts as "animal welfare"? It is to this issue we now turn.

What is the nature of the emerging new ethical thinking that underlies and informs the dramatic social changes just discussed? Although society has always had an articulated ethic regarding animal treatment, that ethic has been very minimalistic, leaving most of the issue of animal treatment to people's personal ethic, rather than to the social ethic. Since Biblical times, that limited social ethic has forbidden deliberate, willful, sadistic, deviant, purposeless, unnecessary infliction of pain and suffering on animals, or outrageous neglect, such as not feeding or watering. Beginning in the early nineteenth century, this set of prohibitions was articulated in the anti-cruelty statutes of the laws in all civilized societies [15]. But even in Biblical and medieval times, the social ethic inveighed against cruelty. The Old Testament injunctions against yoking an ox and an ass together to a plow, or muzzling the ox when it is being used to mill grain, or seething a calf in its mother's milk, all reflect concern with, and abhorrence for what the Rabbinical tradition called tsaar baalei chaiim; the suffering of living things. In the Middle Ages, St. Thomas Aquinas [16], while affirming that, lacking a soul, animals enjoyed no moral status, nonetheless strictly forbade cruelty, on the grounds that permitting such behavior towards animals would encourage its spreading to human beings, an insight buttressed by over two decades of recent research [17]. Numerous serial killers have evidenced early abusive behavior towards animals, as have many of the youths in the U.S. who in recent years wrought massacres on their peers.

For the overwhelming majority of human history, until some four decades ago, the anti-cruelty ethic served as the only socially articulated moral principle for animal treatment. Except for a few sporadic voices following in the wake of Darwin's articulation of human-animal continuity, no one spoke of animals' rights, nor did society have moral concepts for animal treatment that went "beyond cruelty." The obvious question that presents itself is this: What has occurred during the last half century which led to social disaffection with the venerable ethic of anti-cruelty and to strengthening of the anti-cruelty laws, which now make cruelty a felony in almost 40 states.

In a study commissioned by USDA to answer this question, I distinguished a variety of social and conceptual reasons [18]:

1) Changing demographics and consequent changes in the paradigm for animals:

   Whereas at the turn of the century, more than half the population was engaged in producing food for the rest, today only some 1.5% of the U.S. public is engaged in production agriculture[19]. One hundred years ago, if one were to ask a person in the street, urban or rural, to state the words that come into their mind when one says "animal", the answer would doubtless have been "horse", "cow", "food", "work", *etc.* Today, however, for the majority of the population, the answer is "dog", "cat", "pet". Repeated studies show that almost 100% of the pet-owning population views their animals as "members of the family" [20] and virtually no one views them as an income source. Divorce lawyers note that custody of the dog can be as thorny an issue as custody of the children!

2) We have lived through a long period of ethical soul-searching

   For almost 50 years society has turned its "ethical searchlight" on humans traditionally ignored or even oppressed by the consensus ethic—blacks, women, the handicapped, other minorities. The same ethical imperative has focused attention on our treatment of the non-human world—the environment and animals. Many leaders of the activist animal movement in fact have roots in earlier movements—civil rights, feminism, homosexual rights, children's rights, labor.

3) The media has discovered that "animals sell papers"

   One cannot channel-surf across normal television service without being bombarded with animal stories, real and fictional. (A *New York Times* reporter recently told me that more time on cable TV in New

York City is devoted to animals than to any other subject.) Recall, for example, the extensive media coverage a decade ago of some whales trapped in an ice-floe, and freed by a Russian ice-breaker. This was hardly an overflowing of Russian compassion—an oxymoronic notion applied to a people who gave us pogroms, the Gulag, and Stalinism. Rather, someone in the Kremlin was bright enough to realize that liberating the whales was an extremely cheap way to score points with U.S. public opinion.

4) Strong and visible arguments have been advanced in favor of raising the status of animals by philosophers, scientists and celebrities [21,22,23,24,25,26].

5) Changes in the nature of animal use demanded new moral categories

In my view, while all of the reasons listed above are relevant, they are nowhere nearly as important as the precipitous and dramatic changes in animal use that occurred after World War II. These changes were, first of all, huge conceptual changes in the nature of agriculture and second the rise of significant amounts of animal research and testing.

For virtually all of human history, animal agriculture was based foursquare in animal husbandry. Husbandry, derived from the old Norse word "hus/band," bonded to the household, meant taking great pains to put one's animals into the best possible environment one could find to meet their physical and psychological natures which, following Aristotle, I call telos [23], and then augmenting their ability to survive and thrive by providing them with food during famine, protection from predation, water during drought, medical attention, help in birthing, and so on. Thus traditional agriculture was roughly a fair contract between humans and animals, with both sides being better off in virtue of the relationship. Husbandry agriculture was about putting square pegs into square holes, round pegs into round holes, and creating as little friction as possible doing so. So powerful is the notion of husbandry, in fact, that when the Psalmist seeks a metaphor for God's ideal relationship to humans, he seizes upon the shepherd in the 23rd Psalm:

The Lord is my shepherd; I shall not want; He maketh me to lie down in green pastures; He leadeth me beside still waters; He restoreth my soul.

We wish no more from God than what the husbandman provides for his sheep. In husbandry, a producer did well if and only if the animals did well, so productivity was tied to welfare. No social ethic was thus needed to ensure proper animal treatment; only the anti-cruelty designed to deal with sadists and psychopaths was needed to augment husbandry. Self-interest virtually assured good treatment.

After World War II, this beautiful contract was broken by humans. Symbolically, at universities, Departments of Animal Husbandry became Departments of Animal Science, defined not as care, but as "the application of industrial methods to the production of animals" to increase efficiency and productivity. With "technological sanders"—hormones, vaccines, antibiotics, air-handling systems, mechanization—we could force square pegs into round holes, and place animals into environments where they suffered in ways irrelevant to productivity. If a nineteenth century agriculturalist had tried to put 100,000 egg-laying hens in cages in a building, they all would have died of disease in a month; today such systems dominate.

The new approach to animal agriculture was not the result of cruelty, bad character or even insensitivity. It developed rather out of perfectly decent, prima facie plausible motives that were a product of dramatic significant historical and social upheavals that occurred after World War II. At that point in time, agricultural scientists and government officials became extremely concerned about supplying the public with cheap and plentiful food for a variety of reasons. In the first place, after the Dust Bowl and the Great Depression, many people in the U.S. had soured on farming. Second, reasonable predictions of urban and suburban encroachment on agricultural land were being made, with a resultant diminution of land for food production. Third, many farm people had been sent to both foreign and domestic urban centers during the war, thereby creating a reluctance to return to rural areas that lacked excitement; recall the song of the 40's "How are you gonna keep 'em down on the farm now that they've seen Paree?" Fourth, having experienced the spectre of literal starvation during the Great Depression, the American consumer was, for the first time in history, fearful of an insufficient food supply. Fifth, projection of major population increases further fueled concern.

When the above considerations of loss of land and diminution of agricultural labor are coupled with the rapid development of a variety of technological modalities relevant to agriculture during and after World War II and with the burgeoning belief in technologically-based economics of scale, it was probably inevitable that animal agriculture would become subject to industrialization. This was a major departure from traditional agriculture and a fundamental change in agricultural core values—industrial values of efficiency and productivity replaced and eclipsed the traditional values of "way of life" and husbandry.

There is thus no question that industrialized agriculture, including animal agriculture, is responsible for greatly increased productivity. It is equally clear that the husbandry associated with traditional agriculture has changed significantly as a result of industrialization. One of my colleagues, a cow-calf cattle specialist, says that the worst thing that ever happened to his department is betokened by the name change from Animal Husbandry to Animal Science. No husbandry person would ever dream of feeding sheep meal, poultry waste, or cement dust to cattle, but such "innovations" are entailed by an industrial/efficiency mind-set.

In addition, in the mid-twentieth century there arose large scale use of animals in research and testing for toxicity. This too was an unprecedented large-scale use of animals, lacking the fairness of husbandry agriculture.

A moment's reflection on the development of large-scale animal research and high-technology agriculture elucidates why these innovations have led to the demand for a new ethic for animals in society. In a nutshell, these new developments represent a radically different playing field of animal use from the one that characterized most of human history; in the modern world of agriculture and animal research, the traditional anti-cruelty ethic grows increasingly less applicable. A thought experiment makes this clear. Imagine a pie chart that represents all the suffering that animals experience at human hands today. What percentage of that suffering is a result of intentional cruelty of the sort condemned by the anticruelty ethic and laws? When I ask my audiences this question—whether scientists, agriculturalists, animal advocates, or members of the general public—I always get the same response: only a fraction of 1 percent. Few people have ever witnessed overt, intentional cruelty, which is thankfully rare.

On the other hand, people realize that biomedical and other scientific research, toxicological safety testing, uses of animals in teaching, pharmaceutical product extraction from animals, and so on all produce far more suffering than does overt cruelty. This suffering comes from creating disease, burns, trauma, fractures, and the like in animals in order to study them; producing pain, fear, learned helplessness, aggression, and other states for research; poisoning animals to study toxicity; and performing surgery on animals to develop new operative procedures. In addition, suffering is engendered by the housing of research animals. Indeed, a prominent member of the biomedical research community has argued that the discomfort and suffering that animals used in research experience by virtue of being housed under conditions that are convenient for us, but inimical to their biological natures—for example, keeping rodents, which are nocturnal, burrowing creatures, in polycarbonate crates under artificial, full-time light—far exceed the suffering produced by invasive research protocols [27].

Now it is clear that farmers and researchers are not intentionally cruel—they are motivated by plausible and decent intentions: to cure disease, advance knowledge, ensure product safety, provide cheap and plentiful food. Nonetheless, they may inflict great amounts of suffering on the animals they use. Furthermore, the traditional ethic of anti-cruelty and the laws expressing it had no vocabulary for labeling such suffering, since researchers were not maliciously intending to hurt the animals. Indeed, this is eloquently marked by the fact that the cruelty laws exempt animal use in science and standard agricultural practices from their purview. Therefore, a new set of concepts beyond cruelty and kindness was needed to discuss the issues associated with burgeoning research animal use and industrial agriculture.

Society eventually became aware that new kinds of suffering were engendered by modern agriculture. Once again, producers could not be categorized as cruel, yet they were responsible for new types of animal suffering on at least four fronts:

1) Production diseases arise from the new ways the animals are produced. For example, liver abscesses in cattle are a function of certain animals' responses to the high-concentrate, low-roughage diet that characterizes feedlot production. (That is, of course, not the only cause of liver abscesses.) Although a certain percentage of the animals get sick and die, the overall economic efficiency of feedlots is maximized by the provision of such a diet. The ideas of a method of production creating diseases that were "acceptable" would be anathema to a husbandry agriculturalist.

2) The huge scale of industrialized agricultural operations and the small profit margin per animal militate against the sort of individual attention that typified much of traditional agriculture. In traditional dairies 50 years ago, one could make a living with a herd of 50 cows. Today, one needs literally thousands. In the U.S., dairies may have 10,000 cows.

3) Another new source of suffering in industrialized agriculture results from physical and psychological deprivation for animals in confinement: lack of space, lack of companionship for social animals, inability to move freely, boredom, austerity of environments, and so on. Since the animals evolved for adaptation to extensive environments but are now placed in truncated environments, such deprivation is inevitable. This was not a problem in traditional, extensive agriculture.

4) In confinement systems, workers may not be "animal smart"; the "intelligence," such as it is, is in the mechanized system. Instead of husbandmen, workers in swine factories are minimum wage, often animal-ignorant labor. So there is often no empathy with, or concern for, the animals.

These sources of suffering, like the ones in research, are again not captured by the vocabulary of cruelty, nor are they proscribed or even acknowledged by the laws based on the anti-cruelty ethic. Furthermore, they typically do not arise under traditional agriculture and its ethic of husbandry.

A few years ago, I experienced some sharply contracting incidents which dramatically highlight the moral difference between intensive and extensive agriculture. That particular year, Colorado cattle ranchers, paradigmatic exemplars of husbandry, were afflicted by a significant amount of scours. Over two months, I talked to a half dozen rancher friends of mine. Every single one had experienced trouble with scours, and every one had spent more on treating the disease than was economically justified by the calves' monetary value. When I asked these men why they were being what an economist would term "economically irrational," they were quite adamant in their response: "It's part of my bargain with the animal; part of caring for them," one of them said.

It is, of course, the same ethical outlook that leads ranch wives to sit up all night with sick marginal calves, sometimes for days in a row. If the issues were strictly economic, these people would hardly be valuing their time at 50 ¢ per hour—including their sleep time!

Now in contrast to these uplifting moral attitudes, consider the following: One of my animal scientist colleagues related to me that his son-in-law was an employee in a large, total confinement swine operation. As a young man he had raised and shown pigs, keeping them semi-extensively. One day he detected a disease among the feeder pigs in the confinement facility where he works, which necessitated killing them with a blow to the head, since this operation did not treat individual animals, their profit margin being allegedly too low. Out of his long established husbandry ethic, he came in on his own time with his own medicine to treat the animals. He cured them! Management's response was to fire him on the spot for violating company policy! He kept his job and escaped with a reprimand only when he was able to prove that he had expended his own—not the company's—resources. He continued to work for them, but felt that his health has suffered in virtue of what I have called the "moral stress" he experiences every day; the stress growing out of the conflict between what he is told to do and how he morally believes he should be treating the animals. Eventually, he left agriculture altogether. The above-detailed contrasting incidents, better than anything else I know, eloquently illustrate the large gap between the ethics of husbandry and industry. (Many confinement operations are run by accountants, not by animal science or animal husbandry people.)

Given that the old anti-cruelty ethic did not apply to animal research or confinement agriculture, society needed new ethical concepts to express its concern about these new uses. But ethical concepts do not arise ex nihilo.

Plato taught us a very valuable lesson about effecting ethical change. If one wishes to change another person's—or society's—ethical beliefs, it is much better to remind than to teach or, in my martial arts metaphor, to use judo rather than sumo. In other words, if you and I disagree ethically on some matter, it is far better for me to show you that what I am trying to convince you of is already implicit—albeit unnoticed—in what you already believe. Similarly, we cannot force others to believe as we do (sumo); we can, however, show them that their own assumptions, if thought through, lead to a conclusion different from what they currently entertain (judo). These points are well-exemplified in 20th century U.S. history. Prohibition was sumo, not judo—an attempt to forcefully impose a new ethic about drinking on the majority by the minority. As such, it was doomed to fail, and in fact people drank more during Prohibition. Contrast this with Lyndon Johnson's civil rights legislation. As himself a Southerner, Johnson realized that even Southerners would acquiesce to the following two propositions:

1) All humans should be treated equally, and black people were human—they just had never bothered to draw the relevant conclusion.

2) If Johnson had been wrong about this point, if "writing this large" in the law had not "reminded" people, civil rights would have been as ineffective as Prohibition!

So society was faced with the need for new moral categories and laws that reflect those categories in order to deal with animal use in science and agriculture and to limit the animal suffering with which it is increasingly concerned. At the same time, recall that western society has one through almost fifty years of extending its moral categories for humans to people who were morally ignored or invisible—women, minorities, the handicapped, children, citizens of the third world. As we noted earlier, new and viable ethics do not emerge ex nihilo. So a plausible and obvious move is for society to continue in its tendency and attempt to extend the moral machinery it has developed for dealing with people, appropriately modified, to animals. And this is precisely what has occurred. Society has taken elements of the moral categories it uses for assessing the treatment of people and is in the process of modifying these concepts to make them appropriate for dealing with new issues in the treatment of animals, especially their use in science and confinement agriculture.

What aspect of our ethic for people is being so extended? One that is, in fact, quite applicable to animal use, is the fundamental problem of weighing the interests of the individual against those of the general welfare. Different societies have provided different answers to this problem. Totalitarian societies opt to devote little concern to the individual, favoring instead the state, or whatever their version of the general welfare may be. At the other extreme, anarchical groups such as communes give primacy to the individual and very little concern to the group—hence they tend to enjoy only transient existence. In our society, however, a balance is struck. Although most of our decisions are made to the benefit of the general welfare, fences are built around individuals to protect their fundamental interests from being sacrificed to the majority. Thus we protect individuals from being silenced even if the majority disapproves of what they say; we protect individuals from having their property seized without recompense even if such seizure benefits the general welfare; we protect individuals from torture even if they have planted a bomb in an elementary school and refuse to divulge its location. We protect those interests of the individual that we consider essential to being human, to human nature, from being submerged, even by the common good. Those moral/legal fences that so protect the individual human are called rights and are based on plausible assumptions regarding what is essential to being human.

It is this notion to which society in general is looking in order to generate the new moral notions necessary to talk about the treatment of animals in today's world, where cruelty is not the major problem but where such laudable, general human welfare goals as efficiency, productivity, knowledge, medical progress, and product safety are responsible for the vast majority of animal suffering. People in society are seeking to "build fences" around animals to protect the animals and their interests and natures from being totally submerged for the sake of the general welfare, and are trying to accomplish this goal by going to the legislature. In husbandry, this occurred automatically; in industrialized agriculture, where it is no longer automatic, people wish to see it legislated.

It is necessary to stress here certain things that this ethic, in its mainstream version, is not and does not attempt to be. As a mainstream movement, it does not try to give human rights to animals. Since animals do not have the same natures and interests flowing from these natures as humans do, human rights do not fit animals. Animals do not have basic natures that demand speech, religion, or property; thus according them these rights would be absurd. On the other hand, animals have natures of their own and interests that flow from these natures, and the thwarting of these interests matters to animals as much as the thwarting of speech matters to humans. The agenda is not, for mainstream society, making animals have the same rights as people. It is rather preserving the common-sense insight that "fish gotta swim and birds gotta fly," and suffer if they don't.

This new ethic is conservative, not radical, harking back to the animal use that necessitated and thus entailed respect for the animals' natures. It is based on the insight that what we do to animals matters to them, just as what we do to humans matters to them, and that consequently we should respect that mattering in our treatment of use of animals as we do in our treatment and use of humans. And since respect for animal nature is no longer automatic as it was in traditional husbandry agriculture, society is demanding that it be encoded in law. Significantly, in 2004, no fewer than 2,100 bills pertaining to animal welfare were proposed in U.S. state legislatures. 90+ law schools now teach animal law.

With regards to animal agriculture, the pastoral images of animals grazing on pasture and moving freely are iconic. As the 23rd Psalm indicates, people who consume animals wish to see the animals live decent lives, not lives of pain, distress and frustration. It is for this reason in part that industrial agriculture conceals the reality of its practices from a naïve public—witness Perdue's advertisements about raising "happy chickens", or the California "happy cow" ads. As ordinary people discover the truth, they are shocked. When I served on the Pew Commission and other commissioners had their first view of sow stalls, many were in tears and all were outraged.

Just as our use of people is constrained by respect for the basic elements of human nature, people wish to see a similar notion applied to animals. Animals, too, have natures, what I call telos following Aristotle—the "pigness of the pig", the "cowness of a cow". Pigs are "designed" to move about on soft loam, not to be in gestation crates. If this no longer occurs naturally, as it did in husbandry, people wish to see it legislated. This is the mainstream sense of "animal rights".

As property, strictly speaking, animals cannot have legal rights. But a functional equivalent to rights can be achieved by limiting human property rights. When I and others drafted the U.S. federal laws for laboratory animals, we did not deny that research animals were the property of researchers. We merely placed limits on their use of their property. I may own my car, but that does not mean I can drive it on the sidewalk or at any speed I choose. Similarly, our law states that if one hurts an animal in research, one must control pain and distress. Thus research animals can be said to have the right to have their pain controlled. [27]

In the case of farm animals, people wish to see their basic needs and nature, teloi, respected in the systems that they are raised. Since this no longer occurs naturally as it did in husbandry, it must be imposed by legislation or regulation. A Gallup poll conducted in 2003 shows that 75% of the public wants legislated guarantees of farm animal welfare [28]. This is what I call "animal rights as a mainstream phenomenon".

Legal codification of rules of animal care respecting animal telos is thus the form animal welfare takes where husbandry has been abandoned.

Thus, in today's world, the ethical component of animal welfare prescribes that the way we raise and use animals must embody respect and provision for their psychological needs and natures. It is therefore essential that industrial agriculture phase out those systems which cause animal suffering by violating animals' natures and replace them with systems respecting their natures.

REFERENCES AND NOTES

1. McCarthy, C. National Institutes of Health: Bethesda, MD, USA; *Personal communication*; 1988.
2. NCBA: Denver CO USA; *Personal communication*; 1991.
3. HSUS. 2003. Available online: http://www.hsus.org/ace/12031/ (accessed on 29 December 2010).
4. Denver Post, 19 June 1994; P. D1.
5. Animal People, Clinton, WA, USA, March 1999.
6. Laitenschloger, R.A.; Bowyer, R.T. Wildlife management by referendum: When professionals fail to communicate. *Wildlife Soc. Bull.* **1985**, *13*, 564–570.
7. Houston Livestock Show and Rodeo. 1998. Speech delivered in Equine section.
8. Animal legislation. Animals in Scientific Procedures Act. 2003. Available online: http://www.homeoffice.gov.uk/docs/animallegisltation.html/ (accessed on 29 December 2010).
9. Gaskell, G.; Bauer, M.V.; Durant, J. Europe Ambivalent on Biotechnology. *Nature* **1997**, *387*, 845–847.
10. Brambell, F.W.R. *Report of the Technical Committee to Enquire into the Welfare of Animals Kept under Intense Livestock Husbandry Systems*; HMSO: London, UK, 1965.
11. Lohr, S. Swedish Farm Animals Get a Bill of Rights. In *New York Times*; 25; October; 1988; p. 1.
12. *Welfare of Dogs*; EU Health and Consumer Protection Directorate: Luxemburg; June 19; 2001.
13. Putting Meat on the Table. Available online: www.pcifap.org (accessed on 4 January 2011).
14. *Scientific Aspects of the Welfare of Food Animals*; Report #91; Council for Agricultural Science and Technology (CAST): Ames, IA, USA, 1981.
15. Leavitt, E.S. *Animals and their Legal Rights*; Animal Welfare Institute: Washington, DC, USA, 1978.
16. Aquinas, T. *On the Truth of the Catholic Faith*; Doubleday: Providence, NY, USA, 1956.
17. Ascione, F.; Arkow, P. *Child Abuse, Domestic Violence, and Animal Abuse*; Purdue University Press: West Lafayette, IN, USA, 1999.
18. Rollin, B. *Farm Animal Welfare: Social, Bioethical and Research Issues*; Iowa State University Press: Ames, IA, USA, 1995.
19. AMC. 18th Agricultural Machinery Conference, Cedar Rapids, IA, USA, 5–7 May 2003; Available online: http://www.amc-online.org/ (accessed on 29 December 2010).
20. Survey says pets are members of the family. The Acorn 31 January 2002.
21. Singer, P. *Animal Liberation*; New York Review Press: New York, NY, USA, 1975.
22. Rollin, B. *Animal Rights and Human Morality*, 1st ed.; Prometheus Books: Buffalo, NY, USA, 1981.

23. Rollin, B. *Animal Rights and Human Morality*, 2nd ed.; Prometheus Books: Buffalo, NY, USA, 1992.
24. Rollin, B. *Animal Rights and Human Morality*, 3rd ed.; Prometheus Books: Buffalo, NY, USA, 2006.
25. Regan, T. *The Case for Animal Rights*; University of California Press: Berkeley, CA, USA, 1983.
26. Sapontzis, S. *Morals, Reason and Animals*; Temple University Press: Philadelphia, PA, USA, 1987.
27. Wolfle, T. NIH: *Personal Communication*; 1985.
28. Gallup. 2003. Available online: http://www.gallup.com/ (accessed on 29 December 2010).

CHAPTER 15

Pet husbandry and infection control practices related to zoonotic disease risks in Ontario, Canada

Jason W Stull[1,2,4,*], Andrew S Peregrine[1,2], Jan M Sargeant[2,3] and J Scott Weese[1,2]

[1]*Department of Pathobiology, Ontario Veterinary College, University of Guelph, Guelph, Ontario N1G2W1, Canada*

[2]*Centre for Public Health and Zoonoses, University of Guelph, Guelph, Ontario N1G2W1, Canada*

[3]*Department of Population Medicine, Ontario Veterinary College, University of Guelph, Guelph, Ontario N1G2W1, Canada. 4Current address: Department of Veterinary Preventive Medicine, The Ohio State University, College of Veterinary Medicine, 1920 Coffey Rd., Columbus, OH 43210, USA*

ABSTRACT

Background

Many human infections are transmitted through contact with animals (zoonoses), including household pets. Despite this concern, there is limited knowledge of the public's pet husbandry and infection control practices. The objective of this study was to characterize zoonotic disease related-husbandry and infection preventive practices in pet-owning households in Ontario, Canada.

Methods

A self-administered questionnaire was distributed to individuals at two multi-physician clinics in Waterloo, Ontario, Canada during 2010. One adult from each household was invited to participate in the study.

Results

Four hundred one pet-owners completed the questionnaire. Households reported ownership of dogs (68%), cats (48%), fish (13%), exotic mammals (7%), such as hamsters, and reptiles and birds (each 6%). Across all species, individuals at higher risk of infections (i.e. < 5yrs, ≥ 65yrs, immunocompromised) were often (46-57%) present in households. Children < 16 yrs of age had close pet contact, as households reported dogs (13%) and cats (30%) usually slept in a child's bed and dogs often licked a child's face (24%). Household husbandry practices that increase zoonotic disease risk were frequently identified; some fed high-risk foods (i.e. raw eggs, raw meat, or raw animal product treats) to their dogs (28%) or cats (3%); 14% of reptile-owning households allowed the pet to roam through the kitchen or washed it in the kitchen sink. Reported hand washing by children was high for all species (> 76% washed hands sometimes or greater after touching the pet, its feces, or housing), although fewer reported children always washed their hands (3-57%; by species). With a few exceptions, practices were not associated with the presence of higher risk members in the household or recall of having previously received zoonotic disease education.

Conclusions

The results suggest there is a need for education on zoonotic disease prevention practices for pet-owning households with individuals at higher risk of infection and those with high-risk species (e.g., reptiles). Further research is needed to determine the role of education in altering higher risk pet practices.

Keywords: Zoonoses, Pet Education, Infection, Husbandry, Canada

1. BACKGROUND

Pet ownership is common, with the majority of households in many developed countries owning pets [1-8]. A recent study estimated that 56% of Canadian homes have at least one dog or cat, with a minority owning fish (12%), birds (5%), rabbits or hamsters (each 2%), lizards, guinea pigs, snakes, frogs, turtles, ferrets, or gerbils (each 1%) [6]. Pet ownership has been shown to have mental and physical benefits [9], particularly among children, the elderly and immunocompromised individuals [10-16]. However, despite these benefits, there are also potential health hazards associated with pet ownership and contact.

A recent estimate suggests that 14% of all human illness caused by seven common enteric pathogen groups (i.e., Campylobacter spp, Cryptosporidium spp, Shiga toxin–producing Escherichia coli (STEC) O157, STEC non-O157, Listeria monocytogenes, nontyphoidal Salmonella spp., and Yersinia enterocolitica) are attributable to animal contact [17]. Although the proportion of human disease for which pets are specifically responsible is largely unknown, greater than 70 pathogens of companion animals are known to be transmissible to people [18] and pet contact has been clearly established as a risk factor for illness caused by various pathogens, including Campylobacter[19, 20], Salmonella[21], Pasteurella multocida[22] and Capnocytophaga canimorsus[23]. For some pet species, their role in human disease has been more clearly identified; reptiles and amphibians are estimated to be responsible for 6% of all sporadic Salmonella infections (11% among persons less than 21 yrs old) [24]. The emergence and dissemination of multidrug-resistant bacteria in both humans and companion animals has created additional concerns [25, 26].

People can acquire pet-associated zoonotic organisms through the skin and mucous membranes (via animal bites, scratches, or direct or indirect contact with animal saliva, urine and other body fluids or secretions), ingestion of animal fecal material, inhalation of infectious aerosols or droplets, and through arthropods or other invertebrate vectors [27]. Although any exposed person can become infected with a zoonotic pathogen, risks are particularly high for those with a compromised or incompletely developed immune system, such as the young (< 5 yrs), elderly (≥ 65 yrs), pregnant and those with immune function-reducing conditions or treatments (e.g., diabetes, cancer, infection with human immunodeficiency virus (HIV), chemotherapy) [27, 28].

Although incompletely understood, there are many factors that are involved in the occurrence of a zoonotic infection from a pet. Several of these factors likely have important roles in determining if an individual becomes infected, becomes symptomatic once infected, and the degree of illness that occurs. These include: 1) level and dose of exposure to the pathogen, 2) ability of the individual to suppress infection, with immunocompromised and extremes of age less able to successfully respond immunologically to pathogens, and 3) preventative veterinary care, husbandry and infection control practices that remove or reduce the level of the pathogen present and risk of transmission [27]. Existing recommendations stress that those at greatest risk for infectious disease should take the greatest precautions when interacting with pets, and be the most diligent in following infection prevention practices [29-35].

Despite the risk for pet-associated disease in people and role that appropriate infection control and husbandry practices may play in reducing disease risk, the area of household pet husbandry and infection practices has not been thoroughly investigated. The studies that have examined these topics, have noted the frequency of close contact between pets and people (e.g., licking of hands and sleeping in household member beds [36, 37]), general pet ownership patterns by individuals at higher risk of disease [15, 38-43], and general husbandry and infection control practices within pet-owning households (e.g., frequency of preventive veterinary care [40, 44], and occurrence of hand hygiene [36, 37] and husbandry practices [36, 45]). However, previous studies have not analyzed the associations between these factors, which would allow for a better understanding of how the factors may change with differing household disease risks or with prior receipt of zoonotic disease education. The objective of this study was therefore to characterize pet husbandry and infection control practices that relate to zoonotic disease risks in pet-owning households in Ontario, Canada. Additionally, the study aimed to integrate household demographics, including human disease risk status (i.e., extremes of age, immunocompromised) and prior receipt of zoonotic disease education to determine if these factors influenced household practices.

2. METHODS

2.1 Study location and selection of study sites

The site chosen for this study was the region of Waterloo, located in southern Ontario, Canada. The Waterloo Region is composed of three urban and four rural municipalities. This region was selected as it includes both rural and urban settings and its population has similar demographics to the Province of Ontario and Canada as a whole [46]. As of 2006, this region had a population of approximately 478,000 people in 178,000 private households. The median age was 36.4 yrs, with 81% of the population greater than 14 yrs. Median household annual income was $65,000 CAD

[46]. These statistics are similar to those for the Province of Ontario and all of Canada, with the exception of a higher household median annual income ($60,000 for Ontario, $54,000 for Canada) [46].

Medical clinic and participant enrolment was part of a previously reported study that addressed knowledge and attitudes related to pet contact and associated zoonoses. Detailed enrolment methods and inclusion criteria are available elsewhere [47]. In brief, a convenience sample of multi-doctor general practice physician offices located within the Waterloo region was approached to participate in the study. Two practices agreed to participate (clinic A in Kitchener, Ontario and clinic B in Cambridge, Ontario).

2.2 Data collection

During October and November, 2010, adults present in the waiting areas of the participating clinics were individually approached and invited to take part in the study. Individuals were eligible to participate if they were at least 18 yrs of age, able to read and speak English, and no one from their household had previously taken part in the study. When more than one eligible adult for a given household was present in the waiting area, a single household volunteer was requested. Participants were asked to complete an anonymous, confidential, self-administered written general questionnaire. Individuals who currently had pets in their household were also given a self-administered written 5-minute supplemental pet-specific questionnaire (Additional file 1). Household pets were defined as those that were indoor only, outdoor only, and those that spent time both indoor and outdoor. No incentives were offered for participating in the study.

The 7-page supplemental pet-specific questionnaire for this study was developed with guidance from epidemiologists, veterinarians, physicians and zoonotic disease experts. The questionnaire was pre-tested on six members of the public with varying zoonotic disease backgrounds, and revised accordingly. The final questionnaire utilized closed-ended questions, primarily Yes/No or multiple choice (using an ordinal scale). It gathered household-level data on the number and species of pets, pet husbandry and preventive medicine practices, involvement of children less than 16 yrs of age in select husbandry and pet-contact practices, childrens' utilization of infection control practices, and the household's and childrens' emotional attachment to pets. Given the complexity of capturing and assessing flea and endoparasite prevention programs for each household pet, respondents were asked if all household dogs/cats were on intestinal parasite and flea prevention programs, without further defining, verifying or assessing specific components of the programs. Individual supplemental questionnaires were linked by respondent to the general questionnaire [47], allowing select variables collected through the general questionnaire to be utilized for additional analysis in this study. The study was approved by the University of Guelph's Research Ethics Board.

2.3 Data analysis

Data were entered into an Access database (Microsoft Corp., Redmond, WA, USA) and analyzed using Intercooled Stata version 10.1 (Stata Corp., College Station, TX, USA). Descriptive statistics were computed for all variables. Blank and "don't know" responses were excluded from analyses. Household demographic and health data were extracted from the general questionnaire [47] to characterize households as higher risk [one or more individuals < 5 yrs, ≥ 65 yrs, or ever diagnosed with an immunocompromising condition (such as HIV/acquired immunodeficiency syndrome (AIDS), diabetes, cancer)] or not higher risk. Data regarding the respondents' recall of having previously received pet-associated zoonotic disease education (i.e. having ever received information from any source about diseases that can be transmitted from pets or precautions to take with pets) were also extracted from the general questionnaire [47]. For statistical analysis, ordinal levels of each pet husbandry and infection control practice were collapsed to produce dichotomous outcome variables; decisions on how categories were collapsed were made a priori and based on existing recommendations for higher risk households and relevance of comparisons (Table 1) [29-35]. Statistical associations were individually assessed between pet husbandry and infection control practices (outcomes) and 1) households with individuals at higher risk of infections [at least one individual < 5 yrs, ≥ 65 yrs, or immunocompromised (exposed) vs. no one fulfilling these criteria (unexposed)] and 2) receipt of zoonotic disease education [previous receipt of pet-associated zoonotic disease education (exposed) vs. no recollection of having received this information (unexposed)]; Table 1. For households with children under 16 yrs of age, statistical associations were individually assessed between child-specific husbandry and infection control practices (e.g., child's involvement and precautions taken in cleaning pet's cages or feces, level of contact with pets) and households with all children at higher risk of infectious disease [presence of all household children < 5 yrs (exposed) vs. presence of all household children ≥ 5 yrs (unexposed)]. Due to small sample sizes for many of the variables, associations were measured using univariate exact logistic regression with conditional maximum likelihood estimates (CMLEs). When CMLEs were computed to be infinite, median unbiased estimates were calculated [48]. Odds ratios (ORs) and 95% confidence intervals (CI) for ORs were used to estimate the strength of associations. Statistical significance was based on a P-value < 0.05.

Table 1 Outcomes related to animal husbandry and infection control practices tested for association with presence of higher risk individuals in the household and receipt of zoonotic disease education (Ontario, Canada)

I. Outcome variables (yes/no) tested for association with presence of any higher risk individuals in the household and prior receipt of zoonotic disease education[1]	
Any uncontrolled outdoor access (cat)[2]	Dump aquarium water in bathtub, shower, kitchen sink, or bathroom sink (fish)
When outdoors, all always in fenced yard or on leash (dog)	Household very attached (cat, dog, reptile, fish, bird, exotic mammal)
Within past 3 months, any dogs on a leash at a dog park	Any fed raw animal product (cat, dog)[4]
Within past 3 months, any dogs off leash at a dog park (dog)	Any fed raw animal product not including raw animal treats (dog)
All on a flea prevention program (cat, dog)[3]	Any drink from the toilet (cat, dog)
All with outside access on flea prevention program (cat)	Any fed in the kitchen (cat, dog)
All on an intestinal parasite prevention program (cat, dog)[3]	Any used for breeding (dog)
All with outside access on an intestinal parasite prevention program (cat)	Any used for hunting (dog)
All have at least annual veterinary examinations (cat, dog)	Any declawed (cat)
Any allowed to roam freely through house (reptile)	Any ever bitten anyone (cat, dog)
Any allowed to roam in kitchen (reptile)	Any ever scratched anyone (cat)
Any cages or animals washed in bathtub or bathroom sink (reptile)	Clean litter boxes daily (cat)
Cage or litter boxes cleaned every several days, or more frequently (cat, reptile, bird, exotic mammal)	

II. Outcome variables (yes/no) tested for association with all household children being higher risk (less than 5 yrs) and prior receipt of zoonotic disease education[1,5]	
Any children have access to the litter box (cat)	Clean up feces immediately (dog)
Any children clean-up feces (cat, dog)	Never sleep in children's bed (cat, dog)
Any children wear gloves when cleaning feces (dog)[6]	Any children play in same area pet goes to the bathroom (dog)
Children wash hands after cleaning feces (dog)[6]	Any children touch pet (cat, dog)
Ever lick children's faces (dog)	Children never wash hands after touching pet (cat, dog)

[1]Presence of higher risk individuals in the household defined as one or more individuals < 5 yrs, ≥ 65 yrs, or ever diagnosed with an immunocompromising condition (e.g., HIV/AIDS, diabetes, cancer). Receipt of zoonotic disease information defined as recollection of having ever received information from any source about diseases that can be transmitted from pets or precautions to take with pets.
[2]Uncontrolled outdoor access defined as households with one or more outdoor only cats or indoor/outdoor cats not placed on a leash when outdoors.
[3]Respondant asked if all dogs/cats on flea or endoparasite prevention programs; the specific components of the programs were not assessed.
[4]Raw animal products defined as raw meat, raw eggs, or raw animal product treats.
[5]Analysis investigating the presence of higher risk children in the household was only performed for children-specific outcomes and only concerned comparison between children-containing households in which all children were < 5 yrs and those in which all children were ≥ 5 yrs.
[6]Association only assessed with prior receipt of zoonotic disease education.

3. RESULTS

Six hundred forty one of the 853 individuals who were approached (75.1%) completed the general questionnaire [47]. Of these, 408 respondents (63.7%) reported having one or more household pets and 401 (98%) completed the supplemental pet-specific questionnaire. Ownership of dogs, cats, exotic companion animals (such as gerbils and rabbits), reptiles and birds were reported. In addition to having household pets, 16 (4.0%) of the respondents reported they lived on a farm where livestock, horses, or poultry were kept.

3.1 Dog-owning households

Dogs were the most frequently reported owned species, with 68.3% [(274/401); 95% CI: 63.5, 72.9] of respondents reporting one or more household dogs. For those respondents listing the number of dogs in the household (n=254), there was a mean (SD) of 1.3 (0.6) dogs per household (range: 1–6). The location of the dog was provided by 239 households; 6 households (2.5 %) had 1 or more outdoor only dogs, 231 (96.7 %) had one or more indoor/outdoor dogs, and 2 (0.8%) had both outdoor only and indoor/outdoor dogs.

Household demographics, animal husbandry, attachment and infection control practices for dog-owning households are listed in Table 2. The majority (52%) of households surveyed had one or more higher risk individuals. Forty percent of the dog-owning respondents recalled having ever received information about pet-associated zoonotic diseases or precautions to reduce the risk of disease.

Most (76-80%) dog-owning households had all dogs on flea or endoparasite prevention programs, with 92% of respondents taking all dogs to a veterinarian at least annually. High risk feeding or drinking habits were observed in many households. Most households (65%) fed dogs in the kitchen and some (15%) reported dogs drank from the toilet. High-risk foods were fed to dogs in 28% of households with raw animal product treats (e.g. pig ears) fed most frequently (22%), followed by raw meat (4%), and raw eggs (3%).

Potential contact between household children (< 16 yrs) and fecal material was frequently reported, through their involvement in cleaning up fecal material (24%) or playing in the same area where the dog tended to defecate (52%; 58% of these respondents stated feces were removed weekly or less often). Close contact between dogs and children was often reported, as respondents stated the dog slept in a child's bed (26%) or licked a child's face (68%) sometimes or more frequently.

Table 2 Demographics, animal husbandry, attachment and practices of dog and cat-owning households (Ontario, Canada)

Variable	Dog (N=274)		Cat (N=191)	
	N_R	N (%)	N_R	N (%)
Child under 16 yrs living in household	274	122 (44.5)	191	78 (40.8)
Child under 5 yrs living in household	274	44 (16.1)	191	26 (13.6)
All children living in household under 5 yrs	274	25 (9.1)	191	19 (10.0)
All children living in household 5-16 yrs	274	78 (28.5)	191	52 (27.2)
Adult ≥ 65 yrs living in household	274	35 (12.8)	191	28 (14.7)
Individual living in household ever diagnosed with an immunocompromising condition	227	60 (26.4)	162	55 (34.0)
Individual < 5 yrs, ≥ 65 yrs, or ever diagnosed with an immunocompromising condition	237	122 (51.5)	169	90 (53.3)
Previous receipt of pet-associated zoonotic disease education[1]	255	102 (40.0)	172	75 (43.6)
Any ever bitten anyone	269	29 (10.8)	184	59 (32.1)
Any ever scratched anyone	NA	NA	180	113 (62.8)
Any declawed	NA	NA	188	110 (58.5)
All on flea prevention program[2]	263	200 (76.1)	177	98 (55.4)
All on intestinal parasite prevention program[3]	258	205 (79.5)	162	68 (42.0)
Any drink from toilet bowl	267	41 (15.4)	186	36 (19.4)
Any fed in kitchen	269	176 (65.4)	189	107 (56.6)
Any used for breeding	268	15 (5.6)	183	0
Any used for hunting	269	3 (1.1)	NA	NA
All visit veterinarian at least annually	259	237 (91.5)	184	121 (65.8)
Any inappropriately urinate or defecate in house[4]	268		188	
Always		1 (0.4)		7 (3.7)
Often		14 (5.2)		13 (6.9)
Sometimes		74 (27.6)		43 (22.9)
Never		179 (66.8)		125 (66.5)
Litter box(es) cleaned	NA		184	
Daily		NA		73 (39.7)
Every several days		NA		72 (39.1)
Weekly		NA		30 (16.3)
Every 2 weeks or more		NA		9 (4.9)
In the past 3 months, have any been	274		NA	
On leash at a dog park		53 (19.3)		NA
Off leash at a dog park		51 (18.6)		NA
Doggie daycare or boarding		33 (12.0)		NA
		2 (0.7)		NA
When outdoors, how often in a fenced yard or on leash	267		NA	
Always		195 (73.0)		NA

Sometimes		58 (21.7)		NA
Never		14 (5.2)		NA
Fed to any of the pets	264		186	
Commercial canned/dry food		244 (92.4)		185 (99.5)
Raw eggs		9 (3.4)		1 (0.5)
Home cooked pet food		15 (5.7)		3 (1.6)
Raw animal product treats		59 (22.4)		0
Home cooked human food		112 (42.4)		17 (9.1)
Commercial processed pet treats		156 (59.1)		70 (37.6)
Raw meat		11 (4.2)		5 (2.7)
Household's emotional attachment to pet	268		186	
Very attached		247 (92.2)		150 (80.7)
Somewhat attached		21 (7.8)		32 (17.2)
Not very attached		0		2 (1.1)
Not at all attached		0		2 (1.1)
Any children have access to the litter box(es)	NA	NA	74	49 (66.2)
Any children ever clean-up feces[5]	119	29 (24.4)	77	19 (24.7)
Wear gloves (or use scooper for dogs) when cleaning up feces	28	24 (85.7)	18	5 (27.8)
Wash hands after cleaning up feces	26	25 (96.2)	19	19 (100)
Play in area where the pet goes to the bathroom	114	59 (51.8)	NA	
If play in area, feces removed	59		NA	
Immediately		12 (20.3)		NA
Daily		13 (22.0)		NA
Weekly		24 (40.7)		NA
Greater than weekly		10 (17.0)		NA
Any children touch the pet	118	116 (98.3)	77	72 (93.5)
Children wash hands after touching the pet	115		70	
Always		5 (4.4)		2 (2.9)
Usually		31 (27.0)		25 (35.7)
Sometimes		53 (46.1)		26 (37.1)
Never		26 (22.6)		17 (24.3)
Sleeps in one of the children's beds[6]	115		71	
Always		9 (7.8)		11 (15.5)
Usually		6 (5.2)		10 (14.1)
Sometimes		15 (13.0)		21 (29.6)
Never		85 (73.9)		29 (40.9)
Licks one of the children's faces[7]	115		NA	
Daily		7 (6.1)	NA	
Often		21 (18.3)		NA
Sometimes		50 (43.5)		NA
Never		37 (32.2)		NA
Children's emotional attachment to the pet	114		77	
Very attached		82 (71.9)		52 (67.5)
Somewhat attached		25 (21.9)		14 (18.2)
Not very attached		5 (4.4)		5 (6.5)
Not at all attached		2 (1.8)		6 (7.8)

[1]Respondent recalled having ever received information from any source about diseases that they can get from pets or precautions to take with pets to reduce the risk of disease.
[2]If respondent reported they did not know, they were excluded from analysis: dogs (n=5); cats (n=9).
[3]If respondent reported they did not know, they were excluded from analysis: dogs (n=11); cats (n=20).
[4]Cats: Frequency of urination or defection outside of the litter box in the house; dogs: as written. Dogs/cats: Often (at least once per week), sometimes (less than once per week).
[5]Cats: Children ever clean the litter box(es); dogs: as written.
[6]Usually (3–6 nights per week), sometimes (less than 3 nights per week).
[7]Often (several times per week).
N_R: Number responding to question. Blank and "don't know" responses were excluded.
NA: Not asked.

Hand washing was reported after children picked up feces (96%), and sometimes or greater following contact with the dog (77%).

In most cases the presence of higher risk individuals in the household or recalled receipt of zoonotic disease education were not significantly associated with the use of different or additional husbandry or infection control practices compared to other households (Table 3). Households with all children less than 5 yrs of age were less likely to report children cleaned up the dog's feces (OR: 0.05; 95% CI: 0–0.3), but more likely to report a dog ever licked a child's face (OR: 4.1; 1.1-23.5), than households with all children ≥ 5 yrs.

Table 3 Significant univariable associations between outcomes related to animal husbandry and infection control practices and risk factors reported by pet-owning households (Ontario, Canada)

Outcome (species)[1]	All children < 5 yrs (Exposed; N)[2]	All children ≥ 5 yrs (Unexposed; N)[2]	Risk Exposed	Risk Unexposed	Odds Ratio[3]	95% CI[3]	*P*-value[3]
Any children have access to the litter boxes (cat)							
	4	40	25.0%	78.4%	0.1	0.02-0.4	0.0004
Any children clean up feces (cat)							
	0	18	0%	35.3%	0.07*	0-0.5	0.002
Any children touch pet (cat)							
	13	52	72.2%	100%	0.04*	0-0.3	0.001
Never sleep in children's bed (cat)							
	10	14	83.3%	26.9%	12.9	2.4-135.9	0.001
Any children clean-up feces (dog)							
	0	28	0%	36.8%	0.05*	0-0.3	0.0002
Ever lick children's face (dog)							
	20	45	87.0%	61.6%	4.1	1.1-23.5	0.04
Outcome (species)[1]	**Higher Risk Member (Exposed; N)[4]**	**No Higher Risk Member (Unexposed; N)[4]**	**Risk Exposed**	**Risk Unexposed**	**Odds Ratio[3]**	**95% CI[3]**	***P*-value[3]**
Within past 3 months, any off leash at a dog park (dog)							
	15	29	12.3%	25.2%	0.4	0.2-0.9	0.02
Household very attached (dog)							
	106	109	88.3%	97.3%	0.2	0.04-0.8	0.01
Dump aquarium water in bathtub, shower, kitchen sink, or bathroom sink (fish)							
	6	12	25.0%	60.0%	0.2	0.05-0.9	0.04
Household very attached (bird)							
	6	0	54.6%	0%	11.5*	1.3-Inf	0.02

[1]Univariable exact logistic regression used to model the likelihood of each husbandry and infection control outcome as a function of selected risk factors (all children < 5 yrs; higher risk member in household). Each row represents a separate regression model.
[2]Performed for children-specific outcomes. Comparison between households with all children < 5 yrs (exposed) and households with all children ≥ 5 yrs (unexposed).
[3]Odds ratio, 95% confidence interval of the odds ratio (95% CI), and *P*-value computed using exact logistic regression.
[4]Comparision between households with ≥ 1 higher risk members [i.e., < 5 yrs, ≥ 65 yrs, ever diagnosed with an immunocompromising condition (exposed)] and households without higher risk members (unexposed).
*Median unbiased estimate computed since odds ratio calculated to be infinite.

3.2 Cat-owning households

One hundred ninety one respondents (47.6%; 95% CI: 42.7, 52.6) reported owning one or more household cats. For those respondents listing the number of cats in the household (n=178), there was a mean (SD) of 1.7 (1.4) cats per household (range: 1–11). The location of the cat was provided by 176 households; 7 households (4%) had one or more outdoor only cats, 114 (65%) had one or more indoor only cats, and 73 (41%) had one or more indoor/outdoor cats. However, 41 (56%) of households with indoor/outdoor cats kept cats restrained on a leash when outdoors.

Household demographics, animal husbandry, attachment and infection control practices for cat-owning households are listed in Table 2. The majority (53%) of households had one or more higher risk individuals. Forty-four percent of the cat-owning respondents recalled having ever received information about pet-associated zoonotic diseases or precautions to reduce the risk of disease.

Bites and scratches ever occurring by any of the current household cats were frequently reported (32% and 63%, respectively). Approximately half of responding households had all cats on flea (55%) or endoparasite (42%) prevention programs, with 66% of households taking all cats to a veterinarian at least annually. High risk feeding and drinking habits were frequently observed, with many households (57%) having fed cats in the kitchen. High-risk foods were fed to cats in 3% of households with raw meat (3%) and raw eggs (0.5%) reported.

Potential contact between children and fecal material was frequently reported, through their involvement in cleaning up fecal material (25%) or access to the litter box (66%). Twenty-one percent of respondents reported the litter box was cleaned weekly or less often; this percentage was similar for households with and without children. Close contact between cats and children was frequently reported, as 59% of households reported a cat slept in a child's bed. Hand washing was reported after children picked up feces (100%) and sometimes or greater following contact with the cat (76%).

The presence of only children under 5 yrs of age in the household was significantly associated with the occurrence of several infection control practices compared to households with all children 5 yrs of age or older (Table 3). Households with all children less than 5 yrs were less likely to report children cleaned up the cat's feces (OR: 0.07; 95% CI: 0–0.5), less likely to touch the cat (OR: 0.04; 95% CI: 0–0.3), less likely to have access to the litter box (OR: 0.1; 95% CI: 0.02-0.4), and more likely for the cat to never sleep in the child's bed (OR: 12.9; 95% CI: 2.4-135.9), than households in which all children were ≥ 5 yrs.

3.3 Fish-owning households

Fifty-three respondents (13.2%; 95% CI: 10.1, 16.9) reported one or more household fish (median 4 fish per household; range: 1–60). Household demographics, animal husbandry, attachment and infection control practices are listed in Table 4. The majority (53%) of households had one or more higher risk individuals. Thirty percent of the fish-owning respondents recalled having ever received information about pet-associated zoonotic diseases or precautions to reduce the risk of disease.

Table 4 Demographics, animal husbandry, attachment and practices of fish and bird-owning households (Ontario, Canada)

Variable	Fish (N=53)		Bird (N=22)	
	N_R	N (%)	N_R	N (%)
Household Demographics				
Child under 16 yrs living in household	53	34 (64.2)	22	8 (36.4)
Child under 5 yrs living in household	53	10 (18.9)	22	1 (4.6)
Adult ≥ 65 yrs living in household	53	2 (3.8)	22	4 (18.2)
Individual living in household ever diagnosed with an immunocompromising condition	42	15 (35.7)	20	7 (35.0)
Individual < 5 yrs, ≥ 65 yrs, or ever diagnosed with an immunocompromising condition	45	24 (53.3)	21	12 (57.1)
Previous receipt of pet-associated zoonotic disease education[1]	47	14 (29.8)	21	7 (33.3)
Housing cleaned[2]	52		21	
Daily		1 (1.9)		4 (19.1)
Every several days		1 (1.9)		3 (14.3)
Weekly		17 (32.7)		8 (38.1)
Every 2 weeks		12 (23.1)		3 (14.3)
Greater than every 2 weeks		21 (40.4)		3 (14.3)
Dump aquarium water	50		NA	
Toilet		20 (40.0)		NA
Bathtub or shower		3 (6.0)		NA
Kitchen sink		12 (24.0)		NA
Bathroom sink		8 (16.0)		NA
Outside		19 (38.0)		NA
Other		3 (6.0)		NA
Household's emotional attachment to pet	50		21	
Very attached		5 (10.0)		6 (28.6)
Somewhat attached		22 (44.0)		10 (47.6)
Not very attached		19 (38.0)		5 (23.8)
Not at all attached		4 (8.0)		0

Any children ever clean the animal's housing[2]	33	10 (30.3)	8	3 (37.5)
Wear gloves when cleaning	10	0	3	1 (33.3)
Wash hands after cleaning	10	9 (90.0)	3	2 (66.6)
Children touch the pet (or water for fish)	30	12 (40.0)	8	4 (50.0)
Children wash hands after touching	12		4	
Always		4 (33.3)		2 (50.0)
Usually		4 (33.3)		1 (25.0)
Sometimes		3 (25.0)		0
Never		1 (8.3)		1 (25.0)
Children's emotional attachment to the pet	30		8	
Very attached		3 (10.0)		1 (12.5)
Somewhat attached		14 (46.7)		6 (75.0)
Not very attached		12 (40.0)		1 (12.5)
Not at all attached		1 (3.3)		0

[1]Respondent recalled having ever received information from any source about diseases that they can get from pets or precautions to take with pets to reduce the risk of disease.
[2]"Housing" refers to aquaria (fish) and cages (birds).
N_R: Number responding to question. Blank and "don't know" responses were excluded.
NA: Not asked.

Forty-four percent of respondents stated they disposed of used aquarium water by pouring it into a location with direct human or food contact [i.e., kitchen sink (24%), bathroom sink (16%), bathtub/shower (6%)]. Those households with higher risk members were less likely to dispose of aquarium water in this manner than households without these members (OR: 0.2; 95% CI: 0.05-0.9; Table 3). When children were present in a household, they were frequently involved with cleaning the aquarium (30%) and touching the fish or water (40%). Over 90% of households reported that children washed their hands after cleaning, or washed their hands sometimes or greater after touching the fish or water. A large percentage (43-46%) of households reported the household and children (if present) were not very, or not at all, attached to the fish.

3.4 Bird-owning households

Twenty-two respondents (5.5%; 95% CI: 3.5, 8.2) reported one or more household birds (median one bird per household; range: 1–40). Household demographics, animal husbandry, attachment and infection control practices for bird-owning households are listed in Table 4. The majority (57%) of households had one or more higher risk individuals. Thirty-three percent of the bird-owning respondents recalled having ever received information about pet-associated zoonotic diseases or precautions to reduce the risk of disease. Households with higher risk individuals were more likely to be very attached to the birds than households without these individuals (OR: 11.5; 95% CI: 1.3-infinite; Table 3).

3.5 Exotic companion mammal-owning households

Twenty-nine respondents (7.2%; 95% CI: 4.9, 10.2) reported one or more household exotic mammals (median 1 per household; range: 1–13), consisting of hamsters (n=10), rabbits (6), guinea pigs (5), rats/mice (4) and hedgehogs, chinchillas and gerbils (2 each). Household demographics, animal husbandry, attachment and infection control practices for exotic mammal-owning households are listed in Table 5. Many (46%) households had one or more higher risk individuals. Three households (10%) had at least one child less than 5 yrs of age. Forty-one percent of the respondents recalled having ever received information about pet-associated zoonotic diseases or precautions to reduce the risk of disease.

Table 5 Demographics, animal husbandry, attachment and practices of exotic companion mammal and reptile-owning households (Ontario, Canada)

Variable	Exotic Companion Mammal[1] (N=29)		Reptile (N=22)	
	N_R	N (%)	N_R	N (%)
Household Demographics				
Child under 16 yrs living in household	29	19 (65.5)	22	10 (45.5)
Child under 5 yrs living in household	29	3 (10.3)	22	2 (9.1)
Adult ≥ 65 yrs living in household	29	2 (6.9)	22	0
Individual living in household ever diagnosed with an immunocompromising condition	23	8 (34.8)	19	9 (47.4)
Individual < 5 yrs, ≥ 65 yrs, or ever diagnosed with an immunocompromising condition	26	12 (46.2)	20	11 (55.0)
Previous receipt of pet-associated zoonotic disease education[2]	27	11 (40.7)	17	7 (41.2)
Animal Husbandry, Attachment and Practices				
Cage cleaned	29		21	
Daily		2 (6.9)		1 (4.8)
Every several days		6 (20.7)		3 (14.3)
Weekly		18 (62.1)		6 (28.6)
Every 2 weeks		3 (10.3)		5 (23.8)
Greater than every 2 weeks		0		6 (28.6)
Animals allowed to roam through house	NA	NA	22	2 (9.1)
Animals allowed to roam in kitchen	NA	NA	22	2 (9.1)
Animals/cages washed in kitchen sink	NA	NA	22	3 (13.6)
Animals/cages washed in bathtub or bathroom sink	NA	NA	18	5 (27.8)
Household's emotional attachment to pet	29		21	
Very attached		16 (55.2)		3 (14.3)
Somewhat attached		11 (37.9)		11 (52.4)
Not very attached		2 (6.9)		4 (19.1)
Not at all attached		0		3 (14.3)

Questions answered by households with one or more children under 16 yrs				
Any children ever clean the animal's cage	19	12 (63.2)	10	4 (40.0)
Wear gloves when cleaning	12	1 (8.3)	3	0
Wash hands after cleaning	12	11 (91.7)	4	3 (75.0)
Animals roam in children's room	NA	NA	8	0
Children touch the pet	19	19 (100)	9	7 (77.8)
Children wash hands after touching	19		7	
Always		8 (42.1)		4 (57.1)
Usually		8 (42.1)		0
Sometimes		3 (15.8)		2 (28.6)
Never		0		1 (14.3)
Children's emotional attachment to the pet	18		10	
Very attached		11 (61.1)		1 (10.0)
Somewhat attached		6 (33.3)		5 (50.0)
Not very attached		1 (5.6)		3 (30.0)
Not at all attached		0		1 (10.0)

[1]Exotic companion mammals included rabbits, hedgehogs, and rodents (such as gerbils, hamsters, guinea pigs, chinchillas, mice, and rats).
[2]Respondent recalled having ever received information from any source about diseases that they can get from pets or precautions to take with pets to reduce the risk of disease.
N_R: Number responding to question. Blank and "don't know" responses were excluded.
NA: Not asked.

When children were present in a household, they were frequently involved with cleaning the animal's cage (63%) or touching the pet (100%). Most children reportedly washed their hands after cleaning the cage (92%) or washed their hands sometimes or greater after touching the pet (100%). A minority (42%) always washed their hands after touching the pet. Most respondents (> 90%) reported the household and children (if present) were very or somewhat attached to the pets. No statistically significant associations between husbandry/infection control practices and risk factors were detected.

3.6 Reptile-owning households

Twenty-two respondents (5.5%; 95% CI: 3.5, 8.2) reported one or more household reptiles (median one per household; range: 1–11), consisting of lizards (n=9), snakes (5), and turtles (9). Household demographics, animal husbandry, attachment and infection control practices for reptile-owning households are listed in Table 5. Many households (55%) had one or more higher risk individuals, with a high proportion (47%) having been diagnosed with an immunocompromising condition. Two households (9%) had at least one child less than 5 yrs of age. Forty-one percent of the reptile-owning respondents recalled having ever received information about pet-associated zoonotic diseases or precautions to reduce the risk of disease.

Three respondents (14%) reported the pet reptile was allowed to roam through the house, kitchen or washed in the kitchen sink. When children were present in a household, they were often involved with cleaning the animal's cage (40%) or touching the pet (78%). In one household in which children touched the reptile, all children (n=2) were less than 5 yrs. Most respondents reported the children washed their hands after cleaning the cage (75%) or washed their hands sometimes or greater after touching the pet (86%). However, only 57% always washed their hands after touching the reptile. Many respondents reported the household and children (if present) were not very, or not at all, attached to the reptile (33% and 40%, respectively). No statistically significant associations between husbandry/infection control practices and risk factors were detected.

3.7 Farm-owning households

Sixteen respondents (4.0% of those with household pets; 95% CI: 2.3, 6.4) reported they lived on a farm. Respondents most commonly listed horses (n=9), chickens (5), cattle and pigs (3 each) as the farm animals present. Fifty percent (7/14) of households for which the data were provided had one or more higher risk individuals. When children were present [6/16 (38%)], they were often involved in higher risk activities. Most [5/6 (83%)] of such households reported children helped feed and clean-up after the farm animals at least sometimes, and in most households [5/6 (83%)] children usually or always washed their hands after contact with the animals or their housing.

4. DISCUSSION

As pet-human contact frequently occurs and disease transmission is overall rarely reported [47], in most situations such contact should not pose a substantial health risk. However, there are practices that may increase pathogen exposure and individuals (< 5yrs, ≥ 65yrs, or immunocompromised) for whom these risks are known or presumed to be greater. Reported husbandry and infection control practices in this study indicated opportunities for exposure to pathogens. Respondents reported household children had close contact with pets (e.g., face licking, sleeping in beds, cleaning up pet fecal material). These findings are similar to a previous study in which dogs licked their owners' face (50%) and dogs and cats slept in an adult's bed (18%, 30%, respectively) [37]. The oral cavity of the dog and cat contain numerous potentially zoonotic opportunistic pathogens [49] and face licking can result in transmission of bacteria such as Pasteurella spp. and Capnocytophaga canimorsus[50], potentially resulting in life-threatening disease [22]. For these reasons, some discourage young children and immunocompromised individuals from regularly allowing face licking by pets [50]. The disease risk associated with dogs and cats sleeping in a child's bed is largely unquantified, however sleeping with pets has been identified as a risk factor for several diseases, prompting some to discourage this practice by higher risk individuals [50].

Routine preventive veterinary care and husbandry practices are important techniques for reducing zoonotic disease transmission between pets and people [30, 35]. Consistent with previous studies [40, 44], most cat- and dog-owning households had these animals examined at least annually by a veterinarian, which allows for identification and education of potential disease concerns and facilitates implementation of an appropriate preventive medicine program. The differences observed between dog and cat preventative veterinary care, with cats having lesser veterinary contact, has been previously noted [40, 44]. Since cats have less contact with the veterinary healthcare system yet have close interactions with household members, educational approaches that reach beyond the veterinary clinic may be required.

Concerning husbandry practices were observed in all owned species. Feeding of raw meat or egg products are well-established risk factors for salmonellosis in dogs [45, 51] and presumably cats. Raw animal product treats (i.e. pig ears) have historically been a concern for Salmonella infections in pets and pet-owners, with outbreaks identified [52, 53]. Although feeding these items is discouraged, particularly in higher risk households [54], many households reported this practice. These risks are further compounded by feeding pets in the kitchen (frequently reported in this study) as documented in an outbreak investigation of salmonellosis in infants linked to Salmonella-contaminated pet food [55]. As several outbreaks of human salmonellosis have been associated with contaminated dry commercial pet food [55, 56], feeding pets in the kitchen should be discouraged, especially in higher risk households [29]. Concerning husbandry practices were frequently reported in reptile-owning homes (e.g., infrequent hand washing, permitting them to roam freely through a home or kitchen). Due to the high prevalence of Salmonella shedding by healthy reptiles [57] and the high incidence of human salmonellosis attributed to reptile contact [24], the reported practices are strongly discouraged [34, 58].

Hand hygiene serves a critical role in reducing the risk of zoonotic infections. Our results were similar to previous studies in which 96% of dog owners reported they usually or greater washed their hands after picking up feces [36] and 15% always washed their hands after touching the dog [37]. For lower risk circumstances, such as contact with a dog or cat, the level of hand hygiene documented in our study is likely adequate; however, in higher-risk situations, such as following contact with a reptile, hand washing should always occur.

Across all animal species that were owned, many households had an individual at higher risk of infections (< 5yrs, ≥ 65yrs, immunocompromised). Although pet ownership has previously been commonly reported in households with individuals who are immunocompromised [15, 38, 39] and of extremes of age [40, 41, 42, 43], it was concerning to observe this trend with higher risk species such as reptiles, which are discouraged from being kept in higher risk households [29, 31, 32, 33, 34, 58]. However, this finding was not surprising as a previous study involving this respondent group found households with higher risk individuals did not differ from the remaining households regarding their knowledge or perceived risk of pet-associated disease or recall of having received information regarding pet-associated disease risks [47]. Physicians and veterinarians are in an important position to offer guidance on proper pet selection. In order to tailor recommendations, physicians should obtain a history of contact with pets or other animals as part of every wellness evaluation, especially for individuals at higher risk of zoonotic disease, and veterinarians should utilize methods to encourage clients to divulge health information relevant to disease risk status for household members.

Despite evaluating over 50 practices consistent with existing recommendations for individuals at greater risk for infectious disease [29, 30, 31, 32, 33, 34, 35], few significant associations were identified. Most of those identified suggested a lower risk of practice occurrence in households with young children; the extent to which such decisions were purposefully implemented versus simply a consequence of young age is unknown. The finding that young children were at increased risk to be licked on the face by the household dog was expected due to their age and size, but nonetheless concerning because of reports of lick-associated infections in children [22, 50].

The limited use of husbandry and infection control practices by households with higher risk individuals was disappointing but not surprising. Previous research on this respondent group found pet owners were overall unconcerned about pet-associated zoonoses and a minority (36%) recalled previously receiving zoonotic disease education [47]. Most respondents were comfortable with their level of knowledge of, and methods to reduce, zoonotic disease risk [47]. While it is unclear whether participants had reasonable knowledge of those areas, their perceived competency could reduce the impetus to seek out further information.

General zoonotic disease counselling was previously associated with statistically higher zoonotic disease knowledge scores in this respondent group [47], yet no such difference was observed in husbandry or infection control practices in the present study. This discrepancy may have arisen due to the minimally defined nature of this variable (i.e. timing and quality of information provided were not assessed), but it is also possible that although education imparted improved knowledge of zoonotic disease risks, due to a lack of concern for pet-associated zoonoses or competing time commitments, infection control and husbandry practices were unaffected. It is also possible that education regarding diseases was not accompanied by adequate information about preventive measures.

This study identified numerous pet husbandry and infection control practices that may increase disease risk in pet-owning households. In order to determine the relative importance of these practices and educational efforts that should be employed, targeted observational studies are needed. While guidelines for infection-control practices in higher risk households are available, it appears that individuals in such households are no better educated on the topic of pet-associated zoonoses [47], nor do their husbandry or infection control practices differ from other households. Educational efforts are needed for households with individuals at higher risk of infections. Given the multifaceted nature of this topic, with communication barriers among physicians, veterinarians and the public [59, 60, 61], all members of the family healthcare team (i.e., physicians, veterinarians, public health) must work together to reach the common goal of reducing the public's pet-associated disease risks.

Several biases may have been introduced into this study and affected the results. We acquired our data from a convenience sample from the waiting room of general practice physician offices from a limited geographic area. All Canadian residents receive medically necessary healthcare services at no charge [62], reducing the potential that variable access to physicians would result in a biased study population. As previously discussed [47], based on census data, our sample appeared to be representative of the region. Response bias cannot be excluded, however as blank and "don't know" responses were uncommon (<20% for each question with the majority <10%), the magnitude of this bias is likely limited. As many questions addressed health-related actions taken by respondents for which they may have known what answers were ideal, respondents may have misclassified household practices. In these cases, information provided would suggest better husbandry and infection control practices than actually existed. This was perhaps most likely for questions surrounding hand washing [63]. As only a single individual from each household participated in the study, it is possible reported observations and perceptions may not have been representative of the household. Due to the nature of this study, small sample sizes were often encountered, which in some cases prohibited analysis and in others may have limited statistical power. Furthermore, small sample sizes prohibited multivariable analysis and the ability to investigate confounding. A large number of individual regression analyses were conducted and corrections were not made for this multiplicity of testing; estimates are likely to have an increased type I error and fewer significant associations may exist than reported.

5. CONCLUSIONS

This study characterized pet husbandry and infection control practices that relate to zoonotic disease risks in pet-owning households in Ontario, Canada, and identified deficiencies that could result in exposure to pathogens and human zoonotic infections. Individuals at higher risk of infections were frequently present in pet-owning households, regardless of species. With few exceptions, households with members at higher risk for infectious disease and those who recalled having received education on pet-associated disease risks followed similar practices to households without these individuals or education. Targeted educational efforts are indicated for households with individuals at higher risk of infections and those with high-risk species (e.g., reptiles). Healthcare team members, including veterinarians, physicians and public health practitioners, have the potential to serve as valuable resources in reducing pet-associated disease risks, however all must jointly work to address this area and devote more attention toward the public's interaction with animals. Further research is needed to determine the reasoning behind household infection control and husbandry practices and the respective roles of education and perceptions in shaping these practices.

ACKNOWLEDGEMENTS

The authors thank staff and students at the University of Guelph (Departments of Population Medicine and Pathobiology), McMaster University (Department of Pediatrics) and the Children's Hospital of Eastern Ontario (Division of Infectious Diseases) for their input into the questionnaire. They also thank the two medical practices and their patients for their participation. JMS was supported through a Canadian Institutes of Health Research (CIHR) Institute of Population and Public Health/ Public Health Agency of Canada Applied Public Health Research Chair.

REFERENCES

1. Butler JR, Bingham J: Demography and dog-human relationships of the dog population in Zimbabwean communal lands. Vet Rec. 2000, 147 (16): 442-446. 10.1136/vr.147.16.442.
2. Baldock FC, Alexander L, More SJ: Estimated and predicted changes in the cat population of Australian households from 1979 to 2005. Aust Vet J. 2003, 81 (5): 289-292. 10.1111/j.1751-0813.2003.tb12577.x.
3. Downes M, Canty MJ, More SJ: Demography of the pet dog and cat population on the island of Ireland and human factors influencing pet ownership. Prev Vet Med. 2009, 92 (1–2): 140-149.

4. Slater MR, Di Nardo A, Pediconi O, Villa PD, Candeloro L, Alessandrini B, Del Papa S: Cat and dog ownership and management patterns in central Italy. Prev Vet Med. 2008, 85 (3–4): 267-294.
5. Murray JK, Browne WJ, Roberts MA, Whitmarsh A, Gruffydd-Jones TJ: Number and ownership profiles of cats and dogs in the UK. Vet Rec. 2010, 166 (6): 163-168. 10.1136/vr.b4712.
6. Perrin T: The Business Of Urban Animals Survey: the facts and statistics on companion animals in Canada. Can Vet J. 2009, 50 (1): 48-52.
7. Alves MC, Matos MR, Reichmann ML, Dominguez MH: Estimation of the dog and cat population in the State of Sao Paulo. Rev Saude Publica. 2005, 39 (6): 891-897.
8. American Veterinary Medical Association: U.S. pet ownership & demographics sourcebook. 2007, Schaumburg (Illinois): American Veterinary Medical Association
9. Friedmann E, Son H: The human–companion animal bond: how humans benefit. Vet Clin N Am: Small Anim Pract. 2009, 39 (2): 293-326. 10.1016/j.cvsm.2008.10.015.
10. Reaser JK, Clark EE, Meyers NM: All creatures great and minute: a public policy primer for companion animal zoonoses. Zoonoses Public Health. 2008, 55 (8–10): 385-401.
11. Poresky RH, Hendrix C: Differential effects of pet presence and pet-bonding on young children. Psychol Rep. 1990, 67 (1): 51-54. 10.2466/pr0.1990.67.1.51.
12. Melson GF, Schwarz RL, Beck AM: Importance of companion animals in children's lives-implications for veterinary practice. J Am Vet Med Assoc. 1997, 211 (12): 1512-1518.
13. Carmack BJ: The role of companion animals for persons with AIDS/HIV. Holist Nurs Pract. 1991, 5 (2): 24-31.
14. Castelli P, Hart LA, Zasloff RL: Companion cats and the social support systems of men with AIDS. Psychol Rep. 2001, 89 (1): 177-187. 10.2466/pr0.2001.89.1.177.
15. Conti L, Lieb S, Liberti T, Wiley-Bayless M, Hepburn K, Diaz T: Pet ownership among persons with AIDS in three Florida counties. Am J Public Health. 1995, 85 (11): 1559-1561. 10.2105/AJPH.85.11.1559.
16. Raina P, Waltner-Toews D, Bonnett B, Woodward C, Abernathy T: Influence of companion animals on the physical and psychological health of older people: an analysis of a one-year longitudinal study. J Am Geriatr Soc. 1999, 47 (3): 323-329.
17. Hale CR, Scallan E, Cronquist AB, Dunn J, Smith K, Robinson T, Lathrop S, Tobin-D'Angelo M, Clogher P: Estimates of enteric illness attributable to contact with animals and their environments in the United States. Clin Infect Dis. 2012, 54 (Suppl 5): S472-S479. 10.1093/cid/cis051.
18. Weese JS, Fulford MB: Companion Animal Zoonoses. Iowa: Wiley-Blackwell. 2011, 3-298.
19. Friedman CR, Hoekstra RM, Samuel M, Marcus R, Bender J, Shiferaw B, Reddy S, Ahuja SD, Helfrick DL, Hardnett F, Carter M, Anderson B, Tauxe RV: Emerging Infections Program FoodNet Working Group: Risk factors for sporadic Campylobacterinfection in the United States: a case–control study in FoodNet sites. Clin Infect Dis. 2004, 38 (Suppl 3): S285-S296.
20. Tenkate TD, Stafford RJ: Risk factors for Campylobacter infection in infants and young children: a matched case–control study. Epidemiol Infect. 2001, 127 (3): 399-404.
21. Younus M, Wilkins MJ, Davies HD, Rahbar MH, Funk J, Nguyen C, Siddiqi AE, Cho S, Saeed AM: The role of exposures to animals and other risk factors in sporadic, non-typhoidal Salmonella infections in Michigan children. Zoonoses Public Health. 2010, 57 (7–8): e170-e176.
22. Kobayaa H, Souki RR, Trust S, Domachowske JB: Pasteurella multocida meningitis in newborns after incidental animal exposure. Pediatr Infect Dis J. 2009, 28 (10): 928-929. 10.1097/INF.0b013e3181a81f0f.
23. De Boer MG, Lambregts PC, Van Dam AP, Van T, Wout JW: Meningitis caused by Capnocytophaga canimorsus: when to expect the unexpected. Clin Neurol Neurosurg. 2007, 109 (5): 393-398. 10.1016/j.clineuro.2007.02.010.
24. Mermin J, Hutwagner L, Vugia D, Shallow S, Daily P, Bender J, Koehler J, Marcus R, Angulo FJ: Emerging Infections Program FoodNet Working Group: Reptiles, amphibians, and human Salmonellainfection: a population-based, case–control study. Clin Infect Dis. 2004, 38 (Suppl 3): S253-S261.

25. Lin AE, Davies JE: Occurrence of highly fluoroquinolone-resistant and methicillin-resistant Staphylococcus aureus in domestic animals. Can J Microbiol. 2007, 53 (7): 925-929. 10.1139/W07-062.
26. Platell JL, Trott DJ, Johnson JR, Heisig P, Heisig A, Clabots CR, Johnston B, Cobbold RN: Prominence of an O75 clonal group (clonal complex 14) among non-ST131 fluoroquinolone-resistant Escherichia coli causing extraintestinal infections in humans and dogs in Australia. Antimicrob Agents Chemother. 2012, 56 (7): 3898-3904. 10.1128/AAC.06120-11.
27. Mani I, Maguire JH: Small animal zoonoses and immunocompromised pet owners. Top Companion Anim Med. 2009, 24 (4): 164-174. 10.1053/j.tcam.2009.07.002.
28. Abbas AK, Lichtman AH, Pillai S: Congenital and acquired immunodeficiencies. Cellular and molecular immunology. 2007, Philadelphia: Saunders, 463-488. 6
29. Angulo FJ, Glaser CA, Juranek DD, Lappin MR, Regnery RL: Caring for pets of immunocompromised persons. Can Vet J. 1995, 36 (4): 217-222.
30. Glaser CA, Powers EL, Greene CE: Zoonotic infections of medical importance in immunocompromised humans. Infectious diseases of the dog and cat. Edited by: Greene CE. 2012, St. Louis: Elsevier, 1141-1162. 4
31. Centers for Disease Control and Prevention: Guidelines for preventing opportunistic infections among hematopoietic stem cell transplant recipients: recommendations from CDC, Infectious Disease Society of America, American Society of Blood and Marrow Transplantation. MMWR Recomm Rep. 2000, 49 (RR-10): 1-125. CE1-7
32. Centers for Disease Control and Prevention: Guidelines for the Prevention and Treatment of Opportunistic Infections among HIV-exposed and HIV-infected children: recommendations from CDC, the National Institutes of Health, the HIV Medicine Association of the Infectious Diseases Society of America, the Pediatric Infectious Diseases Society, and the American Academy of Pediatrics. MMWR Recomm Rep. 2009, 58 (RR-11): 1-166.
33. Pickering LK, Marano N, Bocchini JA, Angulo FJ: Committee on Infectious Diseases: Exposure to nontraditional pets at home and to animals in public settings: risks to children. Pediatrics. 2008, 122 (4): 876-886. 10.1542/peds.2008-1942.
34. Centers for Disease Control and Prevention: Compendium of Measures to Prevent Disease Associated with Animals in Public Settings, 2011. MMWR. 2011, 60 (RR-4): 1-24.
35. Lappin MR: Pet ownership by immunocompromised people. Compend Contin Educ Pract Vet. 2002, 24 (5A): 16-25.
36. Westgarth C, Pinchbeck GL, Bradshaw JW, Dawson S, Gaskell RM, Christley RM: Dog-human and dog-dog interactions of 260 dog-owning households in a community in Cheshire. Vet Rec. 2008, 162 (14): 436-442. 10.1136/vr.162.14.436.
37. Overgaauw PA, Van Zutphen L, Hoek D, Yaya FO, Roelfsema J, Pinelli E, Van Knapen F, Kortbeek LM: Zoonotic parasites in fecal samples and fur from dogs and cats in The Netherlands. Vet Parasitol. 2009, 163 (1–2): 115-122.
38. Siegel JM, Angulo FJ, Detels R, Wesch J, Mullen A: AIDS diagnosis and depression in the Multicenter AIDS Cohort Study: the ameliorating impact of pet ownership. AIDS Care. 1999, 11 (2): 157-170. 10.1080/09540129948054.
39. Abarca VK, Del Lopez PJ, Pena DA, Lopez GJC: Pet ownership and health status of pets from immunocompromised children, with emphasis in zoonotic diseases. Rev Chilena Infectol. 2011, 28 (3): 205-210. 10.4067/S0716-10182011000300001.
40. Leslie BE, Meek AH, Kawash GF, McKeown DB: An epidemiological investigation of pet ownership in Ontario. Can Vet J. 1994, 35 (4): 218-222.
41. Westgarth C, Pinchbeck GL, Bradshaw JW, Dawson S, Gaskell RM, Christley RM: Factors associated with dog ownership and contact with dogs in a UK community. BMC Vet Res. 2007, 3: 5-10.1186/1746-6148-3-5.
42. Westgarth C, Heron J, Ness AR, Bundred P, Gaskell RM, Coyne KP, German AJ, McCune S, Dawson S: Family pet ownership during childhood: findings from a UK birth cohort and implications for public health research. Int J Environ Res Public Health. 2010, 7 (10): 3704-3729. 10.3390/ijerph7103704.

43. Ramon ME, Slater MR, Ward MP: Companion animal knowledge, attachment and pet cat care and their associations with household demographics for residents of a rural Texas town. Prev Vet Med. 2010, 94 (3–4): 251-263.
44. Volk JO, Felsted KE, Thomas JG, Siren CW: Executive summary of the Bayer veterinary care usage study. J Am Vet Med Assoc. 2011, 238 (10): 1275-1282. 10.2460/javma.238.10.1275.
45. Leonard EK, Pearl DL, Finley RL, Janecko N, Peregrine AS, Reid-Smith RJ, Weese JS: Evaluation of pet-related management factors and the risk of Salmonella spp. carriage in pet dogs from volunteer households in Ontario (2005–2006). Zoonoses Public Health. 2011, 58 (2): 140-149. 10.1111/j.1863-2378.2009.01320.x.
46. Waterloo, Ontario. 2006, : , Community Profileshttp://www12.statcan.ca/census-recensement/2006/dp-pd/prof/92-591/index.cfm?Lang=E,
47. Stull JW, Peregrine AS, Sargeant JM, Weese JS: Household knowledge, attitudes and practices related to pet contact and associated zoonoses in Ontario. Canada. BMC Public Health. 2012, 12 (1): 553-10.1186/1471-2458-12-553.
48. Mehta CR, Patel NR: Exact logistic regression: theory and examples. Statist Med. 1995, 14 (19): 2143-2160. 10.1002/sim.4780141908.
49. Forsblom B, Sarkiala-Kessel E, Kanervo A, Vaisanen ML, Helander M, Jousimies-Somer H: Characterisation of aerobic gram-negative bacteria from subgingival sites of dogs–potential bite wound pathogens. J Med Microbiol. 2002, 51 (3): 207-220.
50. Chomel BB, Sun B: Zoonoses in the bedroom. Emerg Infect Dis. 2011, 17 (2): 167-172. 10.3201/eid1702.101070.
51. Kukanich KS: Update on Salmonella spp. contamination of pet food, treats, and nutritional products and safe feeding recommendations. J Am Vet Med Assoc. 2011, 238 (11): 1430-1434. 10.2460/javma.238.11.1430.
52. Clark C, Cunningham J, Ahmed R, Woodward D, Fonseca K, Isaacs S, Ellis A, Anand C, Ziebell K, Muckle A, Sockett P, Rodgers F: Characterization of Salmonella associated with pig ear dog treats in Canada. J Clin Microbiol. 2001, 39 (11): 3962-3968. 10.1128/JCM.39.11.3962-3968.2001.
53. Centers for Disease Control and Prevention (CDC): Human salmonellosis associated with animal-derived pet treats-United States and Canada, 2005. MMWR Morb Mortal Wkly Rep. 2006, 55 (25): 702-705.
54. Finley R, Reid-Smith R, Ribble C, Popa M, Vandermeer M, Aramini J: The occurrence and anti-microbial susceptibility of Salmonellae isolated from commercially available pig ear pet treats. Zoonoses Public Health. 2008, 55 (8–10): 455-461.
55. Behravesh CB, Ferraro A, Deasy M, Dato V, Moll M, Sandt C, Rea NK, Rickert R, Marriott C, Warren K, Urdaneta V, Salehi E, Villamil E, Ayers T, Hoekstra RM, Austin JL, Ostroff S, Williams IT: Salmonella Schwarzengrund Outbreak Investigation Team: Human Salmonellainfections linked to contaminated dry dog and cat food, 2006–2008. Pediatrics. 2010, 126 (3): 477-483. 10.1542/peds.2009-3273.
56. Centers for Disease Control and Prevention (CDC): Notes from the field: human Salmonella infantisinfections linked to dry dog food - United States and Canada, 2012. MMWR Morb Mortal Wkly Rep. 2012, 61-436.
57. Hydeskov HB, Guardabassi L, Aalbaek B, Olsen KE, Nielsen SS, Bertelsen MF: Salmonella prevalence among reptiles in a zoo education setting. Zoonoses Public Health. 2013, 60 (4): 291-5. 10.1111/j.1863-2378.2012.01521.x.
58. Centers for Disease Control and Prevention (CDC): Reptile-associated salmonellosis-selected states, 1998–2002. MMWR Morb Mortal Wkly Rep. 2003, 52 (49): 1206-1209.
59. Grant S, Olsen CW: Preventing zoonotic diseases in immunocompromised persons: the role of physicians and veterinarians. Emerg Infect Dis. 1999, 5 (1): 159-163. 10.3201/eid0501.990121.
60. Von Matthiessen PW, Sansone RA, Meier BP, Gaither GA, Shrader J: Zoonotic diseases and at-risk patients: a survey of veterinarians and physicians. AIDS. 2003, 17 (9): 1404-1406. 10.1097/00002030-200306130-00021.
61. Hill WA, Petty GC, Erwin PC, Souza MJ: A survey of Tennessee veterinarian and physician attitudes, knowledge, and practices regarding zoonoses prevention among animal owners

with HIV infection or AIDS. J Am Vet Med Assoc. 2012, 240 (12): 1432-1440. 10.2460/javma.240.12.1432.

62. Health Canada: Canada Health Act Annual Report 2009–2010. 2010, Ottawa, Ontario: , 978-1-100-16312-3:4
63. Moret L, Tequi B, Lombrail P: Should self-assessment methods be used to measure compliance with handwashing recommendations? A study carried out in a French university hospital. Am J Infect Control. 2004, 32 (7): 384-390. 10.1016/j.ajic.2004.02.004.

CHAPTER 16

Exploration Feeding and Higher Space Allocation Improve Welfare of Growing-Finishing Pigs

Herman M. Vermeer*, Nienke C. P. M. M. Dirx-Kuijken and Marc B. M. Bracke

Wageningen Livestock Research, P.O. Box 338, Wageningen 6700 AH, The Netherlands

SIMPLE SUMMARY

A lack of exploration materials in pig pens can result in damaging behavior towards pen mates. The objective of our study was to reduce skin and tail lesions by frequently providing small amounts of feed on the floor and by providing more space per pig. Both the so-called "exploration feeding" and the additional space resulted in fewer skin lesions. Finally, this can lead to a more welfare-friendly pig husbandry.

ABSTRACT

Lack of environmental enrichment and high stocking densities in growing-finishing pigs can lead to adverse social behaviors directed to pen mates, resulting in skin lesions, lameness, and tail biting. The objective of the study was to improve animal welfare and prevent biting behavior in an experiment with a 2 × 2 × 2 factorial design on exploration feeding, stocking density, and sex. We kept 550 pigs in 69 pens from 63 days to 171 days of life. Pigs were supplemented with or without exploration feeding, kept in groups of seven (1.0 m^2/pig) or nine animals (0.8 m^2/pig) and separated per sex. Exploration feeding provided small amounts of feed periodically on the solid floor. Skin lesion scores were significantly lower in pens with exploration feeding ($p = 0.028$, $p < 0.001$, $p < 0.001$ for front, middle, and hind body), in pens with high compared to low space allowance ($p = 0.005$, $p = 0.006$, $p < 0.001$ for front, middle and hind body), and in pens with females compared to males ($p < 0.001$, $p = 0.005$, $p < 0.001$ for front, middle and hind body). Males with exploration feeding had fewer front skin lesions than females with exploration feeding ($p = 0.022$). Pigs with 1.0 m^2 compared to 0.8 m^2 per pig had a higher daily gain of 27 g per pig per day ($p = 0.04$) and males compared to females had a higher daily gain of 39 g per pig per day ($p = 0.01$). These results indicate that exploration feeding might contribute to the development of a more welfare-friendly pig husbandry with intact tails in the near future.

Keywords: animal welfare; environmental enrichment; feeding method; pig; stocking density

1. INTRODUCTION

The Dutch Ministry of Economic Affairs would like to see an increased number of pig production chains addressing improved pig welfare. The sustainable pork chain called "De Hoeve" wants to be at the forefront in taking the next step in the area of animal welfare. This step consists of no longer docking the tails of the piglets. Up to now, not docking in conventional pig husbandry increases the risk of tail biting [1,2]. More enrichment, space (low stocking density), stable social groups, and improved management may contribute significantly to healthy curly tails of slaughter pigs [1,3,4].

Enrichment was identified as the main risk factor [5], but also fully and partially slatted floors, more than five animals per feeding space and less than 1 m^2 per animal may increase the risk of tail biting. The European Union welfare regulations require 0.65 m^2 per pig and in The Netherlands 0.80 m^2 per pig up to 110 kg. This project examined how tail and skin damage can be prevented with exploration feeding, a lower stocking density, and separate sexes (males and females). The study was conducted at the demonstration farm of the Dutch sustainable pork chain "De Hoeve". The proximate aim of this project was to reduce biting wounds on the skin and tails by periodically providing small amounts of food on the floor as environmental enrichment.

The objective was to produce slaughter pigs without tail biting in docked tails, skin without lesions, and sound legs, including good performance as a step on the route towards no tail biting in pigs with intact tails.

2. MATERIALS AND METHODS

2.1. Animals

In total 550 pigs in nine batches were followed during the growing finishing period. These were crossbred German Pietrain terminal boar × Topigs20 sow. Half of the pigs were males (n = 275) and half were females (n = 275). The animals entered the growing finishing unit at nine weeks of age. After 108 (89–133) days on average they were ready for slaughter at 116 (91–140) kg body weight. The animals were born and finished at the same farm. At weaning, the littermates were kept together as much as possible, and groups contained both males and females. Weaners were kept in stable groups of 15–20 animals. Mixing happened when the pigs entered the finishing house at 63 days of age, where males and females were raised in separate pens. The animals had a tail of medium length, which was about 8 cm long at the time of slaughter. Tails were docked using electric cauterization at three to four days after birth, combined with an iron injection and receiving an ear tag. The tails were left slightly longer than was routine practice at the farm and males were not castrated.

2.2. Ethical Statement

The study was conducted in accordance with the Declaration of Helsinki, and the protocol was considered not harmful for the animals. The treatments were additional environmental enrichment and regarded as beneficial for animal welfare and the observations were all non-invasive.

2.3. Housing

The animals were housed in a finishing house in nine identical rooms with eight pens of 6.9 m^2 each (Figure 1). Each pen contained either seven or nine animals, resulting in space allowances of 1.0 m^2/pig or 0.8 m^2/pig respectively. The pens were equipped with a chain and plastic ball combination as enrichment material. Starting from the inspection alley, pens had concrete slats (1.20 m), a solid convex lying area (1.40 m), and metal slats (1.05 m). The pens were 1.90 m wide with a total area of 6.94 m². Fresh air entered through the ceiling and was discharged via a fan in the back of the unit. Artificial light was on from 6 a.m. to 10 p.m.

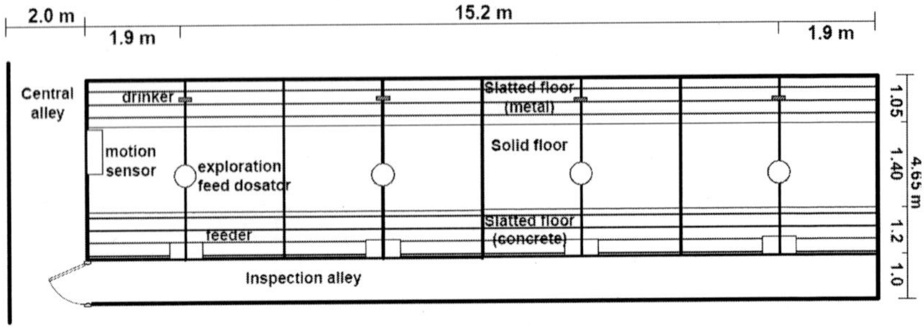

Figure 1. Layout of one of the nine rooms with eight pens with motion sensors and dosators.

The animals were all fed ad libitum using dry pelleted feed. Feed was provided in a dry-wet feeder with one feeding space per pen. Several times a day the feeder was replenished by an automatic feeding system. Drinking water was presented ad libitum in a water bowl in the back of the pen.

2.4. Design

The study had a 2 × 2 × 2 factorial design. The three factors were:

1) Exploration feeding: present or absent.
2) Space allowance/group size: nine pigs at 0.8 m² per pig or seven pigs at 1.0 m²/pig.
3) Sex: males (entire males) or females.

Nine identical rooms were used and each room had eight pens. Pens were randomly assigned to treatment combinations, and every combination was allocated to one pen per room. Males and females were kept in alternate pens. Due to a lack of sufficient animals, 3 pens of the potential 72 pens remained empty, resulting in a total of 69 pens in the dataset.

2.5. Exploration Feeding

In the pens with exploration feeding, the animals received additional small amounts of pelleted feed on the solid floor 25 to 30 times a day during the light period. The feed was provided via a volume dispenser containing several litres of feed (dosator) with an auger. Per two pens, one dosator was mounted with two outflow openings, one per pen (Figure 1). In the pens without exploration feeding the outflow opening was closed. The system was specially designed for this experiment (Coppens Constructie en Stalinrichtingen BV, Westerhoven, The Netherlands).

Provision of the feed was only possible during the light period between 6 a.m. and 10 p.m. The number of seconds that the auger turned determined the amount of feed provided. Times were set using a central process computer in the central corridor, where provided portions were recorded as "pulses". The quantity of feed provided each time was dependent on the age of the animals. During the first three weeks in the finishing house the pigs received feed for 3 s per portion, which resulted in the distribution of 27 g of feed on average. After three weeks, feed was provided for just one second per portion, which resulted in 12 g each time. Supply frequencies depended on the activity of a single pig or group(s) of pigs as registered with a motion sensor. The maximum number of portions was 42 in the first period and 29 in the last period. This infrared sensitive motion sensor was mounted at the front wall of the room where it detected activity especially in the first pens. When activity was registered for three minutes exploration feed was provided; but when within this period a period of more than 15 s of inactivity was registered, the timer was restarted. The minimum waiting time between bouts of exploration feeding was set at 20 min (during the first three weeks) or 30 min (after three weeks).

The first three weeks in the finishing house up to 42 portions per day could be provided (between 6 a.m. and 10 p.m. = 16 h divided by 23 min (20 min + 3 min activity). After three weeks until 100 days after start of the finishing period the maximum number of portions per day was 29 (16 h divided by 33 min). More details are available in the Dutch research report [6].

2.6. Data Collection

All data was collected on the animal level, but for analysis the pen averages per observation were used. From the second week on lesions of the skin, legs, and tails were scored using existing protocols every four to five weeks. Some batches were scored three, others four times, depending on how soon the first pigs were ready for slaughter in the fourth month and the observations stopped.

For skin lesions, separate scores were given for the front, middle, and hind parts of the pigs on a scale from 0 to 5 with increasing level of lesions (where 0 = none, 1 = low, 2 = mild, 3 = moderate, 4 = severe, 5 = very severe). In this visual score fresh, red lesions have a higher weight than older, black scars (crusts). With this protocol, it is not possible to discriminate between lesions caused by aggression due to the initial mixing of the pigs and by competition for resources like feed.

For scoring legs the Welfare Quality®, lameness protocol was used [7], using a scale from 0 to 2 (where 0 = not lame; 1 = non symmetrical walking but using all feet; 2 = not using one of the legs).

Tails were scored according to the protocol of Zonderland [8], where tail length related to starting length is scored in five classes (1: complete tail; 2: three quarters left; 3: half left; 4: one quarter left; 5: less than a quarter of the tail left), tail lesions in three classes (1: no injury; 2: bite marks; 3: one or more wounds), and blood on the tail in four classes (1: no blood; 2: dark brown/black, i.e., dry crust; 3: dark red/brown, i.e., older blood; 4: red, i.e., fresh blood and wet tail tips).

The following production and slaughter parameters were recorded per pig: mortality, starting weight, starting date, slaughter date, cold carcass weight (from which mean end weights and growth rates per pen were calculated), meat percentage, muscle, and fat thickness. We used the Dutch equation live weight = carcass weight × (1.3 + (83 − carcass weight) × 0.0025) to calculate live weight from cold carcass weight [9].

2.7. Statistical Analysis

Skin, leg, and tail scores were analyzed with a generalized linear mixed model for ordinal data in Genstat 18th Edition (VSN International, Hemel Hempstead, UK) [10]. Fixed effects were month after start of batch, exploration feeding, group size, and sex (including two way interactions). Random effects were room, pen number, starting date, and the interaction between starting date and month. The hypotheses were all tested with a significance level of 0.05.

In the ordinal model the K-1 intercept terms or thresholds (αn) are estimated, where K is the number of classes in the class-score, using the equation Logit (γn (x)) = $\alpha n + \beta Tx$, where βTx is a short notation of all model terms. Thresholds, averages for the underlying distribution Z, and variance components for all the terms in the random model were estimated using the method as described in [11], using the procedure IRCLASS [12].

Growth rates were calculated—using the starting weight, number of days in the room, and the calculated live weight—and analyzed using REML (REsidual Maximum Likelihood) in Genstat 18th Edition [10] with exploration feeding, group size, and sex as fixed effects and room and pen as random effects.

3. RESULTS

3.1. Usage of Exploration Feeding

Growing pigs (until three weeks in the finishing house) realized about 30 portions per day, which was 75% of the maximum number of portions. During the rest of the finishing period the pigs realized 23 portions, which was 80% of the maximum portions. Over time, the number of portions slightly decreased, but this is partly caused by the extension of the intervals between the portions. On average, pig pens with exploration feeding received 1250 g of additional feed per animal during the growing-finishing period, this was 12 g per animal per day. Assuming that all the exploration feed was consumed, this implied that exploration feeding contributed about 0.5% to the overall feed intake (which was 2.6 kg/day).

3.2. Skin, Leg, and Tail Scores

In Table 1, the results of the skin scores are summarized for the main factors (exploration feeding, space allowance, and sex). All factors had significant effects on the skin lesion scores. Over time, the skin lesion scores on the front body declined ($p = 0.03$). The skin lesion scores on the middle and hind body were not affected by month (Table 2).

An interaction was found between exploration feeding and sex ($p = 0.022$) for the skin lesion score on the front of the pig. Males had higher skin lesion scores at the front than females, and this difference is smaller with exploration feeding compared to no exploration feeding ($p = 0.022$; Figure 2). Another significant interaction was found between space allowance and month for the skin lesions scores at the hind body part ($p = 0.042$; Figure 3). A higher stocking density resulted in more skin lesions at the hind body part and the difference increased over time.

Of the 2178 observed legs, only 33 (1.5%) scored a 1 or 2 for lameness, with no effect of treatments. Initial tail lengths varied considerably, and the experimental treatments had no effect on the scores for tail length, tail lesion, and blood score. During the trial, hardly any tail was shortened due to tail biting. Occasionally tail biting was observed. The level of tail wounds and blood scores decreased from the first to the third month (Table 3).

Table 1. Mean skin lesion scores (0–5) at the front, middle and back of the pigs related to the main treatment factors with SEM (Standard Error of the Mean) and p value for 64 pens in total.

Lesion Score	Exploration Feeding	No Exploration Feeding	SEM	*p* Value
Front	1.66	1.77	0.039	0.028
Middle	1.34	1.58	0.040	<0.001
Back	1.26	1.5	0.036	<0.001
n pens	33	36		
	1.0 m²/pig	0.8 m²/pig	SEM	*p* Value
Front	1.66	1.78	0.039	0.005
Middle	1.39	1.53	0.040	0.006
Back	1.30	1.46	0.036	<0.001
n pens	33	36		
	Males	Females	SEM	*p* Value
Front	1.86	1.57	0.039	<0.001
Middle	1.54	1.38	0.040	0.005
Back	1.48	1.28	0.036	<0.001
n pens	36	33		

Table 2. Mean skin lesion scores (0–5) at the front, middle and back of the pigs related to month (age) with SEM and p value for 64 pens in total; different superscripts within a row indicate a statistical difference (p < 0.05).

Lesion Score	Month 1	Month 2	Month 3	Month 4	SEM	*p* Value
Front	1.92 [a]	1.89 [a]	1.56 [b]	1.48 [b]	0.034	0.03
Middle	1.64	1.68	1.3	1.22	0.033	0.123
Back	1.32	1.55	1.29	1.42	0.033	0.675
n pens	69	53	63	53		

Table 3. Mean tail length, lesions, and blood scores over time with SEM and p value; different superscripts within a row indicate a statistical difference (p < 0.05).

Month	1	2	3	4	SEM	*p* Value
Tail length	1.028	1.000	1.000	1.002	0.007	no analysis possible *
Tail lesions	1.34 [a]	1.24 [b]	1.10 [c]	1.13 [c]	0.015	0.010
Blood scores	1.20 [a]	1.13 [b]	1.06 [c]	1.09 [b,c]	0.013	0.004

* Too many treatment combinations with only score 1.

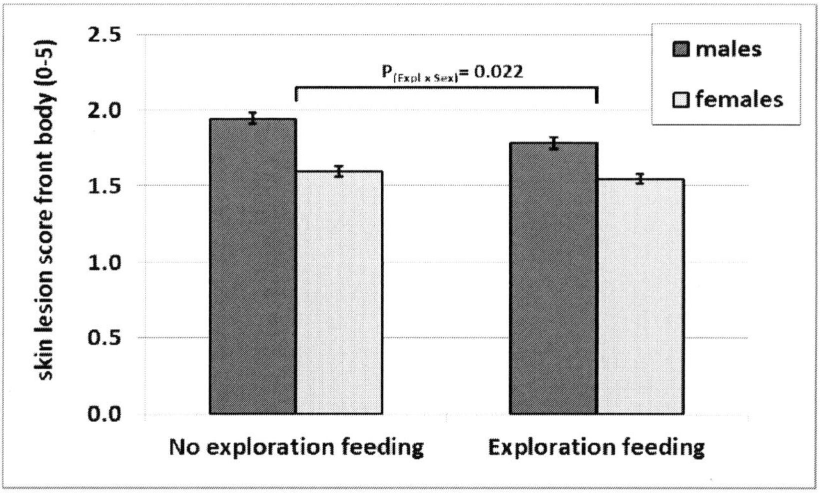

Figure 2. Mean front skin lesion score per treatment (exploration feeding) and sex with SEM.

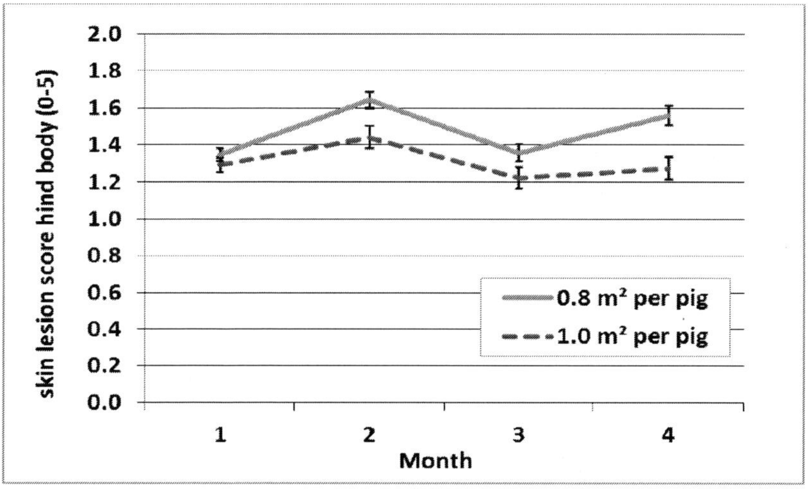

Figure 3. Mean skin lesion scores at the hind body part with SEM in error bars per month for high and low space allowance.

3.3. Performance

Table 4 shows the effects per treatment on performance. Exploration feeding had no effect on performance. Males had a 39 g/d ($p = 0.01$) higher daily gain and a 2 mm ($p = 0.02$) lower muscle thickness than females. A space allowance of 0.8 m^2/pig had a 27 g/d ($p = 0.04$) lower daily gain than 1.0 m^2/pig.

Table 4. Mean performance data per main treatment factor with SEM and p value.

Variable	No Exploration Feeding	Exploration Feeding	SEM	*p* Value
Number of pens	24	21		
Growth (g/pig/d)	808.9	810.1	9.2	0.75
Muscle thickness (mm)	61.3	60.6	0.38	0.42
Fat thickness (mm)	14.3	14.0	0.13	0.29
Lean meat (%)	58.4	58.6	0.11	0.49
	Males	Females	SEM	*p* Value
Number of pens	24	21		
Growth (g/pig/d)	828.9	790.1	9.2	0.01
Muscle thickness (mm)	60.0	62.0	0.38	0.02
Fat thickness (mm)	14.1	14.2	0.13	0.76
Lean meat (%)	58.4	58.6	0.11	0.43
	0.8 m^2/pig	1.0 m^2/pig	SEM	*p* Value
Number of pens	21	24		
Growth (g/pig/d)	796.1	822.9	9.2	0.04
Muscle thickness (mm)	60.7	61.1	0.38	0.60
Fat thickness (mm)	13.9	14.3	0.13	0.21
Lean meat (%)	58.6	58.4	0.11	0.49

4. DISCUSSION

Exploration feeding offers the pigs the opportunity to perform elements of species specific foraging behavior. In this experiment, we measured not the behavior around exploration feeding itself, but features related with health and behavior—like skin lesions, lameness, and tail biting. In this way, the effect of aggression and damaging behavior during the previous weeks is measured with a limited number of observations like described in the Welfare Quality® Pig Protocol [8].

Exploration feeding reduces aggression as measured by the number of scratches (lesions) at the front, middle, and hind body part of the pigs. Although the original objective of reducing tail biting by applying exploration feeding was not achieved in this experiment, the overall skin condition was better in pigs with exploration feeding, implying a reduction of unwanted behavior towards penmates by providing additional nutritional enrichment. In a situation with intact tails, the effect of exploration feeding on tail biting might have been more pronounced. Despite exploration feeding, the pigs in some pens nevertheless showed tail biting. This indicates that tail biting is multifactorial and that exploration feeding alone may not be sufficient to prevent it [1,4,13].

Beattie [14] concluded that pen enrichment is more important than lowering the stocking density to prevent skin damage. In the present study, both exploration feeding and pens with increased space allowance reduced skin lesion scores.

Floor feeding will only work in pens with partly slatted floors, to prevent loss of feed. The pen floors in this study had a 40% convex solid floor measuring 1.4 × 1.9 m. This floor was used both as lying area and for exploration feeding. The convex floor had a downward slope to the slatted floor on both sides, increasing the risk of feed losses. A better design would be a slightly sloping floor with provision of the exploration feed at the highest point reducing the risk of feed loss. The solid floors in pens with exploration feeding remained clean. The pigs clearly separated the functional areas for foraging/lying on one side and excretion on the other side. However, more fouling of the solid floor was (occasionally) observed at the lower stocking densities.

Exploration feeding was controlled by a sensor registering general activity in the front pens in each unit. The pigs received 75%–80% of the maximum number of servings recorded by the process computer. Occasional visual observations showed that the pigs did not react very quickly on a new portion, however hardly any feed was left on the floor at the start of the next serving. The motion sensor only reacted on activity in the first pens of a unit with eight pens on a row, so the sensor was not selective. Providing a sensor in each pen would probably be too costly. However, pig activity within a room is often strongly synchronized [15]. The question remains which moments or types of activity stimulates the sensor the most. Alternatively, the system could work effectively without motion sensors as well, possibly with a limited number of portions with fixed intervals during the light (active) period.

In this experiment, feed was provided ad libitum and exploration feed was the same as the feed provided in the feed trough. Exploration feed might be more interesting when the pigs would be feed restricted or when another type of feed was provided. However, floor feeding as only feeding system is also known to have a higher risk on increased levels of aggression [16], and this would be counterproductive.

Increased space allowance (1.0 versus 0.8 m^2/pig) improved both production and welfare as measured by better skin scores and elevated growth rates. This study does not allow a conclusion as to whether this effect was due to increased space allowance only or also due to reduced group size. The review of D'Eath et al. [2] and the analysis of Gonyou et al. [17] both reported a reducing effect of a lower space allowance on daily gain, especially below 0.8 m^2 per pig. Vermeer et al. [16] even found an improved daily gain with increased space allowances above 1.0 m^2 per pig. This could have an effect through the number of pigs per feeding space. However, both D'Eath et al. [2] and Gonyou et al. [17] considered feeder space allowance as less important. This is supported by our study where feeder space was not limiting with less than 10 pigs per ad libitum feeder. Moinard [5] found similar effects on tail biting of the combined effect of increased space allowances per pig and reduced numbers per feeding space. For an unbiased comparison of stocking densities, the group sizes should be constant and pen sizes could be varied in a future study.

Males generally exhibit more aggression than females. Aggression leads to higher front skin lesion scores. In this experiment, exploration feeding reduced skin scores more in males compared to females. This reduction in male aggression by exploration feeding at the standard space allowance of 0.8 m^2 per pig was not evident at the lower stocking density of 1.0 m^2 per pig. As a consequence, in addition to reducing stocking density, exploration feeding can provide another possible strategy to reduce aggression in the growing EU tendency of raising non-castrated males.

The motion sensor ensured that feed was distributed only when there was activity in the pens. However, only one motion sensor was used for the entire unit with eight pens. The synchronicity of activity in pig rooms ensures that control on a room level will be sufficient. However, there is a possibility that exploration feed was provided in pens where the pigs were not active and as occasionally observed the pigs were not immediately activated by the falling exploration feed. Only when a stockman entered the unit or when the pigs became active at a later moment, they will be focused on the previously spread feed on the floor. This suggests that the pigs were not activated by the servings of exploration feed per se that was also available ad libitum from a dry-wet feeder.

Exploration feeding will be more easily adopted by pig farmers when it is economically beneficial. In this experiment, exploration feeding showed no clear advantage or disadvantage for the performance of the pigs. However, the real advantage should come from a higher financial return in a market concept without castration and tail docking. Exploration feeding could contribute to a conversion towards such a market concept. Detailed costings of the dosage system for large scale application are difficult to make, but a rough estimation is possible. An investment of €10 per pig place with 10% annual costs and three batches per year the costs per pig would be estimated at €0.33. The estimated food spillage is €0.10 to €0.20 per pig, resulting in estimated costs for exploration feeding per pig between €0.45 and €0.50. This implies that a higher welfare level can be reached for €0.005 per kg of carcass weight.

ANIMAL WELFARE IMPLICATIONS

The pig welfare in the exploration feed treatment was improved as concluded from the reduced skin lesion scores and no negative effects compared to the control treatment. Similar welfare and health improvements were found for pigs with a higher space allowance.

5. CONCLUSIONS

From this study on intensively reared, growing-finishing pigs we conclude that:

- Exploration feeding, reduced stocking density, and a lower group size all improved skin lesion scores.
- Males showed more skin lesions than females.
- Exploration feeding reduced the difference between males and females in skin lesion scores at the front of the pigs.
- The difference of skin lesion scores at the hind of the pigs between high and low stocking densities/group sizes increased during the finishing period, with the lowest skin lesion score for the lowest stocking density.
- Tail scores were not affected by one of the treatments, but declined in the course of time.
- The daily gain of the males is higher than the females.
- The daily gain in the low stocking density is higher than in the high stocking density.
- At the low stocking density, the difference in daily gain between males and females is smaller than in high stocking density.

- Exploration feeding has no effect on performance.
- Both exploration feeding and lowering stocking density improve animal welfare.

ACKNOWLEDGMENTS

This project was funded by the Dutch Ministry of Economic Affairs and in kind supported by sustainable pork chain "De Hoeve BV".

REFERENCES

1. European Food Safety Authority. Scientific opinion of the Panel on Animal Health and Welfare on a request from commission on the risks associated with tail biting in pigs and possible means to reduce the need for tail docking considering the different housing and husbandry systems. *EFSA J.* **2007**, *611*, 1–13.
2. D'Eath, R.B.; Arnott, G.; Turner, S.P.; Jensen, T.; Lahrmann, H.P.; Busch, M.E.; Niemi, J.K.; Lawrence, A.B.; Sandøe, P. Injurious tail biting in pigs: How can it be controlled in existing systems without tail docking? *Animal* **2014**, *8*, 1479–1497.
3. Petherick, J.C.; Blackshaw, J.K. A review of the factors influencing the aggressive and agonistic behavior of the domestic pig. *Aust. J. Exp. Agric.* **1987**, *27*, 605–611.
4. Zonderland, J.J. 2010. Talking Tails-Quantifying the Development of Tail Biting in Pigs. Ph.D. Thesis, Wageningen University, Wageningen, The Netherlands, 2010.
5. Moinard, C.; Mendl, M.; Nicol, C.J.; Green, L.E. A case control study of on-farm risk factors for tail biting in pigs. *Appl. Anim. Behav. Sci.* **2003**, *81*, 333–355.
6. Vermeer, H.M.; Dirx-Kuijken, N.; Wisman, A.; Bikker, A. *Effect Van Exploratie Voedering en Hokbezetting op Welzijn van Vleesvarkens (Effect of Exploration Feeding and Space on Welfare of Growing Finishing Pigs)*; Report 546; Wageningen UR Livestock Research: Lelystad, The Netherlands, 2011.
7. Welfare Quality®. *Welfare Quality® Assessment Protocol for Pigs (Sows and Piglets, Growing and Finishing Pigs)*; Welfare Quality® Consortium: Lelystad, The Netherlands, 2009.
8. Zonderland, J.J.; Fillerup, M.; Van Reenen, C.G.; Hopster, H.; Spoolder, H.A.M. *Preventie en Behandeling van Staartbijten Bij Gespeende Biggen. (Prevention and Treatment of Tail Biting in Weaned Pigs)*; RIAH Report 18; Praktijkonderzoek: Lelystad, The Netherlands, 2003.
9. Wageningen UR Livestock Research. *Handboek Varkenshouderij-Praktisch Naslagwerk Voor de Nederlandse Varkenshouderij (Handbook Pig Husbandry)*; Wageningen UR Livestock Research: Wageningen, The Netherlands, 2015; pp. 380–381.
10. Genstat. *Reference Manual*; Release 6.1; VSN International: Oxford, UK, 2002.
11. Keen, A.; Engel, B. Analysis of a mixed model for ordinal data by iterative re-weighted REML. *Stat. Neerl.* **1997**, *51*, 129–144.
12. Keen, A. Procedure IRCLASS. In *Biometris GenStat Procedure Library Manual*, 12th ed.; Goedhard, P.W., Thissen, J.T.N.M., Eds.; Biometris: Wageningen, The Netherlands, 2009; Available online: http://edepot.wur.nl/50492 (accessed on 1 April 2015).
13. Bracke, M.B.M.; Hulsegge, B.; Keeling, L.; Blokhuis, H.J. Decision support system with semantic model to assess the risk of tail biting in pigs: 1. Modelling. *Appl. Anim. Behav. Sci.* **2004**, *87*, 31–44.

14. Beattie, V.E.; Walker, N.; Sneddon, I.A. An investigation of the effect of environmental enrichment and space allowance on the behavior and production of growing pigs. *Appl. Anim. Behav. Sci.* **1996**, *48*, 151–158.
15. Vermeer, H.M.; De Greef, K.H.; Houwers, H.W.J. Space allowance and pen size affect welfare indicators and performance of growing pigs under comfort class conditions. *Livest. Sci.* **2014**, *159*, 79–86.
16. Czermely, D.; Wood-Gush, D.G.M. Agonistic behavior in grouped sows. *Biol. Behav.* **1986**, *11*, 244–252.
17. Gonyou, H.W.; Brumm, M.C.; Bush, E.; Deen, J.; Edwards, S.A.; Fangman, R.; McGlone, J.J.; Meunier-Salaun, M.; Morrison, R.B.; Spoolder, H.; et al. Application of broken-line analysis to assess floor space requirements of nursery and grower-finisher pigs expressed on an allometric basis. *J. Anim. Sci.* **2006**, *84*, 229–235.

CHAPTER 17

Cattle brucellosis in traditional livestock husbandry practice in Southern and Eastern Ethiopia, and its zoonotic implication

Bekele Megersa[1,2*], Demelash Biffa[1,2], Fekadu Niguse[1], Tesfaye Rufael[3], Kassahun Asmare[1] and Eystein Skjerve[2]

[1]*School of Veterinary Medicine, Hawassa University, P.O. Box 05, Hawassa, Ethiopia.*

[2]*Center for Epidemiology and Biostatistics, Norwegian School of Veterinary Science, P.O. Box 8146 Dep., 0033 Oslo, Norway.*

[3]*National AnimalHealth, Diagnostic and Investigation Centre, P.O. Box 04, Sebeta, Ethiopia.*

ABSTRACT

Background

Cattle brucellosis has significant economic and zoonotic implication for the rural communities in Ethiopia in consequence of their traditional life styles, feeding habits and disease patterns. Hence, knowledge of brucellosis occurrence in traditional livestock husbandry practice has considerable importance in reducing the economic and public health impacts of the disease.

Methods

A total of 1623 cattle sera were serially tested using the rose Bengal test as screening and complement fixation test as confirmatory tests. The Stata survey command was used to establish prevalences for the overall and individual variables, while potential risk factors for seropositivity were analyzed using a multivariable logistic regression analysis.

Results

The results showed that 3.5% (95% CI = 2.4, 4.5%) of the animals and 26.1% (95% CI = 18.6, 33.7) of the herds tested had antibodies against Brucella species. Village level seroprevalence ranged from 0% to 100%. A higher seroprevalence was observed in pastoral system than mixed farming although this variable was not significant in the final model. The final logistic regression model identified herd size; with large (odd ratio (OR) = 8.0, 95% CI = 1.9, 33.6) and medium herds (OR = 8.1, 95% CI = 1.9, 34.2) showing higher risk of Brucella infection when compared to small herds. Similarly, the odds of Brucella infection was higher in cattle aged above 4 years when compared to age groups of 1-2 (OR = 5.4, 2.1, 12.9) and 3-4 years (OR = 3.1, 95% CI = 1.0, 9.6). Herd level analysis of the risk factors revealed that large and medium herds as well as herds kept with multiple livestock species were at higher risk of acquiring Brucella infection. Brucellosis in traditional livestock husbandry practices certainly poses a zoonotic risk to the public, in consequence of raw milk consumption, close contact with animals and provision of assistance during parturition. Due to lack of diagnostic facilities and information on its occurrence, human brucellosis is most likely misdiagnosed for other febrile diseases prevailing in the areas and treated empirically.

Conclusions

The results of this study demonstrated that bovine brucellosis is widely prevalent in the study areas particularly in pastoral production system. Hence, the study suggests the need for implementing control measures and raising public awareness on prevention methods of brucellosis.

1. INTRODUCTION

Brucellosis remains widespread in the livestock populations, and represents a great economic and public health problem in African countries. Brucellosis causes abortion which is the major means of spread by infected afterbirth or fetus as well as excretion of excessive organisms which can easily be acquired by susceptible animals. The epidemiology of the disease in livestock and humans as well as appropriate preventive measures are not well understood, and in particular such information is inadequate in sub-Saharan Africa [1]. The epidemiology of cattle brucellosis is complex and influenced by several factors [2]. These can be broadly classified into factors associated with the transmission of the disease between herds, and factors influencing the maintenance and spread of infection within herds. The climatic and agro-ecological diversities of Ethiopia may allow a wide range of livestock production systems, and therefore, different management systems, multiple livestock species per holding, stock density and social organizations to handle livestock may account for the widespread risk factors for maintenance and transmission of cattle brucellosis.

The evidences of Brucella infections in Ethiopian cattle have been serologically demonstrated by different authors [3, 4, 5, 6, 7]. A relatively high seroprevalence of brucellosis (above 10%) has been reported from smallholder dairy farms in central Ethiopia [4] while most of the studies suggested a low seroprevalence (below 5%) in cattle under crop-livestock mixed farming [3, 6, 8, 9]. There is a scarcity of published literature on the status of cattle brucellosis in pastoral areas of the country where large population of cattle are reared. So far, a study carried out in east Showa zone of Ethiopia showed a relatively higher seroprevalence in pastoral than agropastoral system [10].

Most of the previous studies on cattle brucellosis have been carried out in central and northern Ethiopia, and do not provide an adequate epidemiological picture of the disease in different agro-ecological zones and livestock production systems of the country. In particular, there is no information on cattle brucellosis across various livestock production systems of southern and eastern part of the country, which gave impetus to the initiation of this study. The present study was therefore aimed at determining the prevalence of cattle brucellosis and associated risk factors across the two livestock production systems, pastoral and crop-livestock mixed systems, in Southern and Eastern Ethiopia.

2. MATERIALS AND METHODS

2.1 Study area and study animals

The study was carried out in eight administrative zones in southern and eastern Ethiopia; namely Dawro, Sidama, Gedeo, Hadiya, South Omo, Borana, Jijiga and Shinle (Figure 1). A total of 33 districts were selected from these zones, from which 96 villages were chosen for sampling. The study areas are generally characterized by diverse agro-climatic zones with altitude ranging from 370 meters in Dasanach and Omoratte districts (South Omo) to 3175 meters above sea level (m.a.s.l.) in Bulle district (Gedeo). Based on altitude range, the study areas were broadly classified into the traditional agro-climatic classifications of lowland "Kola" (< 1500 m.a.s.l.); midland "Weynadega" (1500 - 2400 m.a.s.l.) and highland "Dega" (> 2400 m.a.s.l.). Geographically, the study areas cover latitude and longitude ranges of 03° 34'21" to 10° 54' 93" East and 36° 01' 50" to 43° 70' 56" North.

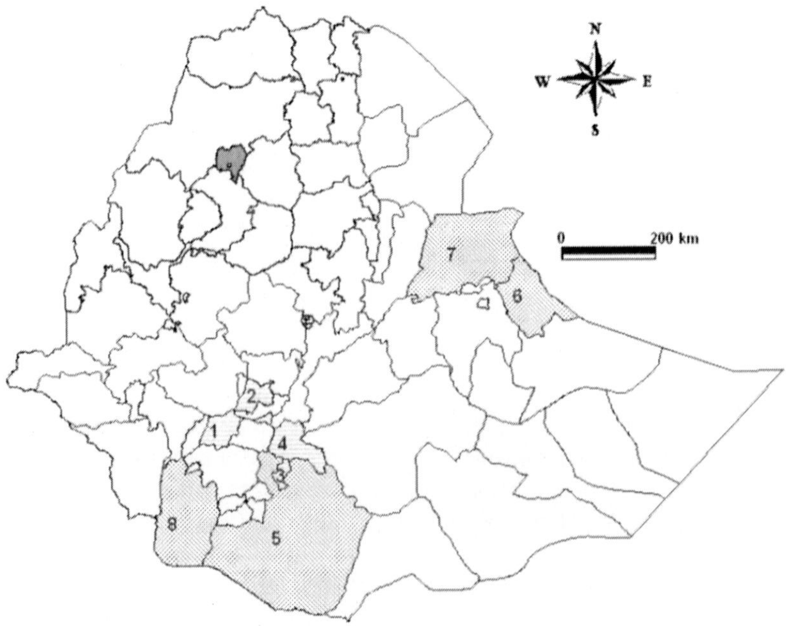

Figure 1 Zonal administrative map of Ethiopia showing the study areas: Zones indicated by numbers 1-4 (1. Dawro, 2. Hdiya, 3. Gedeo, 4. Sidama) are mixed farming while the remaining zones are pastoral (5. Borana, 6. Jijiga, 7. Shinle and 8. South Omo).

Livestock production in the area is dominated by extensive production system, in which indigenous cattle are allowed to graze freely during day time and kept in open enclosures during the night. The extensive production system is further categorized into pastoral and crop-livestock mixed farming systems. Four zones; Borana, Jijiga, Shinle and South Omo are characterized by pastoral system while the remaining four zones practice crop-livestock mixed farming. The number and composition of animal species per holding in the mixed farming is relatively lower than the pastoral system. Table 1 shows the mean herd size and sample proportion over administrative zones. Livestock composition varies from keeping cattle as dominant stock with variable number of small ruminants in crop-livestock mixed farming systems to a camel, cattle and small ruminant composition in pastoral areas. As livestock brucellosis control intervention by immunization has never been attempted in Ethiopia, there is no history of vaccination against brucellosis in the study areas.

Table 1 Mean herd size and sample proportions of the studied herds in each administrative zone of the study areas

Production systems	Administrative Zones	Herd size Mean (95% CI)	Sample proportion (%)
Mixed farming*	Dawro	8.8 (8.4, 9.3)	95.4
	Gedeo	12.6 (11.8, 13.4)	83.3
	Hadiya	15.4 (14.9, 15.9)	87.0
	Sidama	13.1 (12.7, 13.5)	91.0
Pastoral system	Borana	43.6 (42.5, 44.7)	49.4
	Jijiga	20.1 (19.4, 20.9)	77.2
	Shinle	20.7 (19.9, 21.4)	70.0
	South Omo	45.2 (44.0, 46.3)	43.2

* Mixed farming, also known as sedentary farming, is where crop-livestock mixed farming is practiced.

2.2 Study Design and sample size determination

A cross-sectional multi-stage sampling, with zone as highest and herd as lowest sampling stages, district and village in between the two stages, was carried out from October 2007 to March 2008. Selection of the study unit at each stage was based on a mixed design of convenience and random samplings. Zones were conveniently selected based on geographic localities and dominant livestock production system, whereas districts and villages were randomly selected following a randomization of districts and villages when lists were obtained from respective administratives. When this was not the case in pastoral systems, herds were sampled conveniently in consultation with herd owners. As information on prevalence of cattle brucellosis is not available for the study areas, we adopted a sampling technique for detection of disease [11]. Assuming a target prevalence of 5% and district level sensitivity at 95%, we would need to sample a minimum of 59 animals to get at least one positive animal. With available logistics and resources, we managed to sample a total of 1623 animals from 33 districts. The number of animals sampled from each area could vary according to livestock density, access to transportation and availability of logistic facilities. Study animals include all animals aged 1 year and above in a selected herd (while about 50% of the animals in large herds were to be sampled).

2.3 Serum Sample Collection and Testing

From each animal, 10 ml of blood was aseptically collected from the jugular vein using plain vacutainer tubes and clotted at room temperature for 12 hours. Sera were then collected in sterile tubes and transported to the laboratory using ice box where stored at -20°C until tested. Subsequently, the rose Bengal test (RBT), Institut Pourquir, rue de la Galera 34097 Montpellier, France, was carried out by adding an equal volume of antigen (30 μl) and serum onto the glass slide. The antigen and test serum were mixed thoroughly by plastic applicator, shaken for 4 minutes, and degree of agglutination was visually recorded immediately. Complement fixation test (CFT) was performed at the National Animal Health Diagnostic and Investigation Center (NAHDIC), Sebeta Ethiopia, using Brucella antigen and control sera (positive and negative) produced by Veterinary Laboratories Agency (VLA, New Haw Addlestone, Surrey, KT15 3NB, UK). The antigen was standardized at 1:10 dilution. Two-fold dilutions of test sera (1:5, 1:10, 1:20 and 1:40) were prepared in U-shape 96-well micro-titer plates before adding Brucella antigen, guinea pigs complement and 3% sensitized sheep red blood cells. The plates were incubated at 37°C for 30 minutes with agitations (warm fixation) and results were read after the plates have been centrifuged at 2500 rpm for 5 minutes at 4°C. CFT was regarded positive when the reading was as complete fixation (complete inhibition of haemolysis) or nearly complete fixation (25% haemolysis) at 1:10 dilutions. This cut-off point was taken to optimise specificity and ensure that seropositive cases were due to brucellosis. This cut-off is routinely used by NAHDIC in their diagnostic system. An animal was considered positive if tested seropositive on both RBT and CFT in serial interpretation. The test was regarded as valid if the negative control serum showed complete haemolysis and the positive control shows inhibition of haemolysis. The use of RBT/CFT combinations, the most widely used serial scheme, is generally recommended to maximize specificity of the test result by ruling out false positive serological cross-reactions [11].

2.4 Data collection and analysis

Putative biological and environmental factors believed to be associated with the epidemiology of brucellosis were recorded in a Microsoft Excel® Spread Sheet. Data on individual animals such as sex, age, herd size, stock composition, production system and agro-climate were recorded. All the necessary statistical analysis was performed using STATA version 10.0 for Windows (Stata Corp. College Station, TX). The individual positive outcome was defined as any animal with RBT+ and CFT+, while herd or village positivity was any herd or village having at least one seropositive animal.

The prevalence of Brucella antibodies at the individual level was established by the Stata survey command considering village as a primary sampling unit and each variable as a stratum and sampling weight variable. Association of exposure variables with seroprevalence was analyzed at individual animal level using logistic regression following adjustment for sampling weight according to sampled numbers and estimated number of animals in each village. A multivariable logistic regression model was used to identify risk factors associated with Brucella infection, at individual and herd levels, keeping village as the cluster variable. Variables with a p-value lower than or equal to 0.25 (in univariable analysis) were included in the multivariable logistic model. Further selection of variables was based on backward elimination procedure using a LR-test at 0.05 as cut-point. Prior to building a final model, variables were tested for interaction effects using cross-product terms and for multiple-collinearity using the collinearity matrix index. The validity of the model to the observed data was assessed by computing the Hosmer-Lemeshow goodness-of-fit test. Finally, deviant covariate patterns and their influences on parameter estimates of the model were identified.

3. RESULTS

The test results show that 63 of the tested animals were positive for RBT, of which 51 (81.0%) were further confirmed to be seropositive by CFT. The overall seroprevalence records were 3.5% (95% CI: 2.4, 4.5%), 26.1% (95%CI: 18.6, 33.7), and 31.3% (95%CI: 22.4, 41.6) at animal, herd and village levels, respectively. The seroprevalence distribution of Brucella infection at animal, herd and village levels in the study areas is presented by Table 2. The results of herd and village seroprevalence are nearly comparable. This could result from clustering effects at village levels and thus village would be more appropriate unit of the study than herd. Village level seroprevalence ranged from 0% to 100% with higher seroprevalences in pastoral systems. The highest village level seroprevalence (100%) was recorded for Borana, whereas seroprevalences of over 40% were recorded for villages in Jijiga and Shinle pastoral areas of Eastern Ethiopia. The seroprevalence was generally low in mixed farming areas of Sidama and Gedeo zones, while no seropositive case was detected in villages of Dawro zone.

Table 3 presents results of animal level univariable analysis showing the association of the exposure variables and Brucella seropositivity. The results showed that most of the recorded variables showed a high degree of association with seropositivity to Brucella infection.

Variables with a p-value <0.25 from univariable analysis were included in the final multivariable logistic model. Two variables, sex and altitude range that showed collinearity with other variables (sex with age, altitude with production system, livestock composition and herd size), were not included in the multivariable logistic regression model. The rest variables; age, herd size, stock composition and production system were offered to the model. Further selection of variables in the final model was based on stepwise backward elimination procedure.

The final multivariable logistic regression model (Table 4) showed that animals kept in large (OR = 8.1, 95% CI = 1.9, 34.2) and medium (OR = 8.0, 95% CI = 1.8, 35.0) herd sizes were more likely to be exposed to Brucella infections than those maintained in small herds. Similarly, animals above 4 years of age were more likely to acquire infections than those in age groups of 1-2 (OR = 5.4, 2.1, 12.9) and 3-4 years (OR = 3.1, 95% CI = 1.0, 9.6). Herd level analysis of the risk factors identified an increase in herd size and ruminant composition as the major risk factors for herds to acquire Brucella infection. The Hosmer-Lemeshow goodness-of-fit test showed that the model fitted the data well ($\chi^2 = 2.7$, P = 0.61).

Table 2 Distribution of seropositivity (%) to Brucella antigens in indigenous cattle (at different levels) across the study areas

Study areas		Animal level*		Herd level		Village level	
Production system	Zones	No of animals	Prevalence (95% CI)	No of herds	Prevalence (95% CI)	No of Villages	Prevalence (95% CI)
Mixed farming	Dawro	104	0 (-)	13	0 (0)	7	0 (0)
	Gedeo	161	0.5 (0.05, 1.5)	17	5.9 (0.3, 30.8)	10	10 (0.5, 45.9)
	Hadiya	245	3.5 (1.1, 5.8)	20	30.0 (12.8, 54.3)	17	35.3 (15.3, 61.4)
	Sidama	390	1.8 (0.4, 3.0)	37	13.5 (5.1, 29.6)	26	19.2 (7.3, 40.0)
Pastoral system	Borana	271	4.7 (2.1, 7.3)	16	68.8 (41.5, 87.9)	6	100.0
	Jijiga	62	3.0 (1.1, 7.1)	4	50.0 (9.2, 90.8)	4	50.0 (9.2, 90.8)
	Shinle	210	6.6 (3.1, 10.1)	15	40.0 (17.1, 67.1)	14	42.9 (18.8, 70.4)
	South Omo	180	3.4 (0.9, 6.1)	12	33.3 (11.3, 64.6)	12	33.3 (11.3, 64.6)
Total		1623	3.5 (2.4, 4.5)	134	26.1 (19.1, 34.5)	96	31.3 (22.4, 41.6)

* Animal level seroprevalence was calculated following adjustment for sample weight.

Table 3 Prevalences (%) and univariable analysis of the potential risk factors for seropositivity to Brucella antibodies in indigenous cattle (following adjustment for sampling weight)

Variable	Level	No. of Sample	Prevalence (95% CI)	OR (95% CI)	P-value*
Age groups	1 - 2 years	497	1.0 (0.1, 1.8)	1.0 (-)	
	3-4 years	366	3.0 (1.0, 5.0)	3.1 (1.0, 9.3)	0.046
	> 4 years	760	5.1 (3.4, 7.0)	5.4 (2.1, 14.1)	0.001
Sex	Male	485	2.2 (0.7, 3.3)	1.0 (-)	
	Female	1138	4.1 (2.8, 5.4)	2.1 (1.0, 4.3)	0.040
Herd size	< 15 animals	414	0.5 (0.2, 1.2)	1.0 (-)	
	15 - 29 animals	758	3.9 (2.1, 5.6)	8.1 (1.9, 34.7)	0.005
	≥ 30 animals	451	4.3 (2.5, 6.1)	9.1 (2.2, 38.3)	0.003
Agro-climate	Low altitude	435	4.4 (1.8, 7.0)	1.0 (-)	
	Mid altitude	1008	3.3 (2.2, 4.4)	0.7 (0.4, 1.5)	0.391
	High altitude	180	1.0 (0.3, 2.4)	0.2 (0.1, 9.5)	0.043
Production system	Mixed farming	900	1.8 (0.8, 2.8)	1.0 (-)	
	Pastoral system	723	4.5 (2.9, 6.1)	2.6 (1.3, 5.1)	0.007
Species composition	Cattle-small rum	1080	2.4 (1.1, 3.6)	1.0 (-)	
	Cattle-small rum-camel	271	4.7 (2.9, 6.5)	2.0 (1.0, 4.0)	0.041
	Camel-cattle-small rum	272	5.8 (2.2, 9.5)	2.6 (1.1, 6.1)	0.032

* Variables with $p \leq 0.25$ identified as possible risk factors and offered to multivariable model, rum: ruminant.

Table 4 Multivariable logistic regression model of risk factors for Brucella seropositivity in cattle at individual (n = 1623) and herd (n = 134) levels using village as the cluster variable

Variable	level	Odds Ratio	95% CI	P-value
		Individual animal level		
Age	1 - 2 years	1.0	-	-
	3-4 years	3.1	1.0, 9.6	0.045
	> 4 years	5.2	2.1, 12.9	0.000
Herd Size	< 15 animals	1.0	-	-
	15 - 29 animals	8.1	1.9, 34.2	0.005
	≥ 30 animals	8.0	1.8, 35.0	0.006
		Herd level		
Species composition	Cattle-small rum	1.0	-	-
	Cattle-small rum-camel	2.7	1.1, 6.8	0.039
Herd Size	< 15 animals	1.0	-	-
	15 - 29 animals	11.3	2.4, 51.9	0.002
	≥ 30 animals	19.6	3.8, 100.9	0.000

Post-estimation statistics didn't identify any covariate patterns (observations) that showed an outlying distribution and any influence on parameter estimates of the model.

4. DISCUSSION

The study showed that antibodies to Brucella infection were prevalent across the study areas except for Dawro where all tested animals (n = 104) were seronegative (Table 2). The overall animal level seroprevalence of 3.5% was comparable with the findings of other authors in Ethiopia; 3.2% by Berhe et al. [3], 4.6% by Hailemelekot et al. [8], 3.1% by Ibrahim et al. [6], 2.9% by Jergefa et al. [5] and 4.9% by Mekonnen et al. [7]. Similarly, comparable seroprevalences were reported from some other countries: 4.2% in Eritrea [12], 3.3% in Central Africa [13] and 5.8% in Nigeria [14]. Our finding of 26.1% herd seroprevalence is similar to 24.1% reported by Mekonnen et al. [7] whereas most of the other studies in Ethiopia showed a relatively low seroprevalence [5, 6, 9]. Conversely, higher herd level seroprevalences have been recorded by other authors; 62% from Zambia [15], 55.6% from Uganda [16] and 42.3% from Ethiopia [3]. Such contrasting findings could be either related to the overall animal level prevalence status of the disease or number of animals per the studied herds (herd size). The effect of an increased number of animals per herd was also observed in a specific finding of this study (68.8%: higher herd level seroprevalence in Borana than others without much difference in individual animals). This large herd effect reflects the larger numbers of samples in larger herds.

Higher herd and village levels seroprevalences were observed in pastoral production systems, when compared to crop-livestock mixed framings, similar to what has already been demonstrated by earlier researchers [1, 10, 12, 16]. This is mainly attributed to the nature of pastoral production system: high herd mobility, multiple livestock species herding and increased number of animals per holdings. The settlement pattern of pastoral community in Ethiopia is characterized by clustering of households with close proximity of herds in the pastoral camps. Additionally, pastoral households often keep a diverse composite of livestock species as part of a coping mechanism for uncertainties and risks. Such conditions certainly increase aggregation and interaction of different animals at villages, grazing fields and water points, thus, facilitate transmission of the disease. The dynamics and frequent migration of pastoral herds might increase the chance of coming into contact with other potentially infected herds and exposure to geographically limited or seasonally abundant diseases. Mobility also increases the opportunity of interactions with wild animals.

This has already been confirmed by Muma et al. [15] in that herds coming into contact with wildlife had higher likelihood of acquiring infection than those without contact.

The lowest seroprevalence was recorded in mixed farming areas of Sidama and Gedeo zones while no positive case was detected in Dawro zone. These areas are partly cash crop (coffee, fruits, vegetables, spice) growing region of the country where small numbers of animals are kept separately. In some cases, animals are tethered around farmland or homestead and feed on post harvest products of the farms, a condition which decreases mobility and contact between herds. Similar findings of low seroprevalences were reported from crop-livestock mixed farming areas of Eritrea [12] and Ethiopia [3]. Likewise, absence of seropositive animal in Dawro may be due to a small sample size coupled with low prevalence of brucellosis in mixed farming area.

The multivariable logistic analysis identified herd size, age group (animal level) and livestock species composition (herd level) as risk factors for acquiring Brucella infection (Table 4). The higher seropositivity observed in the large herds is in accordance with previous findings [3, 7, 15] and can be explained by the fact that an increase in herd size is usually accompanied by an increase in stocking density, one of the determinants for exposure to Brucella infection especially following abortion or calving [2]. Risks linked to herd size and livestock species composition was observed in final model of herd level analysis. Herds kept with multiple livestock species had higher odds of seropositivity to Brucella infection, suggesting possibilities of cross-species transmission of Brucella infection. Multiple livestock species herding, keeping of small ruminants along with cattle or camels, has been reported as risk factor for seropositivity to Brucella infections [17, 18].

Association of age with seropositivity to Brucella infection is consistent with the findings of earlier studies [3, 4, 6, 7, 8]. Age is one of the intrinsic factors which influences the susceptibility to Brucella infection. Brucellosis appears to be more associated with sexual maturity [19], and higher seroprevalence is repeatedly reported in sexually mature animals. Seroprevalence may increase with age as a result of prolonged duration of antibody responses in infected animals and prolonged exposure to pathogen, particularly in traditional husbandry practice where females are maintained in herds over long period of time. In our data analysis, the fact that females showed higher seropositivity than male animals, and this variable (sex) showed collinearity with age may also substantiate this fact. In the study areas, female animals are maintained in herds over extended time period thus, have ample time for exposure to the pathogen and being source of infection for other animals. Hence, practice of culling breeding females with reduced reproductive performances and old age could reduce the risk of within herd spread of brucellosis and its zoonotic hazard to human.

Although developed countries have successfully controlled brucellosis, many developing countries such as Ethiopia, have not been able to react adequately and the disease continues to be a major public and animal health problem. Control and eradication of brucellosis is almost exclusively based on the serological testing of animals and the subsequent culling of those that are seropositive for antibodies to Brucella species [20, 21]. As no single serological test is appropriate in all epidemiological situations, the use of two tests applied serially is usually recommended for maximal specificity and ruling out false positive cross-reactions [20, 21]. A combination of RBT and CFT tests is the most widely used serial testing scheme. Selection of RBT as screening test is based on cost, easy performance and high sensitivity, especially in endemic areas [15]. The second test, CFT is selected due to its high specificity to discriminate between false positive cross-reactions and Brucella infections [20]. When test specificities are conditionally independent of each other, the resulting expected specificity of serial testing is said be higher than the corresponding individual specificities of each test [11]. Conversely, serial testing using pairs of specificity-correlated serological tests (RBT, CFT, c-ELISA) has been argued, in favor of a highly specific single test such as i-ELISA, to have lower specificity than expected when applied to disease free population [22]. When such test is applied to a low disease prevalence (below 1%) or disease free population, the predictive value of the test drops closer to zero and increased proportion of non-infected animals are classified as seropositive [21, 22]. Therefore, consideration should be given to all factors that have impact on the relevance of the test method and test results to a specific diagnostic interpretation or an epidemiologic situation.

Adherence to traditional farming practices, preference for fresh dairy products and contact with animals have been reported to be risk factors for human exposure [23, 24, 25, 26]. In our study area, close intimacy with livestock, low awareness on zoonotic importance of brucellosis, tradition to consume raw milk and pattern of the disease in animals may certainly increase the risk of human exposure to Brucella infections. Despite the widespread distribution of brucellosis in animals and ample exposure factors for humans in Ethiopia, only scanty published information is available regarding human brucellosis. According to these studies, there are large number of undiagnosed cases of febrile diseases, neurological complications, joint problems and certain generalized complications in rural communities that might be associated with brucellosis [9, 24, 25, 26]. Seroprevalences of 34.9% and 29.4% have been reported from patients with fever of unknown origin in Borana and South Omo (Hamar) pastoral communities, respectively [24]. Similarly, a seroprevalence of 5.3% has been reported from limited number of animal health professionals, occupationally risk group, in Sidama zone of Southern Ethiopia [9]. These suggest that large number of undiagnosed cases with fever, neurological complications and other generalized complications in rural and pastoral communities are misdiagnosed and treated empirically as malaria or fever of unknown origin.

In conclusion, our study revealed that bovine brucellosis is widely prevalent in cattle herds of most villages of the study areas with higher seroprevalence in pastoral than mixed farming areas. Animals aged above 4 years, large herd size and herds kept mixed with more livestock species are at increased risk of acquiring Brucella infection. Hence, the need for implementing control measures and raising public awareness on zoonotic transmission of brucellosis are recommended.

ACKNOWLEDGEMENTS

The study was supported by the research and extension office of Hawassa University and the National Animal Health Diagnostic and Investigation Center (NAHDIC), in addition to the willingness and cooperation of animal owners. All contributions are gratefully acknowledged. Drs. Ajebu Nurfeta and Sandip Banerjee are also thanked for helpful suggestions and comments.

REFERENCES

1. McDermott JJ, Arimi SM: Brucellosis in Sub-Saharan Africa: epidemiology, control and impact. Veterinary Microbiology. 2002, 20: 111-134. 10.1016/S0378-1135(02)00249-3.
2. Crawford RP, Huber JD, Adams BS: Epidemiology and Surveillance. Animal brucellosis. Edited by: Nielsen K, Duncan JR. 1990, CRC Press Inc., Florida, 131-148.
3. Berhe G, Belihu K, Asfaw Y: Seroepidemiological investigation of bovine brucellosis in extensive cattle production system of Tigray region of Ethiopia. International Journal of Applied Research in Veterinary Medicine. 2007, 5: 65-71.
4. Kebede T, Ejeta G, Ameni G: Seroprevalence of bovine brucellosis in smallholder dairy farms in central Ethiopia (Wuchale-Jida district). Revue de' Elevage et Medicine Veterinaire des Pays Tropicaux. 2008, 159: 3-9.
5. Jergefa T, Kelay B, Bekana B, Teshale S, Gustafson H, Kindahl H: Epidemiological study of bovine brucellosis in three agro-ecological areas of central Oromia, Ethiopia. Revue Scientifique et Technique de l Office International des Epizooties. 2009, 28: 933-943.
6. Ibrahim N, Belihu K, Lobago F, Bekana M: Seroprevalence of bovine brucellosis and its risk factors in Jimma zone of Oromia region, South-western Ethiopia. Tropical Animal Health and Production. 2010, 42: 35-40. 10.1007/s11250-009-9382-z.
7. Mekonnen H, Kalayou S, Kyule M: Serological survey of bovine brucellosis in Barka and Arado breeds (Bos indicus) of Western Tigray, Ethiopia. Preventive Veterinary Medicine. 2010, 94: 28-35. 10.1016/j.prevetmed.2009.12.001.

8. Hailemelekot M, Kassa T, Tefera M, Belihu K, Asfaw Y, Ali A: Seroprevalence of brucellosis in cattle and occupationally related humans in selected sites of Ethiopia. Ethiopian Veterinary Journal. 2007, 11: 85-100.
9. Asmare K, Prassad S, Asfaw Y, Gelaye E, Ayelet G, Zeleke A: Seroprevalence of brucellosis in cattle and high risk animal health professionals in Sidama Zone, Southern. Ethiopian Veterinary Journal. 2007, 11: 69-84.
10. Dinka H, Chala R: Seroprevalence study of bovine brucellosis in pastoral and agro-pastoral reas of East Showa zone, Oromia Regional State, Ethiopia. American-Eurasian Journal of Agricultural and Environmental Science. 2009, 6: 508-512.
11. Dohoo I, Martin W, Stryhn H: Veterinary epidemiologic research. 2003, AVC Inc. Charlottetown, Prince Edward Island, 32-120.
12. Omer MK, Skjerve E, Holstad G, Woldehiwot Z, MacMillan AP: Prevalence of antibodies to Brucella species in cattle, sheep, horses and camels in the state of Eritrea: influence of husbandry system. Epidemiology and Infection. 2000, 125: 447-453. 10.1017/S0950268899004501.
13. Nakoune E, Debaere O, Koumand-Kotogne F, Selenkon B, Samory F, Talarmin A: Serological surveillance of brucellosis and Q fever in cattle in the Central African Republic. Acta Tropica. 2004, 92: 147-151. 10.1016/j.actatropica.2004.06.007.
14. Cadmus SIB, Ijagbone IF, Oputa HE, Adesoken HK, Stack JA: Serological survey of brucellosis in livestock animals and workers in Ibadan Nigeria. African Journal of Biomedical Research. 2006, 9: 163-168.
15. Samui KL, Oloya J, Munyeme M, Skjerve E: Risk factors for brucellosis in indigenous cattle reared in livestock-wildlife interface areas of Zambia. Preventive Veterinary Medicine. 2007, 80: 306-317. 10.1016/j.prevetmed.2007.03.003.
16. Faye B, Castel V, Lesnoff M, Rutabinda D, Dhalwa J: Tuberculosis and brucellosis prevalence survey on dairy cattle in Mbarara milk basin in Uganda. Preventive Veterinary Medicine. 2005, 67: 267-281. 10.1016/j.prevetmed.2004.11.002.
17. Al-Majali AM, Talafha AQ, Ababneh MM, Ababneh MM: Seroprevalence and risk factors for bovine brucellosis in Jordan. Journal of Veterinary Science. 2009, 10: 61-65. 10.4142/jvs.2009.10.1.61.
18. Kaoud HA, Zaki MM, Shimaa ARD, Nasr A: Epidemiology of brucellosis among farm animals. Nature and Science. 2010, 8: 190-197.
19. Radostits OM, Gay CC, Blood DC, Hinchcliff KW: Veterinary medicine: a textbook of diseases of cattle, sheep, goats, pigs and horses. 2000, W.B. Saunders Company Ltd, London, 867-882. 9
20. Office International des Epizooties (OIE): Bovine brucellosis. Manual of diagnostic tests and vaccines for terrestrial animals. 2009, OIE, Paris, 409-435.
21. Godfroid J, Saegerman C, Wellemans V, Walravens K, Letesson JJ, Tibor A, McMillan A, Spencer S, Sanna M, Bakker D, Pouillot R, Garin-Bastuji B: How to substantiate eradication of bovine brucellosis when a specific serological reactions occur in the course of brucellosis testing. Veterinary Microbiology. 2002, 90: 461-477. 10.1016/S0378-1135(02)00230-4.
22. Mainar-Jaime RC, Muñoz PM, de Miguel MJ, Grilló MJ, Marín CM, Moriyón I, Blasco JM: Specificity dependence between serological tests for diagnosing bovine brucellosis in Brucella-free farms showing false positive serological reactions due to Yersinia enterocolitica O:9. Canadian Veterinary Journal. 2005, 46: 913-916.
23. Meky EA, Hassan FA, Abd-Elhafez AM, Aboul-Fetou AM, El-Gazali SMS: Epidemiology and risk factors of brucellosis in Alexandria governorate. Eastern Mediterranean Health Journal. 2007, 13: 677-685.
24. Regassa G, Mekonnen D, Yamuah L, Tilahun H, Guta T, Gebreyohannes A, Aseffa A, Abdoel TH, Smits HL: Human brucellosis in traditional pastoral communities in Ethiopia. International Journal of Tropical Medicine. 2009, 4: 59-64.

25. Tolosa T, Regassa F, Belihu K, Tizazu G: Brucellosis among patients with fever of unknown origin in Jimma University Hospital, Southwestern Ethiopia. Ethiopian Journal of Health Sciences. 2007, 17: 1-6.
26. Kassahun J, Yimer E, Geyid A, Abebe P, Newayeselassie B, Zewdie B, Beyene M, Bekele A: Sero-prevalence of brucellosis in occupationally exposed people in Addis Ababa, Ethiopia. Ethiopian Medical Journal. 2006, 44: 245-252.

CHAPTER 18

The role of the companion animal veterinary surgeon in behavioural husbandry

Kendal Shepherd

Address: 16 Church Street, Finedon, Wellingborough, Northants, NN9 5NA, UK

An American Veterinary Medical Association mission statement regarding the human-animal bond reads, "The veterinary surgeon's role in the human-animal bond is to maximise the potential of this relationship between people and animals." To fulfil this aim and implicit obligations, as well as enhancing the welfare of both human and animal companion, it is impossible to ignore the emotional and behavioural aspects of this relationship in favour of purely physiological illness and disease. Rather than only addressing defects in the dog-human relationship which have already occurred – the client complains about destruction in the home, urinating indoors or being snappy towards the children for example – the onus is now on all veterinary surgeons to be rather more pro-active in their approach. It must be appreciated that a relationship does not exist merely between a particular pet and members of their family, but also between that animal and all the humans it may happen to come into contact with elsewhere, including those inhabiting veterinary surgeries. The dog which becomes progressively harder to handle in the surgery over time, although frequently labelled as 'difficult', 'aggressive' or simply 'vicious', is in reality simply one whose behavioural needs and emotional welfare have been inadvertently damaged in the process of maintaining physical health or attending to physical disease. On the other hand, a dog who is genuinely pleased to enter a veterinary surgery, and whose behaviour and emotions have been nurtured in the same way as its physical needs, is a tribute to the attending veterinary surgeon as well as its owner.

At present, lack of adequate and appropriate behavioural management is routinely compromising much of the treatment vets try to deliver to their patients, both medical and surgical. In the UK, the new Animal Welfare Act (2007) stipulates that all those in contact with an animal have a 'duty of care' towards it, which encompasses both physical and mental needs, health and well being. It is therefore a legal, if not also moral, obligation of veterinary surgeons in the UK to ensure that necessary, effective and humane behavioural management is in place, bearing in mind these requirements. It would however often seem that farm animals have the better deal in this regard. Management systems and environment are routinely considered to have a bearing on the cow, sheep or pig's physical health and the development of disease, and are therefore automatically included in the large animal veterinary surgeon's remit. Not so for the domestic dog, which may be skidding uncomfortably on a fashionable laminate floor, subjected to the impact of noise from a multitude of household gadgets, or contained and punished with electric shock, and yet its small animal veterinary surgeon be completely unaware of its plight. How can we truly say such animals are 'under our care' if this is the case?

This poster presentation will summarise some of the very simple means whereby not only appropriate behaviour, but also accompanying emotions, can be generated and maintained in dogs in all veterinary contexts. Using basic knowledge of learning theory and its practical application, the behavioural repertoire of a dog can be manipulated and cared for in such a way as to ensure emotional welfare and maximum success of medical and surgical veterinary interventions. Not only will such care have benefits within the context of the surgery, as a dog's need to become aggressive will be diminished, but veterinary surgeons and nurses will be presented as 'role models' to be emulated by their clients in all aspects of the animal's life. It will help prevent iatrogenic behavioural damage, allow 'best practice' in terms of clinical care, and result in much improved client loyalty and compliance. Above all, the field of behaviour must not be viewed as time-consuming and financially inefficient, only to be tackled on a 'case-by-case' basis. Instead, it must become second nature for any veterinary surgeon to consider both external behaviour and underlying emotions as part of his or her daily clinical repertoire and demonstrate routine care for them. This is the basis of 'behavioural husbandry'.

CHAPTER 19

The current state of welfare, housing and husbandry of the English pet rabbit population

Nicola J Rooney, Emily J Blackwell, Siobhan M Mullan, Richard Saunders, Paula E Baker, Jenna M Hill,Clare E Sealey, Matthew J Turner and Suzanne DE Held

[1]*Department of Clinical Veterinary Science, Animal Welfare and Behaviour Group, University of Bristol, Langford, BS40 5DU Bristol, UK*

ABSTRACT

Background

The welfare of pet rabbits is an area of growing interest in Europe and the UK. This study analyses questionnaire results from a diverse population of 1254 rabbit owners from three different geographical areas in England with the aim of providing an accurate representation of how pet rabbits are currently housed and cared for and key aspects of their health and welfare.

Results

Rabbits were kept in a variety of different housing types, the most common being a traditional hutch/cage (59%). Although the majority had additional exercise areas, access was often unpredictable, or ill-timed, which may compromise welfare. Only 41.9% of owners kept their rabbit with conspecifics, limiting their ability to engage in social behaviour. Of those rabbits housed with a companion, although many were reported to be amicable and to engage in positive interactions, over a quarter were reported to fight at least occasionally (25.3%), whilst 22.7% guarded resources and 27.1% avoided one another. Whilst low levels of some of these behaviours may be a normal part of social interaction, the relatively high levels reported here suggest that not all cohabiting pairs of rabbits are compatible, which is potentially a significant welfare issue.

Although the vast majority of owners fed hay for over 10% this was less than daily. Pelleted foods were very popular (71.4% at least daily) compared to commercial muesli mixes (32.6%). As in previous studies, dental problems were commonly reported (12.2% of rabbits); however, so were eye problems (12.9%), digestive problems (11.5%) and parasites (11.3%). A large proportion of rabbits (58%) were thought to be fearful of loud noises, and 61% were not reported as calm when handled by their owner, which may be a significant concern for this species.

Conclusion

This study has confirmed and expanded on previous findings: many pet rabbits were found to be in good health, had compatible companions and were provided with enriched living areas. However, it also found numerous welfare issues that affect large numbers of pet rabbits. We suggest further studies are required exploring the accuracy of owner reports (which possibly under-report many problems) and prioritising the issues raised here.

Keywords: Pet, rabbit Survey, Husbandry, Companionship, Health Behaviour, Diet, Housing, Welfare

1. BACKGROUND

Rabbits are common pets in many western European countries [1]. They are the third most popular mammalian pet in the UK [2], with an estimated 1.7 million kept in 4% of UK households. However, given their popularity, surprisingly little is known about the way pet rabbits are kept and cared for [1], or how their housing and husbandry conditions affect their health, behaviour and general wellbeing. Three recent studies have suggested that traditional housing and husbandry practices may detrimentally affect rabbit welfare. Schepers et al. [1], in a survey of 912 owners and direct observations of their rabbits, identified several potential welfare concerns, including small hutches, solitary housing, poor socialisation, inadequate and inappropriate diets, and a lack of veterinary care. However, their study was conducted in the Netherlands and it is not known how similar their population was to the UK pet rabbit population.

UK studies have to date been less extensive. Edgar and Mullan [3] surveyed 52 owners at point of sale, whilst Mullan and Main [4, 5] took direct measurements on 102 rabbits in their homes in the South West of England. They found that 20% were kept in hutches smaller than 0.54 m^2, the minimum cage area recommended for laboratory and farmed rabbits (up to 6kg: Home Office [6]); nearly 50% were housed alone, and the most common health problem was dental disease, of which the majority of owners were unaware. The studies by Mullan and Main likely targeted particularly keen rabbit owners by recruiting via advertisements in local shops and newspapers and it is not known how the relatively small populations studied reflect the wider English population. However, they do highlight several potential welfare issues.

In 2011, (and repeated annually) the PDSA (People's Dispensary for Sick Animals e.g. [2, 7]) conducted a UK-wide internet survey reaching 1,132 rabbit owners and reporting welfare concerns relating to *each* of the five welfare needs [8]. While the sample size was larger than Mullan and Main [4, 5], the survey was distributed only to owners with time and skills to access the internet which represents a skewed demographic. Also since it included questions about three species (dogs, cats, rabbits), the amount of information regarding rabbits was limited.

The present study aimed to provide a representative point sample of the current state of care and welfare of English pet rabbits, against which subsequent surveys could be compared, and change assessed.

A questionnaire was devised consisting of over seventy questions and was presented in three different forms: a) written form, distributed via pet care outlets, veterinary surgeries and schools; b) verbally usually over the telephone, and c) via the internet, recruited in multiple ways. Diversity of recruitment methods aimed to overcome some of the sampling bias inherent in past surveys. A prize draw was used to incentivise owners to take part.

Owners were asked about the way they housed and cared for their rabbits, collecting information on many factors hypothesised to affect their welfare, including housing type and size, exercise and grazing provision, husbandry and cleaning regimes, diet, companionship and veterinary care (e.g. vaccination, health checks).

Increasingly, animal welfare is measured in terms of outcome- rather than input- based measures (e.g. [9]). Therefore, we collected data on a range of potential indicators of rabbit welfare status; including indicators of physical health and disease, and behavioural indicators of welfare; data on in-cage behaviour, general temperament, and responses to potential stressors including handling and common fear-provoking stimuli. We asked owners about the occurrence of specific behaviours including those which we later classified as "positive" and "negative". Positive behaviours were classified as those whose expression is known to be important for physical health (e.g. rearing up for musculo-skeletal health), and those which are thought to indicate positive affective states such as locomotory play behaviour [10] (e.g. "binkying" (running and hopping or twisting in the air)). Negative behaviours were those whose expression can be symptomatic of an underlying welfare issue (e.g. chewing the cage, head swaying). We categorised chin rubbing (usually territory marking) [11] and throwing objects as neutral. Throwing could be considered as negative, but we think it ambiguous as it could also be motivated by frustration, investigation or play.

Past estimates of longevity have been based on the age of rabbits at the time of surveys and have raised concerns about potentially low life expectancy (e.g. 2.2 years in [1]). To avoid underestimating longevity here, we asked respondents about the age at death of their last rabbit. We present the data collated for 1254 rabbit owners responding to our survey. The aim is to provide a description of current housing and husbandry practices for English pets and to highlight potential welfare issues.

2. METHODS

2.1 Questionnaire content

The questionnaire included 76 questions divided into eight sections:

Section A Respondent - gender, age, details of any children, number of rabbits currently owned, main carer;

Section B Rabbit - name, sex, neuter status, age and who it was obtained for;

Section C Companionship - whether rabbit lives with another animal, length of cohabitation and frequency with which they show specific social behaviours;

Section D Housing - size and details of main living space, any attached and separate runs; information on the length, height and also the area of the home enclosure. The diversity of accommodation in which rabbits are kept has made some past survey findings difficult to interpret. Therefore we categorised the main living space as the area to which the rabbit had permanent access, but we also asked questions about any "separate" and "attached" runs to which the rabbit had intermittent access, with and without the necessity to be moved.

Section E Husbandry - exercise and cleaning routines, in summer and winter;

Section F Diet - frequency of feeding 12 different foodstuffs, and whether rabbit exhibits selective feeding;

Section G Health - whether ten specific symptoms of ill health had ever, occasionally, or often been observed; whether the rabbit suffered from 15 common veterinary complaints, currently or in the last year, and whether it required veterinary treatment. Estimates of current teeth and nail condition.

Section H Behaviour - owner's opinion of how the rabbit reacts to a number of situations, including handling by themselves, children and other adults, approaching their living space, and five potentially fear-provoking stimuli e.g. loud noises and open spaces, and similarly the rabbits' most common activity.

To avoid bias created by owners of multiple rabbits choosing to give information about all or the healthiest or most interesting rabbit, respondents were instructed to answer only about the rabbit whose name came first alphabetically. They were then given the opportunity to provide extra details on the social groupings of all additional rabbits they owned, and the age at death of their most recently deceased rabbit.

Many of the questions included component parts, hence a maximum of 326 variables were collected from each respondent.

2.2 Questionnaire distribution

The same questionnaire was presented via three forms; a printed version, one delivered verbally during telephone interviews and an internet version. Each included the same questions presented in the same order, but the formatting was slightly modified for the internet software system available.

All three questionnaires were distributed in rural and urban locations within three target regions:

South West – centring around Bristol and North Somerset;

North West – centring around Manchester and the Wirral;

Eastern – centring around Norwich and Eastern Norfolk.

2.3 Written questionnaire

A total of 1419 written questionnaires, accompanied by prepaid return envelopes and advertising posters were placed in 82 different prominent locations, in veterinary surgeries, pet shops, RSPCA (Royal Society for the Prevention of Cruelty to Animals) clinics, rehoming centres, and other pet-related outlets. Regular checks were made at each outlet, to replenish supplies. Questionnaires were also given to rabbit-owning families in four schools, and to friends and family of the research team. In total, 343 owners responded, giving a response rate of 24.2%.

2.4 Telephone/verbal questionnaire

Veterinary surgeries in each target area were contacted, and ten (including both surgeries located within pet stores and independent practices) agreed to contact their rabbit-owning clients to recruit survey respondents. In addition, sign-up lists were placed in three PDSA clinics and all rabbit-presenting clients were asked to participate. Recruitment lists were displayed at a very large rabbit show, postcards handed out at an education stall in Manchester and by RSPCA inspectors on routine calls and were sent to owners who had recently acquired a rabbit from a re-homing centre. In total, 484 owners volunteered. They were contacted by telephone at a convenient time, and 335 interviews taking between 5 and 80 minutes, were conducted (mean = 27.8 ± 9.1 mins).

2.5 Internet questionnaire

The survey was linked to the University of Bristol homepage and publicised in each key region via radio, newspapers, fliers at agricultural shows, local shops, libraries and community centres, emails to staff of several companies and to members of the Rabbit Welfare Association and Fund; alerts on University of Bristol payslips, postings on Facebook and twitter pages, money-saving websites, and general pet and rabbit forums. A total of 1490 rabbit owners responded.

2.6 Data handling

Responses from the internet questionnaire were downloaded and exported to Excel, telephone and written responses were entered into the same spreadsheet. All those respondents from outside the key areas were excluded. Descriptive analysis was carried out using SPSS v19.

Research was approved by the approved the University of Bristol ethics committee and was performed in accordance with the Declaration of Helsinki.

3. RESULTS

3.1 The sample

Between January and July 2011, 2169 completed surveys were returned (Table 1), 915 were from outside the key areas so were excluded leaving a sample of 1254 owners (Table 2).

3.2 Respondents

Respondents were mainly female (89.1%) in the age range 30–49 (56.5%). At the time, 48.3% had no children in the household, whilst 45.6% had children permanently and 6.1% visited at least once every two weeks. More respondents had girls, either under (63.2%) or over ten (54.5%), than boys of the same ages (58.6 and 40.4%).

Table 1 Composition of the complete sample of returned questionnaires – reported as number of respondents

	Verbal	Written	Internet	TOTAL
South West	130	136	305	571
North West	103	129	157	389
Eastern	100	77	117	294
Other	2	2	911	915
Total	335	344	1490	2169

For those 877 owners who stated how many rabbits they owned, answers ranged from one to 37, two rabbits was most common (44.4%) and only 3.3% of owners owned more than six. Most were cared for by the person answering the questionnaire (53.1%), whilst 24.7% of respondents shared care with another adult and 16.4% with a child.

Table 2 Recruitment sources of the1254 completed questionnaires from within the three key recruitment areas

	Verbal	Written	Internet	Total
Veterinary surgeries	243	150	3	396
Facebook	0	0	159	159
Word of mouth	1	0	100	101
Newspapers	0	0	90	90
Pet shops	1	90	0	91
RSPCA or PDSA clinics	42	5	12	59
RSPCA mailing or website	0	0	45	45
Rabbit Welfare Association and Fund	0	0	44	44
Sign up sheet, show or vets	35	2	1	38
Postcards	9	0	23	32
Friends and family	2	22	5	29
University of Bristol website	0	0	25	25
Schools	0	21	0	21
Rabbit focussed website or forum (not RWAF)	0	0	21	21
Other animal centred outlet e.g. kennels	0	11	7	18
Other charities e.g. PDSA	0	12	1	13
University employee payslip	0	0	11	11
Flier picked up in shops etc.	0	1	7	7
Twitter	0	0	6	6
Agricultural show	0	0	4	4
Work email (not University)	0	0	4	4
Non- animal websites e.g. money saver	0	0	8	8
Radio	0	0	2	2
General pet websites	0	0	1	1
Unknown	0	28	0	28
Total	333	342	579	1254

3.3 Rabbits

In total, 58.5% of the rabbits were male, 40.9% female and 0.2% of unknown sex (0.4% of owners failed to answer). A reported 64.9% of males and 51% of females (overall 59.1%) were neutered, for 0.5% the neuter status was unknown. Rabbits ranged in age from 2 months to 12. 7 years, (mean = 3.2 ± 2.3 years). The mean length of ownership was 2.9 (±2.2) years.

The most common breed/type were Lops, whilst Lionheads, Netherland dwarfs and mixed breeds were also popular (Figure 1). A total of 28.4% of rabbits were described as dwarf, 51.5% as small, 17.3% as large and 2.6% as giant.

The most popular source of rabbits was a pet shop or garden centre (39.1%), followed by rehoming centre (17.6%), breeder (15.8%), friend or neighbour (15%), advert (4%), whilst 2.1% were found as strays. Most rabbits were obtained for the respondents (49%), but 27.6% were mainly for a child.

3.4 Housing

The majority of owners described their rabbit's main living space as a hutch or cage (59.1%), whilst 29.9% were described as house rabbits, although most of these (82%) had a cage/hutch, but 5.5% of rabbits lived in the house, without any cage at all (Figure 2). A total of 59.5% of the rabbits lived mainly outdoors, 27.6% predominantly indoors and 12% in a shed, garage or outbuilding. For 19.5% location varied with season.

Most hutches/cages were on a single level (53.5%), 42.3% were over two and 4.2% had three or more levels. The surface area of the main hutch or cage, (including multiple levels if applicable, and the attached run or exercise area only if the rabbit had continual access to it), ranged from 0.1 m^2 to 150503 m^2, with a median size of 1.8 m^2 (25th percentile = 0.9 m^2, 75th percentile = 3.78 m^2). When the number of animals in the enclosure was taken into consideration, the average space was 1.27 m^2 per animal (0.72 m^2, 2.57 m^2), but 8.2% of the population (95 rabbits) were provided with less than 0.54 m^2.

Six cages were recorded as over 10 m tall (presumably roofless). When these were removed the median height was 0.9 m (0.6, 1.2), but for 8.3% of the rabbits their main living space (with permanent run included only if they had constant access to it) was less than 0.5 m high throughout and for 42.1% it was less than 0.75 m. Within their home cage, 86.7% were reported to have a separate area for denning/sleeping, 71.1% had shelters, tunnels or boxes, 65.9% platform(s), and 59.2% had continual access to toys.

The vast majority of rabbits (97.5%) had some exercise area outside their home-cage, but the frequency and duration of access varied greatly. In total, 43.4% had an attached run (area they could reach without being moved by the owner); the average area being 198.59 (±406.3) m^2. Of these 42.1% (23.5% of population) had continual access; 43.3% were shut away at night and for 8.8%, run use varied with the weather.

Separate runs were more popular; 62.9% of rabbits having access in summer, most commonly every day (25.2% of all rabbits), and for 4–8 hours (30.9%). However for 42.2% of rabbits, the frequency of access differed with the season; in winter only 49.9% had any access, with occasional (16.7%) 1–2 hour sessions being most frequently reported (11.1%). The mean run size was 30.56 (±90.63) m^2, and whilst inside, 84.6% of rabbits had access to shelter, tunnel or box; 87.9% had grass to eat and 74% had ground to dig.

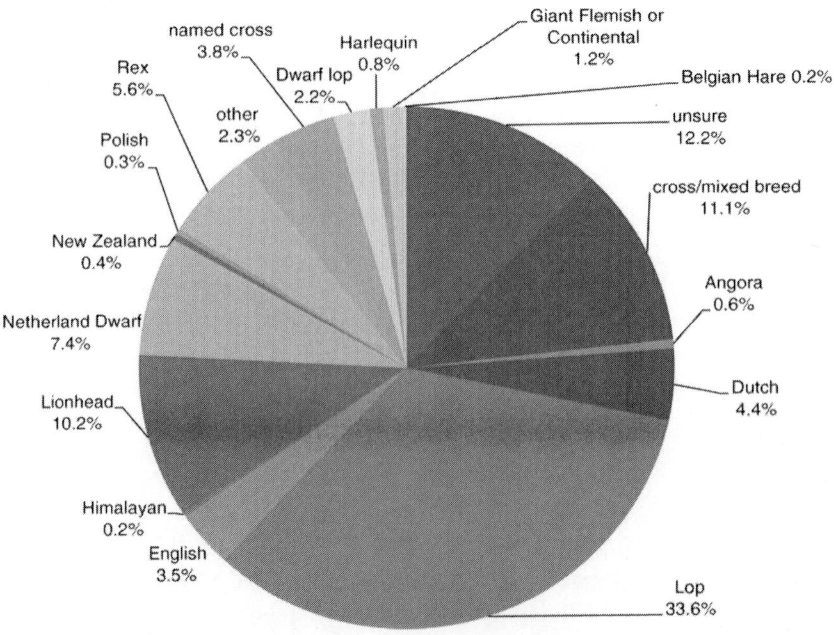

Figure 1 Reported breeds/types of the 1254 rabbits in the questionnaire survey.

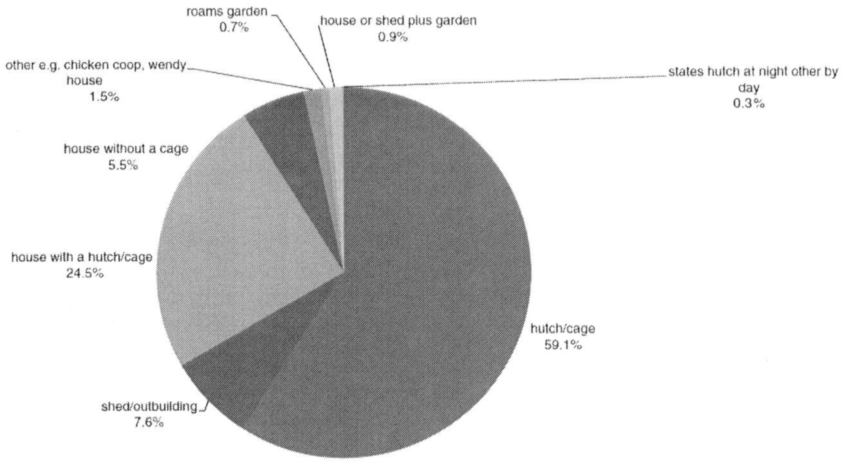

Figure 2 Reported description of main living space.

A reported 62.2% of rabbits were allowed to run loose in the garden, although only 50% in winter. For 22.8%, access was daily, or several times daily 8%; for the remainder, sessions were less predictable. Free running in the house was reported by 59% of owners.

3.5 Companionship

Although 43.5% of rabbits were described as living alone, only 41.9% lived with other rabbit(s), whilst others lived with guinea pigs (4.3%), cats, dogs, quail, chickens, chameleons and even kune kune pigs. Rabbits living with conspecific companions were reported to show a variety of social behaviours (Table 3). Resting in contact, grooming and playing were very common, but considerable numbers of rabbits were also reported to mount, pull fur out and fight, at least on occasion.

3.6 Husbandry routines

In the summer, most owners cleaned the cage weekly (44.6%) or more (27.2%), and took the soiled material out daily (50.2%) or more than once a week (27.2%). However 2.6% reported never thoroughly cleaning their rabbit's living space. When cleaning, 45.2% sometimes (29.7%) or always (13.8%) left bedding inside, most used disinfectant occasionally (42.2%) or always (41.3%).

Most rabbits were handled at least weekly (Table 4), but less than half were handled daily. Many rabbits were groomed weekly, but a large proportion were never groomed (Table 4).

3.7 Feeding

The most popular food types were root vegetables (e.g. carrots; 95.7) and hay (98.3%; Table 5). However, 10.6% of respondents reported feeding hay less than daily. A total of 74.1% of owners fed pelleted-foods at least daily compared to 32.5% feeding commercial muesli mixes. Of rabbits fed mixes, 52% were reported to leave specific parts (e.g. the pellets). Grass (freshly picked or growing) was fed at least weekly by 66.9% of owners. The most popular "other foodstuffs" were bread/crackers (10%) human breakfast cereals (3.8%), biscuits (2.3%) and herbs (3.2%).

Table 3 Percentage of pair and group -housed rabbits reported to show social behaviours at differing frequencies

	Never	Occasionally	Often
Playing with each other	5.3	31.8	62.9
Fighting (biting or scratching)	74.7	22.9	2.4
Resting in contact with each other	1.8	4.7	93.5
Grooming each other	4.4	12.1	83.5
Pulling out fur from each other	62.6	29.4	8.0
Mounting each other	34.4	51.6	14.0
Guarding items or areas from each other	77.3	19.0	3.7
Avoiding each other	72.9	25.8	1.3
Circling around each other	53.6	35.6	10.8
Chasing each other	22.5	54.8	22.7

3.8 Health care

Nail inspections most commonly occurred weekly whilst teeth inspections were most commonly monthly (Table 4). In total, 70.8% of rabbits were vaccinated, but only 11.7% were insured.

3.9 Indicators of physical health

The majority of owners judged their rabbit's nails and teeth, to currently be "fine" (nails: 87.1% teeth: 91.2%), a small proportion were unsure (2.9% and 0.4% respectively), and some admitted that they were possibly (8%, 2.7%) or definitely (1.1%, 0.7%) overgrown.

Of the presented list of clinical signs, indicative of underlying health issues, dirty bottoms were the most common; being reported as occurring at least occasionally by 30.1% owners (Table 6). Swollen body parts/lumps were the least common (3.5%). Caecotrophs (defined as smaller, sticky, often darker droppings- the caecal faeces usually re-ingested for nutritional purposes) were occasionally seen by 53.6% of owners and often by 9.4%.

Of the range of presented veterinary conditions (Table 7), digestive problems were the most commonly occurring over the last year (6.6%) whilst dental disease, and being overweight were the most common currently (4.8% and 7.5% respectively). Flystrike (88.5%), ear problems (88%) and dental disease (88%) were most commonly referred to a vet, whilst only 52% of overweight and 50-1% of rabbits with sore hocks or inside legs were taken to a vet. When asked specifically about their rabbit's body condition, 12.1% were described as overweight (0.2% very), 86.5% were thought to be the correct weight and only 1.5% underweight.

3.10 Behavioural indicators of welfare

In-cage behaviour Most rabbits were described as spending most time resting alone (Figure 3; 35%) or resting with a companion (26.9%), moving and eating were relative common but a surprising 12% were believed to spend most time drinking.

Of the positive behaviours listed, most (Figure 4) (with the exception of rolling on their backs) were reported to be shown by the vast majority of rabbits, but at varying frequencies. However, the potentially negative behaviours of thumping hind limbs, grunting, guarding and digging on hard surfaces were also common.

Table 4 Percentage of respondents reporting handling (by various) people, health checking and grooming at differing frequencies

	N	Never	Less than once a month	Approx monthly	Approx weekly	Most days	Daily
Picked up and handled by respondent	1246	2.2	6.7	5.6	18.0	21.3	46.3
Picked up and handled by other adults	1239	24.1	15.1	9.6	20.0	15.3	16.0
Picked up and handled by children under 10	1241	72.0	9.2	3.5	7.4	3.2	4.8
Picked up and handled by children 11-18	1241	65.7	7.2	4.8	8.5	5.4	8.5
Groomed	1243	23.9	15.2	17.5	26.6	93	7.6
Inspected to check the length of nails	1242	5.3	12.9	32.4	35.4	7.5	6.4
Inspected to check the length of teeth	1232	11.5	25.4	29.6	26.3	3.9	3.9

3.11 Human-directed behaviour

Most rabbits were thought likely to approach in a friendly manner (83.2%) when people approached their living area, but ignoring (25.2%), retreating (10.2%) or hiding (5.4%) were also common (some owners reported multiple responses). However, 1.7% of rabbits were thought likely to move towards the owner in an aggressive manner, and 2.6% to thump their hind legs or vocalise (2.2%).

When rating their own confidence at handling their rabbit 73% of owners scored 5 (very confident), 13% scored 4, 8% 3, 4% 2 and 2% scored 1, (not at all confident).

In total 39% of rabbits were described as calm when handled, but 61% of handled rabbits showed various signs of fear, (Figure 5); only 25% were described as calm when handled by other adults. The most commonly reported fear-provoking stimuli were loud noises (Figure 6).

3.12 Longevity

Overall 830 owners provided data on the age at death of their last rabbit, which ranged from one month to 12 years with a mean of 5.6 (±0.1) years.

4. DISCUSSION

The survey shows the diversity in ways rabbits are kept and cared for and highlights a number of potential welfare issues in the English pet rabbit population. By recruiting in diverse ways and targeting less keen owners, we aimed to reach a more diverse sample of rabbit owners than have some past detailed surveys (e.g. [4]). Our sample is likely more representative of the general rabbit-owning population, although undoubtedly it is still skewed towards self-selected, keen rabbit owners and in particular by the large proportion of house rabbit owners (29%) who responded, and so the prevalence or severity of some problems may be underestimated.

Our survey shows that although the traditional hutch or cage remains the main living space of most English pet rabbits, rabbits are kept in diverse ways, including in Wendy houses, rooms and sheds. Patterns of access to living space also vary greatly between individuals. It is therefore important to understand how design and complexity, combined with size, affect welfare if guidance regarding optimal housing design is to be offered to rabbit owners, and this should be the focus of future research.

Table 5 Percentage of respondents reporting feeding each of thirteen foodstuffs at differing frequencies

	Percentage of those responding reporting to feed with each frequency						
	Never	Occasionally	Once a week	More than once a week	Daily	Several times a day	Constant access
Commercial rabbit mix (cereal pieces of different shapes and colours)	60.0	5.7	0.5	1.3	23.4	1.8	7.3
Pellet feed (pieces all same shape and colour)	17.6	5.5	0.3	2.5	47.4	8.9	17.8
Green vegetables e.g. cabbage, broccoli	5.1	7.5	7.0	22.7	46.9	7.9	3.0
Salad e.g. lettuce	57.2	15.7	6.5	10.3	7.8	1.5	1.1
Garden/wild plants e.g. dandelions, brambles	13.0	28.7	7.9	20.4	22.1	1.9	6.0
Grass freshly picked or growing	14.8	20.4	7.0	16.4	29.2	2.1	10.1
Root vegetables e.g. carrots	4.3	13.5	9.0	31.0	36.6	3.4	2.2
Grass clippings	77.3	14.6	1.9	2.6	2.2	0.4	1.1
Hay/dried grass	1.7	2.1	2.3	4.5	23.5	1.3	64.6
Fruit e.g. apples, pears	17.9	34.7	11.6	20.2	13.3	1.0	1.2
Rabbit treats	23.6	41.8	9.9	11.4	11.7	0.9	0.6
Gnawing blocks	8.7	16.6	4.2	5.3	9.7	0.6	54.9
Other	80.0	7.3	2.7	3.4	5.2	1.0	0.5

Table 6 Percentage of rabbits reported to show each of 10 ill health symptoms at differing frequencies

	N	Never	Occasionally	Often
Runny eyes	1236	85.4	11.7	2.9
Runny nose	1235	92.4	6.9	0.7
Lack of appetite/not eating	1230	80.7	19.0	0.3
Wet fur around mouth and/ or chest (drooling)	1231	95.9	3.9	0.2
Dirty bottom or droppings in fur	1241	69.9	25.7	4.4
Intense scratching	1233	94.7	5.0	0.2
Matted or soiled fur	1232	86.5	11.5	1.9
Swollen parts of body or lumps	1232	96.5	3.0	0.5
Limping/problems moving	1233	96.0	3.1	0.9
Hay or fur in droppings	1225	91.7	7.4	0.9

The results also show that although the majority of rabbits have additional exercise areas, less than a quarter of them are given continual access to these areas. We propose that the value of additional access areas varies with the timing and predictability of access, since unpredictable, uncontrollable routines have been shown to compromise welfare in numerous species (e.g. [12]). Patterns of access were often erratic (i.e. less than daily), and unpredictable, especially in winter, and environmental enrichment in this space is not universal. Since rabbits are naturally crepuscular [13], the timing of access to exercise areas (i.e. whether it occurs when the animal would naturally be most active) is likely to be integral to their welfare value. Many exercise areas contained no shelters or tunnels (15.4%). Since rabbits commonly show fear of open spaces (e.g. [14]), this may limit their value. Although the median cage size was considerably larger than minimum laboratory standards and similar to the PDSA survey [2]; and the number of rabbits living in small cages was less than that recorded by Mullan and Main [4], there were still nearly 10% with less than 0.54 m^2 space available. It has been demonstrated that cages of 0.88 m^2 limit the behaviours which rabbits can exhibit [15] and 27.5% of rabbits in this survey were kept in cages less than this area. Recommendations for pets include that cages should be sufficiently long to allow the animal to carry out three unrestricted hops, and to lie outstretched (e.g. [16]), many of the cages do not meet this suggestion.

Table 7 Percentage of rabbits reported to have experienced each of 15 veterinary conditions, over differing time periods

	Over a year ago	In the past year	Currently	% of affected animals taken to vet
Ear problems	2.29	1.31	0.33	88
Eye problems	4.56	4.32	399	74
Digestive problems (e.g. gut blockage)	4.27	6.64	0.82	75
Overweight	3.20	2.87	7.46	52
Underweight	1.23	1.31	1.15	74
Dental/teeth problems	3.76	3.59	4.82	88
Respiratory/breathing problems	1.23	2.13	1.31	80
Skin problems	2.54	1.96	0.65	70
Parasites (e.g. mites /fleas)	5.98	4.10	1.23	77
Fly strike (i.e. maggots)	1.15	0.74	0.00	88
Fight wounds	2.95	1.97	0.41	64
Urine or kidney infections	1.23	1.15	0.33	73
Soreness on inside of legs	0.25	0.57	0.33	50
Sore hocks (reddened areas or bleeding of underside of heels or feet)	1.31	0.82	0.66	51
Neurological problems (e.g. head tilt)	0.58	0.25	0.74	80

Previous research has shown that rabbits sit and rear more in pens 75 cm high or with no ceilings, and the average height needed to fully 'rear-up' is 52.6 cm [15]. Platforms can encourage rearing and climbing and Table 7 Percentage of rabbits reported to have experienced each of 15 veterinary conditions, over differing time periods Table 7 Percentage of rabbits reported to have experienced each of 15 veterinary conditions, over differing time periods possibly improve musculoskeletal fitness, but 34.1% of rabbits were not provided with such devices. Large numbers of the surveyed rabbits would therefore be unable to rear regularly and consequent musculoskeletal health problems may arise. However rabbits also show behavioural preferences for enclosed areas [17], so the lack of tunnels (in 28.9% of cases) and denning areas (13.3%) in some enclosures is likely to also be a welfare concern.

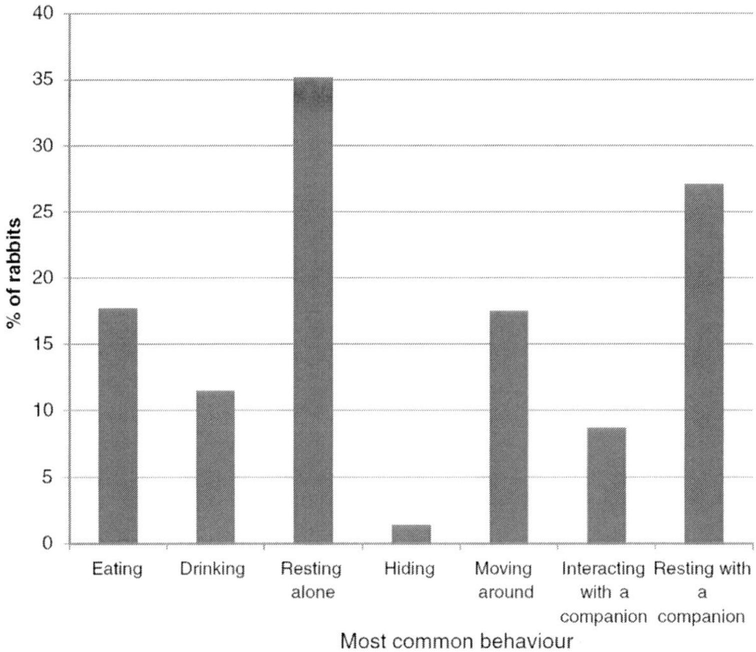

Figure 3 Percentage of rabbits reported to spend the majority of time in each of seven activities.

Less than half of rabbits (42%) in this survey, and an even lower proportion of house-rabbits, lived with a conspecific. This figure is slightly higher than the recent PDSA [7] survey (35%), and highlights solitary housing as a potential welfare issue [1, 18, 19]. Rabbits are strongly motivated to gain social contact [20]. Solitary living precludes their ability to engage in normal social behaviour and negates one of the five basic needs, laid down in the Animal Welfare Act [8]. However, importantly, of those rabbits housed with a companion, over a quarter were reported to at least occasionally fight, whilst many guarded resources or avoided one another at least on occasion. Regular chasing and mounting were also common. Low levels of these behaviours may be a normal part of social interaction [21], but the relatively high levels reported here suggest that not all cohabiting pairs are compatible. Hence, to ensure welfare is improved, rather than compromised by living with a conspecific, it is essential that compatible pairings are selected and introduced appropriately, and adequately sized and structured living space is provided to allow rabbits to avoid one another if they so choose. The effects of mixing and incompatible rabbit combinations have been investigated within production systems (e.g. [22]), yet even the most comprehensive Code of Practice for pet rabbits [16] lacks a definition of "appropriate companion", or details of the signs of compatibility as compared to incompatibility. Studies to determine the most compatible pet pairings, and optimal methods of introduction are required, as is advice to owners on how to monitor compatibility.

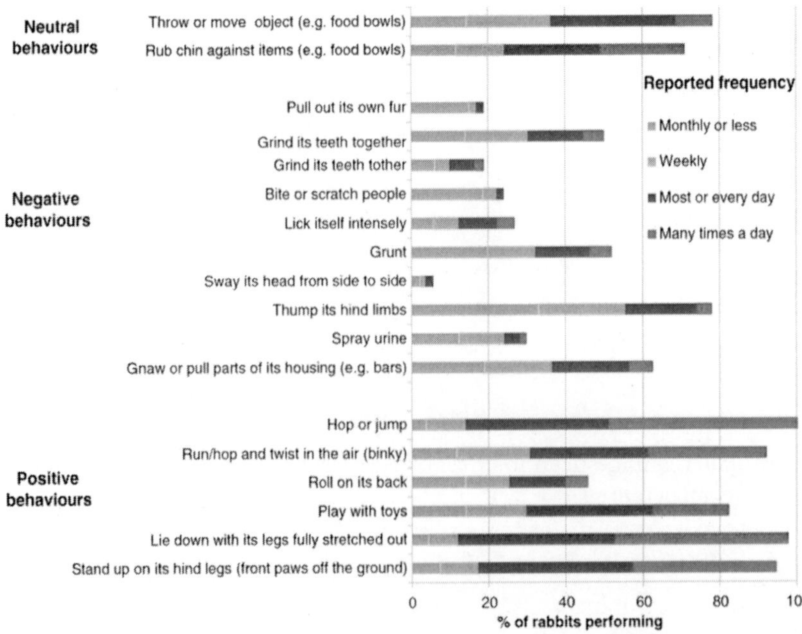

Figure 4 Reported frequency of occurrence of a range of behaviours.

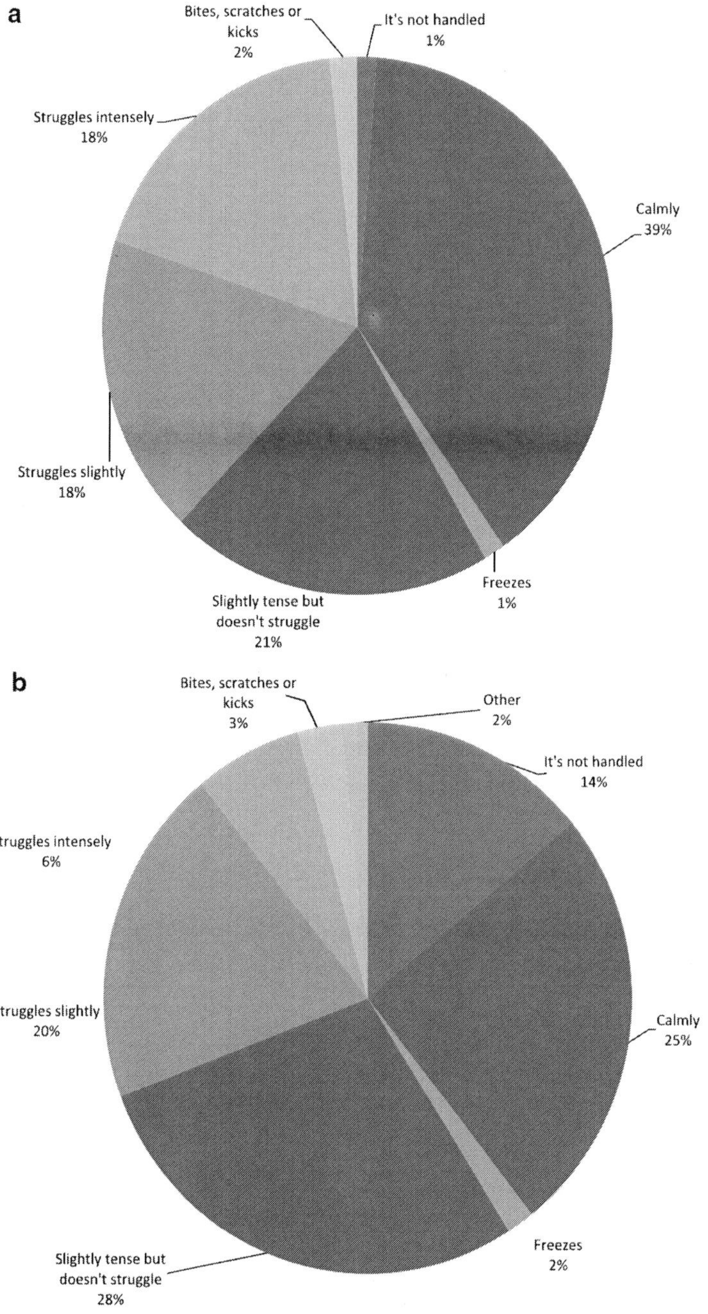

Figure 5 Reported response to a) owner and b) other adults when handled.

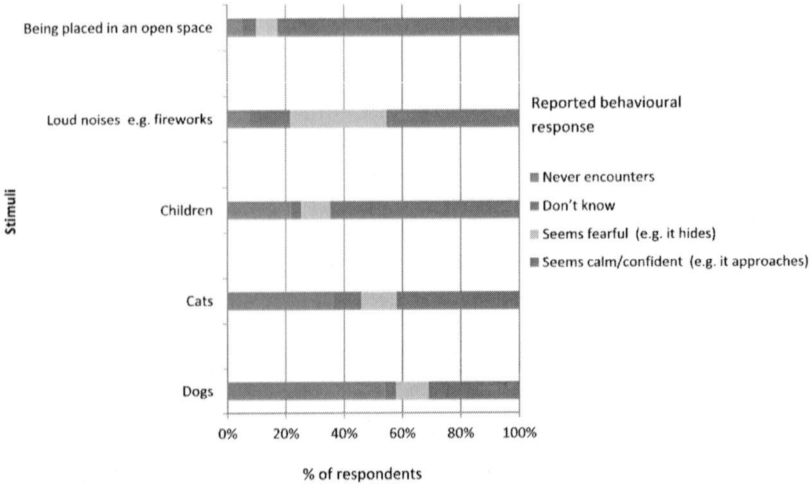

Figure 6 Reported reactions to a range of potentially fear provoking stimuli.

Our results show a marginal difference in food provision compared to past surveys as 40% of rabbits were reported to currently be fed commercial muesli mix, compared to 44% previously reported by Mullan and Main [4]. This could possibly indicate that past education campaigns warning of the potential links between muesli and obesity, dental disease and reduced water intake (e.g. [2, 16, 23]) may have had a small effect. However, over half of those feeding mix reported their rabbits selectively feeding on specific components, meaning they are unlikely to receive a balanced diet. For some rabbits, overfeeding of sugary foods such as root vegetables, treats and fruits (fed at least daily by 42.2, 13.2 and 15.5% of owners respectively) is a potential problem.

When grazing opportunities are limited, it is recommended that rabbit diets should consist primarily of high fibre forage, dried grass and hay. This is known to maintain intestinal physiology [24], to aid digestion, and to allow grazing behaviour which in wild counterparts occupies a large proportion of the day. Provision of forage is therefore believed to also prevent stereotypic behaviours [25, 26]. In total, 14.8% of rabbits were never fed grass, and 1.7% were never given hay or dried grass. This equates to 18 rabbits within this self-selected population, and a further 4.4% only received it weekly or less. Lack of, or inadequate (quantity or quality) of forage may be a contributory factor to a number of the veterinary and behavioural issues (e.g. dental disease; [27]; caecotroph appearance, obesity; [23]; repetitive behaviours) seen in this survey. Ongoing owner education about optimal rabbit feeding regimes is therefore still required.

Whilst this survey suggests that pet rabbits on average live longer than was previously reported (5.6 compared to 4.2 years in [1]), morbidity levels are high. Frequent symptoms of ill health were reported with perineal soiling (reported as "dirty bottoms") being most common as has been previously suggested [28]. As seen in previous studies, dental problems were very common (12.2% reported to have suffered in this survey), but owners also described lack of appetite, runny eyes, digestive problems and many reported seeing caecotrophs, which may also be symptomatic of dental problems (or obesity), preventing the rabbits from consuming caecotrophs. The common occurrence of veterinary problems may, at least in part, be due to recruitment including distribution through veterinary surgeries. However dental disease was similarly found to be very common and under-reported by Mullan and Main [4] and our findings suggest that other issues may also be high and potentially unrecognised or underreported. Confirmation of causal factors, via additional research and improved education of carers regarding diet and health care is evidentially required.

Owners reported behaviours which were categorised as potentially indicative of "positive" welfare in the vast majority of rabbits, but with varying frequencies, for example although binkying was reported in 88%, only a quarter of rabbits were seen to do this many times per day. Behaviours like binkying, rearing up and lying stretched out may be prevented in some living environments (e.g. due to inadequate space or height) and they may also be under-reported in specific environments where owners are less likely to watch their rabbits, e.g. in runs compared to indoors. Potentially negative behaviours such as thumping hind limbs, gnawing housing, grunting and digging on hard surfaces were also common, which could be cause for concern. The relationship between the occurrence of these behavioural indicators and the rabbits' environment and care warrant further research.

Within this sample, although most owners reported "picking up and handling" their rabbit at least weekly, a small minority of rabbits (2.2%) were never picked up and handled. It is however plausible that some of these were handled and stroked whilst on the floor, which can be less stressful to the rabbit and is generally good practice (e.g. http://www.bio.miami.edu/hare/firstrabbit.html). Hence we would recommendthat future surveys make this distinction between "picking up" and "handling".

Owner reports of behaviour suggest that most rabbits do not respond calmly when handled either by their owner (61%) or other adults (75%). Combined with the facts that 27% of owners do not describe themselves as "very confident" when handling their rabbit, yet the majority of rabbits are handled at least weekly; this could potentially represent a significant source of stress and suggests that appropriate handling protocols are essential to ensure this is not aversive, and that fearful animals do not become further sensitised to handling. The value of early positive handling has been demonstrated for laboratory rabbits (e.g. [29]) but further attention and research is needed to determine and promote optimal protocols for pets.

Over half of the pet rabbits were reported to be fearful of loud noises, suggesting that celebrations, for example involving fireworks, could be a significant welfare concern for this species, especially as they are more likely to be housed outdoors than other domestic pets. Research into ways to best habituate rabbits to and treat developed fears of loud noises would be very valuable.

Finally, it is worth noting that unlike in Mullan and Main [4], most rabbits in our survey had been obtained from pet shops or garden centres. Since Edgar and Mullan [3] noted that the conditions observed at the point of sale affected the way owners proposed to house their rabbits, it is likely that welfare education campaigns which engage pet shops and improve the way they exhibit rabbits will also positively affect how the public keep their pet rabbits.

5. CONCLUSION

This large-scale survey gives us up-to-date in-depth information on the ways in which pet rabbits are housed and cared for in England.

Owner-reported data presented here suggest that whilst some rabbits have good health, compatible companions and are provided with enriched living areas, large numbers of rabbits experience potential stressors and welfare issues on a daily basis. The accuracy of the owners' reports is yet to be tested, and it is likely that some of these issues are even more prevalent than this survey has suggested. Many rabbits are kept in ways which do not meet common recommendations (e.g. Wales [16]; Northern Ireland [30]).

This study has identified a large number of areas in which further research is required, for example prioritisation of the range of welfare issues highlighted by this survey, and to confirm postulated associations between aspects or housing and husbandry and rabbit welfare status. Future research also needs to focus on overcoming common issues, such as determining optimal protocols for habituating rabbits to human handling and to loud noises, which have been shown to be common causes of fear.

COMPETING INTERESTS

The authors declare that they have no competing interests.

AUTHORS' CONTRIBUTIONS

The study was designed and planned by NR, SH, EB, SM and RS. The questionnaire was constructed by NR, EB, PB, MT, JH and CS, with input from SH, SM and RS. The questionnaires were distributed and verbal interviews conducted by PB, JH, CS, MT and NJ was instrumental in constructing the internet version. Data entry was performed by PB, MT, JH and CS. Data analysis and initial manuscript drafting was by NR, with significant input for SH. All authors contributed to the critical review and revision of the manuscript, and have approved the final version.

ACKNOWLEDGEMENTS

We are extremely grateful to the many veterinary practices, pet shops, rehoming centres, retails outlet, Bristol Zoo Gardens, and schools which facilitated questionnaire distribution. We are very thankful to the very amenable veterinary practices which contacted their clients in order to recruit our verbal questionnaire respondents: Companion Care, Speke, Bristol and Norwich, Langford Veterinary Services, Highcroft (Bristol), Ashleigh Veterinary Surgery (Manchester) and the PDSA centres in Kirkdale, Manchester, Huyton and Bristol, and RSPCA at Brent Knoll who encouraged clinic attendees to take part in the verbal survey.

We are grateful to the Rabbit Welfare Association and Fund and in particular Rae Todd and Susie Hughes (RSPCA) who helped in innumerable ways and to all the owners who took the time to complete the survey. Thank you also to the four anonymous reviewers for their very valuable comments.

This work was commissioned by the RSPCA by a competitive tendering process. We thank them for their financial support

REFERENCES

1. Schepers F, Koene , Beerda B: Welfare assessment in pet rabbits. Anim Wel f. 2009, 18: 477-485.

2. PDSA: PDSA Animal Wellbeing Report. 2011, http://www.pdsa.org.uk/pet-health-advice/pdsa-animal-wellbeing-report. Accessed 11/01/2015
3. Edgar JL, Mullan SM: Knowledge and attitudes of 52 UK pet rabbit owners at the point of sale. Vet Rec. 2011, 168: 353.
4. Mullan SM, Main DCJ: Survey of the husbandry, health and welfare of 102 pet rabbits. Vet Rec. 2006, 159: 103-109. 10.1136/vr.159.4.103.
5. Mullan SM, Main DCJ: Behaviour and personality of pet rabbits and their interactions with their owners. Vet Rec. 2007, 160: 516-520. 10.1136/vr.160.15.516
6. Home Office: Code of Practice for the Housing and Care of Animals Used in Scientific Procedure. 1989, London: The Stationery Office, https://www.gov.uk/government/uploads/system/uploads/attachment_data/file/228831/0107.pdf, Accessed 12/01/2015
7. PDSA: PDSA Animal Wellbeing Report. 2013, http://www.pdsa.org.uk/pet-health-advice/pdsa-animal-wellbeing-report. 2013. Accessed 01/03/2014
8. Department of Environment Food and Rural Affairs (DEFRA) Archive Animal Welfare Act: 2006, http://archive.defra.gov.uk/foodfarm/farmanimal/welfare/act/. Accessed 12/01/2015. Crown Copyright 2006. The Animal Welfare Act. http://www.legislation.gov.uk/ukpga/2006/45/contents
9. Vasseur E, Pellerin D, de Passille AM, Winckler C, Lensink BJ, Knierim U, Rushen J: Assessing the welfare of dairy calves: outcome-based measures of calf health versus input-based measures of the use of risky management practices. Anim Welf. 2012, 21 (1): 77-78. 10.7120/096272812799129439.
10. Held SDE, Spinka M: Animal play and animal welfare. Anim Behav. 2011, 81: 891-899. 10.1016/j.anbehav.2011.01.007.
11. Hoffman KL, Hernandez Decasa DM, Beyer Ruiz ME, Gonzalez-Mariscal G: Scent marking by the male domestic rabbit (Oryctolagus cuniculus) is stimulated by an object's novelty and its specific visual or tactile characteristics. Behav Brain Res. 2010, 207 (2): 360-367. 10.1016/j.bbr.2009.10.021.
12. Carlstead K, Brown JL, Strawn W: Behavioural and physiological correlates of stress in laboratory cats. Appl Anim Behav Sci. 1993, 38: 143-158. 10.1016/0168-1591(93)90062-T.
13. Jilge B: The rabbit: a diurnal or nocturnal animal. J Exp Anim Sci. 1991, 34 (5–6): 170-183.
14. Perez SS, Jacksic D, Meriggi A, Vidus RA: Density and habitat use by eth European wild Rabbit (Oryctolagus Cuniculus) in an agricultural area of Northern Italy. Hy Strix – Italian. J Mammal. 2008, 19 (2): 63-76.
15. Dixon LM, Hardiman JR, Cooper JJ: The effects of spatial restriction on the behavior of rabbits (Oryctolagus cuniculus). J Vet Behav. 2010, 5 (6): 302-308. 10.1016/j.jveb.2010.07.002.
16. Welsh Assembly Government: Code of Practice for the Welfare of rabbits. 2009, http://wales.gov.uk/docs/drah/publications/091109rabbitcodeen.pdf. Accessed 12/01/2015.
17. Princz Z, Radnai , Biro-Nemeth E, Matics Z, Gerencser Z, Nagy I, Szendro Z: Effect of cage height on the welfare of growing rabbits. Appl Ani Behav Sci. 2008, 114: 284-295. 10.1016/j.applanim.2008.01.006.
18. Held SDE, Turner RJ, Wootton RJ: Choices of laboratory rabbits for individual or group-housing. Appl Anim Behav Sci. 1995, 46: 81-91. 10.1016/0168-1591(95)00632-X.
19. Seaman SC, Waran NK, Mason G, D'eath RB: Animal Economics: assessing the motivation of female laboratory rabbits to reach a platform, social contact and food. Anim Behav. 2008, 75: 31-42. 10.1016/j.anbehav.2006.09.031.
20. Huls WL, Brooks DL, Bean-Knudsen D: Response of adult New Zealand White rabbits to enrichment objects and paired housing. Lab Anim Sci. 1991, 41 (6): 609-612.
21. Mykytowycz R, Dudziński ML: Aggressive and Protective Behaviour of Adult Rabbits Oryctolagus cuniculus (L.) towards Juveniles. Behav. 1972, 431 (4): 97-120.
22. Graf S, Bigler L, Failing K, Wurbel H, Buchwalder T: Regrouping rabbit does in a familiar or novel pen: Effects on agonistic behaviour, injuries and core body temperature. Appl Anim Behav Sci. 2011, 135 (1–2): 121-127.

23. Prebble J, Meredith A: Food and water intake and selective feeding in rabbits on four feeding regimes. J Anim Phys Anim Nut. 2014, 98 (5): 991-1000. 10.1111/jpn.12163.
24. Kamphues J: Die artgerechte Fu¨ tterung vo Kaninchen in der Heimtierhaltung. Deutsche tiera¨ rztliche Wochenschrift. 2001, 108: 131-135.
25. Berthelsen H, Hansen LT: The Effect of hay on the behaviour of caged rabbits (Oryctolagus Cuniculus). Anim Welf. 1999, 8 (2): 149-157.
26. Lidfors L: Behavioural effects of environmental enrichment for individually caged rabbits. Appl Anim Behav Sci. 1997, 52: 157-169. 10.1016/S0168-1591(96)01141-0.
27. Jekl V, Redrobe S: Rabbit dental disease and calcium metabolism - the science behind divided opinions. J Small Anim Pract. 2013, 54: 481-490. 10.1111/jsap.12124.
28. Sayers I: Approach to preventative health care and welfare in rabbits. In Practice. 2012, 32: 190-198.
29. Zucca D, Redaelli V, Marelli SP, Bonazza V, Heinzl E, Verga M, Luzi F: Effect of handling in pre-weaning rabbits. World Rabbit Sci. 2012, 20 (2): 97-101.
30. Department of Agriculture and Rural Development (DASRD): Welfare Code of Practice for Rabbits. 2013, http://www.dardni.gov.uk/rabbits-code-of-practice.pdf. Accessed 12/01/ 2015

CHAPTER 20

Animal Welfare in Relation to Standards in Organic Farming

Karl-Erik Hammarberg

Hudiksvalls distriktsveterinärstation, Malnvägen 2, S-824 50 Hudiksvall

The new EU-regulations on organic farming (1804/1999) are also influencing the animal welfare. A lot of positive regulations is to find, but also regulations that seen to mind more about the general public and customer and their view on organic farming, than the health and welfare of the animals.

The paper specially focus on the impact of the regulations and the recommendations that phytotherapeutic essences and homeopathic products take precedence over the so called chemically-synthesised allopatic veterinary medical products, and that the use of the same is prohibited for preventive treatments. Key questions here are the lack of scientific evidence concerning homeopathy in animals, and that Swedish veterinarians are not allowed to work with homeopathy.

Differences in interpretation of the regulations between animal owners and veterinarians will also be discussed. What is a disease that needs treatment? Who is to decide about the treatment? Parasitic infections are discussed as an illustrative example.

Other consequences of the regulations concerning the animal welfare are problems in certain geographical zones, for instance subarctic areas where necessary crops are impossible to grow.

Animal transports and splitting mother-offspring are briefly discussed as future problems to be handled in the regulations, and the paper ends by presenting the need of regulated herd health control programs in organic husbandry, which can detect and focus on welfare and production problems. The organic movement is not static, and must not be so.

1. INTRODUCTION

Ever since our human ancestors began to abandon the life as hunter gatherers and enter into an existence based on permanent settlements living from agriculture, animal husbandry has been a natural part of producing food. Through animals, nutrients, indigestible for humans, e.g. grass and leaves, could be converted to edible proteins, energy, fat and minerals. In our colder climate zones these nutrients could be stored in the animals until they were needed. It is no exaggeration to state that animal husbandry has been a necessity for man in order to be able to survive in cold climate zones.

Agriculture developed to begin with slowly. Initially by slash and burn methods, later with hay making agriculture leading on to farming with crop rotation. During the last 80 years, in Sweden especially during the last 50 years, the use of artificial fertilisers which now, with increased political support and a maturing view on agriculture, is changing towards organic farming. The main reason for this is the last decades focus on the damage caused by artificial fertilisers and the impacts of the circulation of nutritive substances on the environment. With the increase in industrial agriculture followed a large-scale animal farming, focusing on maximum production output, not always favourable to the animals.

Personally I grew up on a farm and later worked as a veterinarian in rural practice. This has given me the opportunity to follow the rapid development in animal farming during the last 50 years. I have even been a part in this "progress". To partake in this development does not necessarily mean that one sympathises with all that is seen, or with all that has been done. It has, especially during the last decade, become increasingly clear to me that a part of my work has been to act as a "cleaning lady" after the advisers and companies that has led the development in animal farming. Technicians have introduced ventilation systems and constructed cubicles and pens. I have had to deal with calf pneumonia and mastitis. Feeding advisers have created food for animals and I have had to take care of disturbed stomachs, enteritis and distorted immune systems. Economists have recommended large-scale farming and the spreading of contagious diseases have been left for me to attend.

To become aware of the fact that I, by hopefully being a skilful vet, have supported this development, and maybe even helped to maintain this practice of which parts should have been abandoned a long time ago, has increasingly frustrated me.

With this background I see organic animal farming as one of several possible ways to leave this form of animal husbandry behind us, a husbandry that does not put the animals and their needs in focus.

Organic animal farming has in itself been in existence all the time parallel to this development, but it was not until the 1970:s that it began to be formalised. Organic products had to be certified, in one way or another, and give the customer information and knowledge of what it meant to buy "organic". Gradually also the politicians began to catch the organic trend and started to support it with financial subsidies. Targets were set up. One was the 10% level. This meant that by the year 2000, 10% of Swedish farming was to be organic.

The rest is well known to you all. An international umbrella organisation, IFOAM (International Federation of Organic Agriculture Movements), was set up with criteria to define, among other things, organic animal husbandry. Control work began in every participating country supervised by its central authorities. In my country the Swedish Board of Agriculture has appointed KRAV (Kontrollföreningen för ekologisk odling) and the Demeter organisation to set the directive for organic production and to control the suppliers.

2. EU-REGULATION 1804/1999 AND ANIMAL WELFARE

Organic animal husbandry is not static. It is under continuous development. Many of the initial blunders and mistakes have in co-operation with others, among them the vets, been altered and corrected and work continues. In the year 1999, EU decided that certain common rules were to be in force within the whole common market. I am referring to Regulation 1804/1999. It is from the basis of this regulation that I want to give some veterinary comments and thoughts on organic animal farming.

We must presume that the regulation has been introduced to foster and improve the life of the animals. This is, however, not obvious to me. The same thoughts have been put forward in some papers from the Network for Animal Health and Welfare in Organic Agriculture, NAHWOA. No, the decisions behind some of the directives in the regulation are more to satisfy people's expectations and wishes on organic products and to keep the confidence in organic produced products high. No chemicals, no medical drugs.

Many of the regulations are clearly to the advantage of the animals and their general welfare, e.g. demands of prolonged access to milk and increased possibility to live in an environment which favours natural behaviour. These and other EU regulations are to be praised, but others really have to be debated.

As a veterinarian I believe I can do better for animal welfare. Not only by giving assistance with setting up regulations to foster animal welfare, but also by bringing to your attention the weaknesses and discrepancies I see through my veterinary coloured glasses.

We should also bear in mind that some people start organic farming from a subjective image of an organic ideal and not from sound knowledge of agriculture. This does perhaps not lead to any serious consequences when it involves plant growing, but when it is associated with animal farming the consequences can be serious.

There is nothing in the organic regulation that stipulates any knowledge on how to run and manage an animal farm. It does not even state to strive for as natural a life as possible for the animals. The real nature is far too tough for that.

So let us take a look at some of these new regulations where, as a vet, I am of the opinion that animal welfare might be affected in a negative way.

"The use of chemically synthesised allopathic veterinary medicinal products or antibiotics for preventive treatments is prohibited." [1]

Here I would like to debate the fear of using chemicals or medical drugs. I find among some organic farmers that this regulation tends to support their subjective opinion. The Swedish animal welfare legislation, which is alleged to be the toughest in the world, claims that animals have to be protected against diseases. My interpretation of the EU-regulation is, however, that you are not allowed to protect animals from diseases when there are no other ways left but using chemical "allopathic" products. In my view the regulation here leaves the welfare of animals in the hands of the general public's and the customer's opinion on organic farming. Let me give you some examples:

In the North of Sweden attacks from gnats and mosquitoes can be deadly to animals. When these insects flutter about, humans, even the organic farmer protects himself with chemical repellents, or tight fitting clothes. Animals on conventional farms are protected with chemicals, e.g. spot on's, but animals on organic farms are denied this protection, they are unable to run away from their paddock or "cover themselves with clothes". When you see animals attacked by gnats and mosquitoes, you realise they do suffer badly. I have on many occasions seen animals killed in this way.

Another example is tick born diseases. There are chemical protectors on the market, but it is unclear if they can be used prophylactically on organic farms without losing the organic status. Some tick born TBE-viruses can even spread via the milk from goats to humans if the milk is used in a non pasteurised state. [2]

The same thing can be said about Paraphilaria. There is no organically acceptable way to protect cattle against such fly attacks if you follow the EU-regulation. [1]

And, what about Haemonchosis in sheep and scabies in pigs? The eradication programme in conventional farming might leave only the animals bred in organic farming to suffer from these parasites and spread them.

There is also a need to make clear if vaccines are seen as allopathic drugs. This includes the vaccines where its manufacturing is a result of genetic engineering. In my opinion vaccination must be permitted and used in animal disease control and welfare programmes.

Another thing that concerns me and many Swedish veterinarians and farmers is the ban on the use of synthetic vitamins in ruminant food. These additives are necessary and important in our Nordic climate with short grazing seasons and long, dark winters. The result might be lack of vitamins in high producing animals that can, among other things, affect their resistance against diseases. It can also have an affect on the human customer who prefers to drink organic milk. What about the vitamin E and selenium additives that we find is needed in selenium deficient grazing and harvesting areas? How about pain killers to animals in pain? What signals do parts of this EU-regulation send to the consumers?

For me as a veterinarian it is in this situation more important to protect the animals than the customers and consumers.

A consequence of the regulation, that does not immediately spring to mind, is that organic farms can serve as reservoirs for diseases which can be kept under control by the prophylactic use of allopathic drugs. Paraphilaria has, for instance, been mentioned. In conventional animal farming you can control the multiplication and spreading of parasites by using ear-tags with repellents, but an organic farm is not allowed to use these repellents. The parasite can thus survive and multiply there, with the result that the farms neighbouring such an organic farm have to continue to use chemical ear-tags more then if the organic farm had not been there. The result might be that a regulation aiming to reduce chemicals will lead to increased use when you look at the total effect. What will be the customer's reaction to organic farming when they hear that the organic animals are the only ones suffering from parasite problems?

3. SICK OR INJURED ANIMALS MUST BE TREATED

The EU regulation gives an example on how to maintain good and healthy conditions among the animals. This is great. But what happens when an animal gets sick or hurt regardless of this? What does the regulation say then?

"If, despite of the above preventive measures, an animal becomes sick or injured it must be treated immediately, if necessary in isolation, and in suitable housing". [1]

So sick and diseased animals are to be treated. So far so good, but we still face a problem here. When is an animal sick and what should the treatment be? I am not so sure that what a veterinarian claims to be a disease or a welfare problem is judged in the same way by the animal owner. In Sweden KRAV is absolutely clear on this matter. The veterinarian's point of view is valid. But we still face problems here. Let me just mention sheep suffering from endoparasites. When are they sick and what is natural?

We face the same problem when it comes to the treatment. The animal owner might prefer an alternative treatment to the one the veterinarian in a given situation believes to be the best, most effective and even necessary.

4. HOMEOPATHIC TREATMENT AND WELFARE

When it comes to treatment of sick animals the new EU-regulation puts restrictions on the vet's alternatives to act. I am referring to the following:

"Phytotherapeutic (e.g. plant extracts [excluding antibiotics], essences, etc.), homeopathic products (e.g. plant, animal or mineral substances) and trace elements and products listed in Part C, section 3 of Annex 11, shall be used in preference to chemically-synthesised allopathic veterinary medicinal products or antibiotics, provided that their therapeutic effect is effective for the species of animal, and the condition for which the treatment is intended". [1]

The first part, about phyto-therapy and homeopathy has of course as a background that animals must be kept in a way which reduces disease problems and that the use of chemically-synthetic drugs is to be reduced as much as possible.

One interpretation of the regulation, at least in Sweden, is that anyone except a veterinarian is preferred to administer to sick animals on an organic farm, as Swedish vets are not allowed to use homeopathy and phyto-therapy. We must work only with methods based on science and proven experience. The mentioned therapies, especially homeopathy, do not satisfy this demand. There are thousands of reports on homeopathy, most of them bad or very bad, when seen through the eyes of science. However, there is still no proof that the homeopathic fundamentals "alike cures alike" and "potensation increases the effect" are true.

Funnily enough, you can interpret the regulation in quite a different way. By concentrating on the second part of it, *"provided that their therapeutic effect is effective for the species of animal, and the condition for which the treatment is intended"*. I have not yet found any reports or tests giving me necessary information on this.

There are also other EU directives which can be interpreted that homeopathic drugs must be registered the same way as ordinary medical drugs in order to be allowed to be used in food producing animals [4].

I have in many different ways tried to clarify the situation to myself. What does it mean to me as a veterinarian and to the farmer? But so far I have been unsuccessful. Many things, also those concerning animal welfare, are still unclear to me. Those who believe in homeopathy understand and read published reports in their favour. The non-believers have another interpretation. But face the facts! There are no scientific indisputable reports that can support the positive and controllable effects from homeopathy in medical treatment of animals.

A consequence hereof has already been mentioned. Organic animal farming gets a reputation of not treating sick animals in an appropriate way. This can influence the consumer's view. Should this be so, the result of this regulation will be directly opposite to what is was intended.

In reality, I do not think the regulation will have any greater effect on the individual farmer. Those who are believers in homeopathy will continue to believe in it and treat sick animals as before. Those who do not believe in it will not change their minds as a result of the regulation. However, in the eyes of Swedish veterinarians and zoologists the EU-decision is more than astonishing. We have already now examples from both the organic and the conventional farming world where animals have suffered and even died when the farmer has chosen homeopathic treatment before traditional veterinary medicine [3]. And we have seen the spreading of a contagious disease, BVD, through an homeopath. I believe that sooner or later, once reported to the police, this matter will end up as a court case where the charge will be that homeopathic treatment leads to undue suffering of animals. We will then finally get a judicial verdict on what is valid.

I really want to underline that I do not wish to glorify the use of chemical drugs and medicines with this paper. We shall of course strive for animal husbandry built on optimal environmental conditions. Feeding, genetics and care with the minimum use of drugs, but when the need arises they shall be used. This need shall, in my opinion, be decided by a veterinarian who has full knowledge of organic animal farming. With this I mean a vet who can see animal problems against the background of environmental influences, organic feeding, breeds, genetics, animal behaviour and other considerations. In other words a vet with an holistic view. We need a vet who can see further than just infectious matters as a background to the problem. But sometimes even infectious agents have to be considered.

5. MAXIMUM NUMBER OF TREATMENTS

The regulation states that animals loose their organic status if they are treated 3 times or more with chemical-synthetic allopathic drugs or antibiotics within one year. I have still no experiences of the consequences. Of course I see the risk that organic animal owners, when facing the choice between another allopathic treatment with possible loss of organic status and all its consequences, or no allopathic treatment at all, hope that the animal will get well anyway. Or will choose another treatment. Painkillers have been mentioned. But I hope that the animal owner who has already treated an animal twice during this period thinks more of the animal than the organic status. As I have understood it, the last word on this matter has still not been spoken.

6. SEPARATING MOTHERS FROM THEIR OFFSPRING

I can see more than just medical problems when I consider animal welfare in the organic farming. I can also give you some ethological aspects on what I see, aspects I hope will be taken into consideration in a future regulation. Let me give you some examples:

In both conventional and organic dairy farming it is common to separate mother and calf already after a few days. This separation is very unnatural. Everybody who have heard a mother and calf calling for each other after a separation knows we should try to find a system where it would be possible for calves to stay longer with their mothers, stepmother or other adult cows. Research at the Agricultural University in Skara, indicates that such a system is feasible. The reason for separating mother and calf is, as you know, that the mother's milk is to be used for other purposes. However, the research mentioned, shows that it is possible to have the calf with its mother longer without any considerable loss of bulkmilk. The cow produces more milk when the calf suckles. My view is that we should start planning for systems where the calf has increased access to its mother (or stepmother) for longer and more often than is common today. The calf could be together with its mother a couple of times every day, or spend parts of the day or night with her.

Another example we can look at is the common separation of herds of sheep into groups, the purpose being to be able to feed every animal after their needs. A sheep herd actually consists of a number of small herds with mothers and daughters sticking together. To be able to feed the animals right we break up these social groups and put the animals in different pens. In nature this works without the groups being separated. I believe it is possible to find feeding systems where we do not need to separate the family groups as we do today. I have seen an organic sheep farmer with 35 ewes solving this problem brilliantly. They use a feeding system where the animals have free access to good hay and silage plus concentrate feeding in a way that gives also the younger ewes and lambs plenty of space to eat. It is worth a try to find similar systems for larger flocks.

7. TIED UP COWS

We need a further debate with regards to the tying up of cows. According to organic regulations the cows shall be given possibility to move, but there are situations when such motion causes more stress to the animal than when they are tied up. Consequently I do not recommend tied systems, just that we still do not have all relevant information and know-how about free walking systems.

8. GIVE THE ANIMALS SOMETHING TO DO

We have over the years learnt a lot about how to put animals in houses, fences, pens, cubicles and feedlots. But still we do not give them anything to do, something for their curiosity, motoric training and playfullness. I wish that in the organic regulation there were some demands on this, for instance some differences in heights in their area, something to play with for instance balls or tires hanging from the ceiling, drag some bushes or cut down trees into their pens, or give them the possibility to really search for food. Give them something to do in a boring life situation. The organic animal farming should be forerunners here.

This is what I call animal welfare.

9. GEOGRAPHICAL RESTRICTIONS

I have above in this paper, in connection with gnat attacks, mentioned cold climate zones. A substantial part of the fodder, on average about 50%, has to be harvested on the home farm. This means that farms geographically located outside what can be called the farming limit are excluded from organic farming, despite the fact that mountain farming of, for instance, cattle and sheep can be very friendly to the animals and the environment.

It can also be very difficult to grow grain for pigs and to get straw. The demand on outdoor living, or as close to outdoor living as possible, can cause problems with, for instance, frostbites on pigs or lamb ears in the sub arctic parts of Scandinavia. The regulation should be more flexible and adapt to local situations but still be based on an organic thinking. The regulation can, as it stands today, cause geographical restrictions on organic animal farming. If the new EU-regulation had been written by individuals with knowledge of organic matters and interested, practising skilful farmers from northern thinly populated areas of Europe, it would surely have looked different in many aspects.

10. TRANSPORTS TO SLAUGHTER

And so the time comes for slaughtering. According to general organic regulation, animals shall be slaughtered in approved and controlled abattoirs or slaughterhouses.

Against the present background of the closing down and centralisation of slaughterhouses, organically farmed animals have to be transported longer and longer distances, (but not exceeding the Swedish legal maximum of 8 hours transport) sometimes bypassing non approved slaughterhouses. In many cases this is neither organic nor animal friendly. To me it goes without saying that organically farmed animals should be slaughtered at the nearest slaughterhouse. Most of all I would like to have them slaughtered on the farm or at a nearby approved small farm slaughterer.

To favour such a scheme I wish that central authorities would increase the number of animals accepted for slaughter in such small farm slaughterers. This is possible within the EU-regulation. Such a decision would shorten the transit time and perhaps also improve the economy for small farm slaughterhouses.

Another way to go forward, is to accept the same veterinary inspection of the animals as when inspecting deer in deer farms at slaughter, or the inspection of game (moose, bears) for hunters. However, other statutes in the EU-regulation might not accept this 'modus operandum'.

11. ANIMAL HEALTH CARE SERVICE

There are organic farmers who believe that the natural life is always best for the animal. That is not true. Nature is cruel. The natural death rate is in nature very much higher than the one acceptable to normally controlled animal farms. The aim is to keep diseases and production problems, which have turned out to be the cause of serious problems in organic farming, under control. I wish that organic farms had some sort of animal health care service within their regulation. This can be carried out either with the help of already existing animal health services, or by establishing a new health control programme within the organic organisations, with the help of veterinarians. Such a health care programme can look at environmental factors, parasite control, production results, slaughter results, post mortems etc. All with the end purpose of finding and determining the factors and weaknesses which influences animal health.

An animal health care programme must have a built in prophylactic work facility with effective methods to solve problems when other avenues have been exhausted. This must include the ability to use chemical-synthetic allopathic drugs, for instance, effective de-worming methods and vaccines. This is one of the fundamental aspects of animal health care. Strategic use of prophylactic measures is an advantage for the whole herd in stead of just being bound to a late treatment of single, sick individuals. A clarification with perhaps an alteration to the regulation in order to adapt to reality is desired. A positive consequence of an organised animal health care service within the organically run farming industry is that the buying of animals, when necessary, would be safer from an animal health point of view. Only buy from farms on the same or a higher level within the health care service program.

12. THE ORGANIC MOVEMENT MUST NOT REMAIN STATIC

After having said all this, there might be listeners who judge me as negative to the concept of organic farming. This is wrong. Critical on some points maybe, but basically very positive to an organisation opening doors to an animal husbandry more on animal terms than what we very often see in conventional farming.

More and more reports show that the animal health status on most organically run farms is satisfactory and thereby making it possible to maintain good production. There are, however, some exceptions, which perhaps will be discussed here by other speakers. The organic movement must seriously continue to tackle these situations.

I also respond very positively to the politicians who with their decisions open the possibilities for organic farming to compete and survive in a world where mass production is an economic advantage. If organic farming is not given the incentive, support and the economic capability to compete with the big mass producing organisations, the positive trend for organic farming will soon be broken.

The criticism I present here is meant to be positive. My talk here is meant to highlight a situation where we, unless we start working on them today, might face difficulties in the future. There are still bits and pieces to look at and work for, before we find a form of animal husbandry that is primarily based on the animal's needs. But I am optimistically hopeful as long as I can see the direction in which the organic movement in general is going today. The organic movement is not static, and must not be so.

REFERENCES

1. EU-Regulation 1804/1999.
2. Personal communication, State Veterinary College, Lithuania.
3. Ekesbo I: Förekomst och konsekvenser av homeopati på husdjur. Svensk Veterinärtidning. 2000, 52: 871-880.
4. EU-regulation 1992/74

CHAPTER 21

Factors Influencing Effective Use of Information and Communication Technology (ICT) Tools By Various Animal Husbandry (A.H.) Organizations in Andhra Pradesh

Dr. P.Sireesha1 , Dr.B.Sudhakar Rao2 , Dr.D.Thammi Raju3

1 Assistant Professor, Department of Veterinary & Animal Husbandry Extension, College of Veterinary Science, Korutla, Andhra Pradesh, India

2 Retired Professor & Head, Department of Veterinary & Animal Husbandry Extension, College of Veterinary Science, Rajendranagar, Andhra Pradesh ,India

3 Principal Scientist, Education Systems Management Division, NAARM, Hyderabad, Andhra Pradesh, India

ABSTRACT

Data was collected from 33 organizations which were actively involved in Animal Husbandry activities and elicited through a questionnaire following an exploratory research design on "Utilization of Information and Communication Technology (ICT) tools by various organizations in Animal Husbandry - A study in Andhra Pradesh." Majority of the organizations agreed that ICT tools are functional at all times, sufficient number of ICT tools are available in the organization, the connectivity is good and access to the ICT tools is always available. As regards to affordability of ICT tools majority indicated that the organization is able to afford the ICT tools, has sufficient funding for purchase and maintenance of ICT tools, has adequate technical support for maintenance of ICT tools. Responses were varied for the statement that organization is providing loans for the purchase of ICT tools. Regarding the knowledge of computer applications, majority of the users have medium level of knowledge.

Keywords: ICT, Affordability of ICT, infrastructure availability of ICT tools, Animal Husbandry organizations

1. INTRODUCTION

Livestock sector is one of the allied sectors which has proved to be an integral part of livelihood system of rural families and continues to be a major source of rural economy. There is rapid growth in the overall livestock production of our country which can be attributed to the technological revolution in A.H. and the successful adoption of recommended technologies by the rural farmers through different channels of information dissemination by the extension agencies. Apart from the conventional methods of information dissemination like traditional extension methods, electronic and print media, the recent foray into information gateway is the initiative of ICT tools in Agriculture and allied sectors.

In order to transform livestock sector into information driven, modern and competitive sector, the role of Information and Communication Technology (ICT) cannot be overruled. They provide faster and newer ways of delivering and accessing information. The people working in the livestock sector are least equipped with proper tools to deal with rapidly changing livestock production scenario and international competitive environment. At present, the ratio of the farmers to the extension workers is 1000:1(Kumar, 2005), also the existing transfer of technology mechanisms and extension programmes run by the government departments are slow and in many cases ineffective in view of the vast gaps between the research and farmers linkages. This is partly due to inadequate use of new areas of information dissemination in various development programmes. The introduction of ICT helps in upgrading the information at least cost.

Information technology has demonstrated its utility and advantages in all elements of human life and especially in vocations of agriculture and allied sectors enhancing the productivity either directly or indirectly. The extension activities in Indian scenario are traditional in nature and suffer from certain drawbacks such as expensive in nature, irrelevancy of the content, inadequate coverage, one way information system, erosion of messages in passing through different layers, extension personnel's ineffective communication, neglect of extension in livestock development etc. (Sasidhar and Sharma, 2006). Information technology can help to overcome the barriers of traditional extension system to a greater extent. Earlier the mindset of extension system is of transferring technology packages, now with the initiation of ICT tools, the mindset has changed to transferring knowledge or information packages. By this, extension system will become more knowledge intensive and demand driven, and thus more effective in meeting farmer's information needs.

Without communication capabilities, access to information, analyze and share the information in the livestock sector remains at the mercy of global market forces. At this juncture, a rural friendly ICT policy that recognizes the Government's fibre network as a national asset and offers concession to those who extend the network to rural India will certainly prove to be a big boost to the rural GDP in the coming years. Complementing the new ICT revolution, it is important for the Government to recognize the need for the establishment of Community Information centers at both village and block level that have the potential to reach out to rural communities, which will extend the Government's audiovisual and publicity efforts on various entitlements and programmes and information on technical, socio-economic and health related issues, while empowering the rural community to communicate and develop structure for a knowledge intensive rural production system.

1.1 Objectives of the study

1) To study the qualitative indicators of infrastructure availability of ICT tools of the Animal Husbandry organizations.
2) To find out the affordability of ICT tools utilized by the Animal Husbandry organizations.
3) To know the knowledge about computer applications of Animal Husbandry organizations utilizing the ICT tools

2. METHODOLOGY

The study was purposively conducted in Hyderabad and Rangareddy districts of Andhra Pradesh as most of the Animal Husbandry organizations are located in these two districts. A total of 33 organizations which are actively involved in Animal Husbandry activities were selected for the purpose of study. Several organizations involved in the livestock development either directly or indirectly in Andhra Pradesh were ascertained in consultation with various sources viz. consultation with experts, secondary sources such as reports, literature, documents etc. The questionnaire for the organizations was developed in consultation with experts in the field of Animal Husbandry extension and those involved in computer applications. Rapport with the representatives of organizations is very essential so as to get accurate responses. In the present study the investigator established initial rapport with the organizations selected for the study and got acquainted with the officials personally. Later after the development of the questionnaire the investigator handed over the questionnaire to the concerned officials of the selected organizations and elicited their responses through the questionnaire. Questionnaire was prepared in English and data was collected from 33 organizations. It was made sure that all the questions in the questionnaire were self explanatory. Statistical tools used for analysis of data included frequency and percentages.

3. RESULTS AND DISCUSSION

A total of 33 organizations which are actively involved in Animal Husbandry activities were selected for the purpose of study. Based on the nature of work, functions, type of administration, the organizations considered for the study were grouped into five categories for better presentation of the results and subsequent discussion. They are:

1) Service Providers
2) Financial Organizations
3) 3.Educational and Research Organizations
4) Non Governmental Organizations and Cooperative Organizations
5) Private Organizations

The different organizations were categorized into the above five categories are detailed below.

3.1 Service Providers

1) State Animal Husbandry Department
2) Commissionerate of Rural Development -Rural Livelihood project
3) Andhra Pradesh Livestock Development Agency (APLDA)

4) Andhra Pradesh Meat Development Corporation (APMDC)

5) State Management Institute for Livestock Development in Andhra Pradesh (SMILDA)

3.2 Financial Organizations

1) National Bank for Agriculture and Rural Development (NABARD)

2) Andhra Pradesh Cooperative Bank (APCOB)

3) State Bank of India -Agri business unit (SBI)

4) Bharatiya Samruddhi Investments and Consulting Services Limited (BASIX)

3.3 Educational and Research Organizations

1) National Institute of Rural Development (NIRD)

2) National Academy of Agricultural Research Management (NAARM)

3) Project Directorate on Poultry (PDP)

4) Central Research Institute for Dryland Agriculture (CRIDA)

5) National Research Centre on Meat N(NRC on Meat)

6) Sri Venkateswara Veterinary University -College of Veterinary Science, Hyderabad (SVVU)

7) Veterinary Biologicals Research Institute (VBRI)

8) Indian Immunologicals

3.4 Non Governmental Organizations and Cooperative Organizations

1) JK Trust Gram Vikas Yojana

2) ANTHRA

3) Watershed Support Services and Activities Network

4) InterCooperation Social Development

5) Andhra Pradesh Dairy Development Cooperative Federation (APDDCF)

6) National Egg Coordination Committee (NECC)

3.5 Private Organizations

1) Venkateswara Hatcheries
2) Suguna Poultry
3) IndBro Research and Breeding farms
4) Poshak Feeds Private Limited
5) Vimala feeds Private Limited
6) Miracle Feeds Private Limited
7) Intas Pharmaceuticals Limited
8) Neospark drugs and chemicals Limited
9) Sarabhai Zydus Animal Health Limited
10) Natural Remedies Private Limited

4. QUALITATIVE INDICATORS OF INFRASTRUCTURE AVAILABILITY OF ANIMAL HUSBANDRY ORGANIZATIONS

The quality of infrastructure was measured by ascertaining their response on a three point continuum i.e. agree, undecided, disagree for four statements reflecting on accessibility, functional status, connectivity and sufficiency of ICT tools. The response for each statement was rated on a three point continuum i.e. agree, undecided, disagree with weightages of 3, 2 and 1 respectively. The maximum score an organization could get was 12 and minimum score could be 4. Frequencies and percentages were calculated.

4.1 Qualitative indicators of infrastructure availability of Service Providers

From Table 1, it was clear that cent percent of the organizations agreed that access to the ICT tools is always available and that ICT tools are functional at all times was agreed by 60% and 40% of the users were undecided where as 80% agreed that connectivity is good and the same was undecided by 20% of the users. The statement on availability of sufficient number of ICT tools in the organization was agreed by 60% and an equal percent of

20% of the users were undecided and disagreed. This indicated that majority of the organizations were satisfied with the infrastructure availability made for ICT.

Table 1: Qualitative indicators of infrastructure availability of Service Providers.

n= 5

S.No.	Statements	Agree		Undecided		Disagree	
		f	%	F	%	f	%
i.	Access to the ICT tools is always available	5	100	–	–	–	–
ii.	ICT tools are functional at all times	3	60	2	40	–	–
iii.	Connectivity is good	4	80	1	20	–	–
iv.	Sufficient number of ICT tools are available in the organization	3	60	1	20	1	20

4.2 Qualitative indicators of infrastructure availability of Financial Organizations

From Table 2 it was inferred that all of the Financial Organizations agreed that access to the ICT tools is always available, functional at all times, connectivity is good, sufficient number of ICT tools are available in the organization. This indicated that all the Financial Organizations possessed all the important ICT tools and used for the required purpose and as such satisfied with the infrastructure provided to it.

Table 2: Qualitative indicators of infrastructure availability of Financial Organizations

n= 4

S.No.	Statements	Agree		Undecided		Disagree	
		f	%	F	%	f	%
i.	Access to the ICT tools is always available	4	100	–	–	–	–
ii.	ICT tools are functional at all times	4	100	–	–	–	–
iii.	Connectivity is good	4	100	–	–	–	–
iv.	Sufficient number of ICT tools are available in the organization	4	100	–	–	–	–

4.3 Qualitative indicators of infrastructure availability of Educational and Research Organizations

From Table 3, it was revealed that cent percent of the Educational and Research Organizations agreed that access to the ICT tools is always available which are functional at all times while 87.5 % and 12.5 % indicated they agreed and undecided that connectivity is good. 75% of the users agreed that sufficient number of ICT tools are available in the organization as against 25% who were undecided on the above statement. This trend revealed that majority of the organizations had good infrastructure facilities for ICT and satisfied with its requirement.

Table 3: Qualitative indicators of infrastructure availability of Educational and Research Organizations

n= 8

S.No.	Statements	Agree		Undecided		Disagree	
		f	%	F	%	f	%
i.	Access to the ICT tools is always available	8	100	--	--	--	--
ii.	ICT tools are functional at all times	8	100	--	--	--	--
iii.	Connectivity is good	7	87.5	1	12.5	--	--
iv.	Sufficient number of ICT tools are available in the organization	6	75	--	--	2	25

4.4 Qualitative indicators of infrastructure availability of Non Governmental and Cooperative Organizations

A cursory look at the Table 4 revealed that cent percent agreed that access to the ICT tools is always available and ICT tools are functional at all times, 66.7% agreed and 33.3% disagreed that connectivity is good while 83.3% and 16.6 % showed their agreement and disagreement to the statement that sufficient number of ICT tools are available in the organization.

4.5 Qualitative indicators of infrastructure availability of Private Organizations

The findings from Table 5 showed that cent percent of them agreed that ICT tools are functional at all times. 90% of them agreed and 10% were undecided to the statement that the connectivity is good and access to the

ICT tools is always available while 70% of them agreed and 30% were undecided that sufficient number of ICT tools are available in the organization.

Table 4: Qualitative indicators of infrastructure availability Non Governmental and Cooperative Organizations

n= 6

S.No.	Statements	Agree		Undecided		Disagree	
		f	%	f	%	f	%
i.	Access to the ICT tools is always available	6	100	--	--	--	--
ii.	ICT tools are functional at all times	6	100	--	--	--	--
iii.	Connectivity is good	4	66.7	--	--	2	33.3
iv.	Sufficient number of ICT tools are available in the organization	5	83.3	--	--	1	16.6

Table 5: Qualitative indicators of infrastructure availability of Private Organizations

n= 10

S.No.	Statements	Agree		Undecided		Disagree	
		f	%	f	%	f	%
i.	Access to the ICT tools is always available	9	90	1	10	_	_
ii.	ICT tools are functional at all times	10	100	_	_	_	_
iii.	Connectivity is good	9	90	1	10	_	_
iv.	Sufficient number of ICT tools are available in the organization	7	70	3	30	_	_

5. AFFORDABILITY OF ICT TOOLS BY ANIMAL HUSBANDRY ORGANIZATIONS

It was operationally defined as the opinion of an organization to pay for possession of ICT tools and or accessing the information through various ICT tools.

It consisted of 4 statements and the response for each statement was obtained on a three point continuum i.e. agree, undecided and disagree with weightages of 3, 2 and 1 respectively. The maximum score an organization could get was 12 and the minimum score could was 4. Frequencies and percentages were calculated according to the response.

5.1 Affordability of ICT tools by Service Providers

Table 6 brought out that among Service Providers, 60% and 40% of the users agreed and undecided that the organizations are able to afford ICT tools and that the organization has sufficient funding for the purchase of ICT tools. 20% of the users agreed, 60% were undecided while 20% disagreed that the organization has adequate technical support for maintenance of ICT tools. 20% agreed and 80% disagreed that the organization is providing loans for purchase of ICT tools. It is a good sign that majority of the organizations shifting from traditional methods to modern methods and also majority stated that organization has sufficient funding for the purchase of ICT tools. Majority of the organizations are quite regularly using ICT tools such as computers, mobile phones, fax etc.

Table 6: Affordability of ICT tools by Service Providers

n= 5

S.No.	Statements	Agree		Undecided		Disagree	
		f	%	f	%	f	%
i.	Organization is able to afford ICT tools	3	60	2	40	--	--
ii.	Organization has sufficient funding for purchase and maintenance of ICT tools	3	60	2	40	--	--
iii.	Organization has adequate technical support for maintenance of ICT tools	1	20	3	60	1	20
iv.	Organization is providing loans for purchase of ICT tools	1	20	--	--	4	80

5.2 Affordability of ICT tools by Financial Organizations

The results presented in Table 7 showed that 100% of the Financial Organizations are able to afford ICT tools, has sufficient funding for purchase and maintenance of ICT tools, has adequate technical support for maintenance of ICT tools and the organization is providing loans for purchase of ICT tools. It is a general notion that the output oriented financial organizations does give importance to infrastructure that facilitates the maximum productivity from its employees. So the results emerged out from the study were corroborating with current trend with reference to the affordability.

Table 7: Affordability of ICT tools by Financial Organizations

n= 4

S.No.	Statements	Agree		Undecided		Disagree	
		f	%	f	%	f	%
i.	Organization is able to afford ICT tools	4	100	_	_	_	_
ii.	Organization has sufficient funding for purchase and maintenance of ICT tools	4	100	–	–	–	–
iii.	Organization has adequate technical support for maintenance of ICT tools	4	100	–	–	–	–
iv.	Organization is providing loans for purchase of ICT tools	4	100	–	–	–	–

5.3 Affordability of ICT tools by Educational and Research Organizations

Table 8 brought that cent percent of the Educational and Research Organizations agreed that organizations are able to afford ICT tools while 75% and 25% of them agreed and disagreed that organization has sufficient funding for purchase and maintenance of ICT tools. 75% of the respondents agreed, 12.5 % were undecided and 12.5% showed disagreement that organization has adequate technical support for maintenance of ICT tools while the statement that organization is providing loans for purchase of ICT tools was agreed by 67.5%, undecided by 12.5% and disagreed by 25% of the users. It is a good sign that majority of the organizations shifting from traditional methods to modern methods and funds are also allocated for establishment of required infrastructure for the benefit of their employees.

Table 8: Affordability of ICT tools by Educational and Research Organizations

n= 8

S.No.	Statements	Agree		Undecided		Disagree	
		f	%	f	%	f	%
i.	Organization is able to afford ICT tools	8	100	_	_	_	_
ii.	Organization has sufficient funding for purchase and maintenance of ICT tools	6	75	–	–	2	25
iii.	Organization has adequate technical support for maintenance of ICT tools	6	75	1	12.5	1	12.5
iv.	Organization is providing loans for purchase of ICT tools	5	67.5	1	12.5	2	25

5.4 Affordability of ICT tools by Non Governmental Organizations and Cooperative organizations

Table 9: Affordability of ICT tools by Non Governmental Organizations and Cooperative Organizations

n= 6

S.No.	Statements	Agree		Undecided		Disagree	
		f	%	f	%	f	%
i.	Organization is able to afford ICT tools	6	100	–	–	–	–
ii.	Organization has sufficient funding for purchase and maintenance of ICT tools	6	100	–	–	–	–
iii.	Organization has adequate technical support for maintenance of ICT tools	5	83.3	1	16.6	–	–
iv.	Organization is providing loans for purchase of ICT tools	1	16.6	2	33.3	3	49.9

It was inferred from Table 9 that cent percent of the Non Governmental Organizations and Cooperative organizations agreed that organizations can afford ICT tools and have sufficient funding for purchase and maintenance of ICT tools while 83.3% and 16.6% of the users agreed and undecided that the organization has adequate technical support for maintenance of ICT tools. 16.6 %, 33.3% and 49.9% agreed, undecided and disagreed to the statement that the organization is providing loans for the purchase of ICT tools. This positive attitude on ICT indicate that the organizations have realized that ICT tools are advantageous than existing traditional methods.

5.5 Affordability of ICT tools by Private Organizations

From the overall impressions of various organizations, it can be concluded that there is no dearth of funds to provide the required infrastructure for better efficiency of its employees and provision of loans for purchase of these ICT gadgets for their personal use. All the organizations had perceived the advantages of ICT tools and realized the importance of use of ICT tools in the effective functioning of organizations. A positive attitude was observed in harnessing the ICT tools by various organizations.

6. KNOWLEDGE ABOUT COMPUTER APPLICATIONS

It was operationally defined as the knowledge of the organizations on the various basic applications of computer.

Table 10: Affordability of ICT tools by Private Organizations

n=10

S.No.	Statements	Agree		Undecided		Disagree	
		f	%	f	%	f	%
i.	Organization is able to afford ICT tools	9	90	1	10	_	_
ii.	Organization has sufficient funding for purchase and maintenance of ICT tools	8	80	1	10	1	10
iii.	Organization has adequate technical support for maintenance of ICT tools	9	90	_	_	1	10
iv.	Organization is providing loans for purchase of ICT tools	7	70	_	_	3	30

A questionnaire was developed for measurement of knowledge of the organizations, which had 8 questions on computer applications. The responses were obtained through dichotomous pattern and fill up the blanks. A score of 1 was given for correct answer and 0 was given for wrong answer. Thus the maximum score and minimum score could be 8 and 0 respectively and the respondents were grouped into three categories as low, medium and high based on their means and standard deviation.

a. Low – Below (Mean – SD)

b. Medium – Between (Mean ± SD)

c. High – Above (Mean + SD)

From the Table 11, it was evident that majority of the organizations (51.5%) had medium level of knowledge about computer applications followed by low level (30.3%) and the remaining 15.1% had high level of knowledge about computer applications which reflects that about 67% of the users have good knowledge on computer applications with mean and standard deviation values of 5.6 and 1.477.

From the above finding it can be inferred that the users in the organization had fair knowledge on computer applications but to make them more knowledgeable, the organizations have to make their staff exposed to training in the latest applications of computer so that the relevance of use of ICT tools is justified.

These findings are in conformity with Salampsis (2004).

Table 11: Distribution of the organizations based on the knowledge about computer applications

n=33

S.No.	Category	f	%
1	Low	11	33.30
2	Medium	17	51.5
3	High	5	15.10
$\bar{X}$ = 5.60			Σ = 1.477

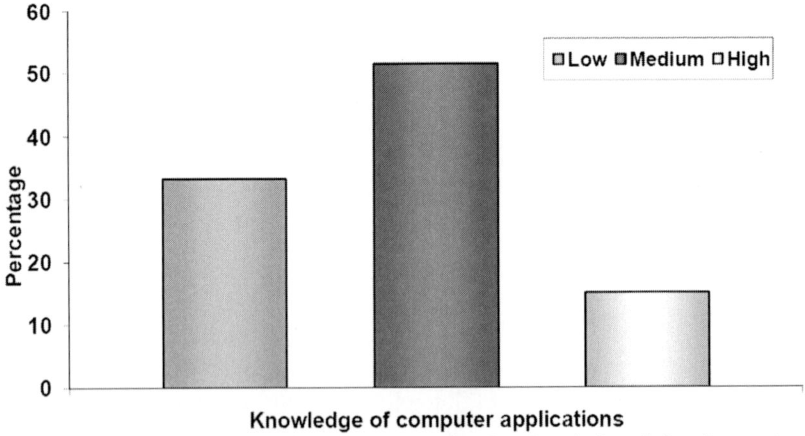

Fig. 1: Distribution of the organizations based on the knowledge of computer applications

Figure 1: Distribution of the organizations based on the knowledge of computer applications

REFERENCES

1. Bheenick K J and Brizmohun R 1999 The scope of information technology applications in agricultural extension in Mauritius. Proceedings of the Third Annual Meeting of Agricultural scientist-Reduit, Mauritius, 17-18 November-1998 : 23-40.
2. Heeks 1999 Development Informatics.Working Paper Series. Working paper No.5. University of Manchester, UK. 'Information and Communication Technologies' Poverty and Development. http://www.sed.manchester.ac.uk.idpm/research/publications/wp/di/documents/di_wp05.pdf.
3. Cecchini S and Scott C 2003 Can information and communication technology applications contribute to poverty reduction? Lessons from rural India. Information Technology for Development 10(2): 73-84 ; ref 33.

4. Maru A 2003 Information and Communication Technology Use in Agricultural Extension in India. http://www.cta.int/observatory2003/case_studies/case_study_India.pdf.
5. Meera N S, Jhamtani A and Rao D U M 2004 Information and Communication Technology (ICT) as an Alternative Extension approach: A critical analysis of ICT projects in India. Proceedings of National Seminar on Information and Communication Technologies for Agriculture and Rural Development held at NAARM Hyderabad on September 9 and 10, 2004, pp 129-140.
6. Shenoy S and Banerjee P N 2004 Knowledge Networking in Information and Communication Technologies (ICTs) for women in Agriculture and Rural Development. Manage Extension Research Review 5 (1) : 85-101
7. Kumar D 2005 Information and Communication Technology in Indian Agriculture Dessiminating Information to Farmers. http://129.3.20.41/eps/get/papers/0503/0503002.pdf
8. Jain D K, Sharma A K and Ruhil A P 2005 Role of information technology in dairy science: a review. Indian Journal of Animal Sciences 75(8): 985- 991.
9. Salampsis M, Batzoi Ch, Samathris V, Adroulidakis S and Adroulidaki M 2005 Use and Impact of Internet in the Greek Agricultural Sector: final results of a of website owners, www.efita.net/apps/accessbase/dbsomnaire.asp?d=539&t=0&identobj=bruqNwgy&uid=57305290&sid=57&idk=1.
10. Maningas V R 2006 Mainstreaming Farmers and Intermediaries into Information and Communication Technology (ICT): A strategy towards adopting ICT for rural development and agricultural extension, Computers in Agriculture and Natural Resources,4th World Congress Conference, Colorado, Florida, USA, Proceedings of the 24-26 July, 2006.
11. Sasidhar P V K and Sharma V P 2006 Cyber livestock outreach services in India: a model framework. www.cipav.org.co/Irrd/Irrd18/1/sasi18002.htm.

INDEX

A

animal husbandry, 2, 168, 169, 186, 230, 251, 258, 296, 297, 372, 373, 377, 381, 382

animal welfare, 2, 5, 9, 10, 11, 12, 13, 25, 42, 98, 199, 208, 373, 374, 377, 378, 379

anti-cruelty ethic, 262, 269, 272, 274

B

Biosecurity, 97, 101, 103

biotelemetry, 1, 2, 3, 4, 5, 6, 9, 10, 11, 12, 13

boar taint, 125, 126, 144, 146, 147, 148, 149, 234

Bovine, 24, 45, 151, 152, 165, 338

C

carcass quality, 18, 22, 24, 44, 53, 121, 152

Companionship, 344, 346, 353

D

Disease surveillance, 97, 100

double muscling, 18, 19, 20, 21, 22, 23, 50

E

environmental enrichment, 2, 4, 9, 13, 15, 223, 256, 257, 307, 308, 309, 321, 359, 369

ethical review, 2, 4

F

feather pecking, 208, 210, 222, 223, 224, 226, 229, 230, 231, 233, 234, 246, 253, 257, 258

foot sole lesions, 208, 220, 226

G

Goat production, 97

H

Hanwoo, 111, 112, 113, 114, 115, 116, 117, 118, 119, 120, 122, 123, 124

helminths, 189, 190, 191, 196, 199, 204, 205

I

Intestinal Tract, 76

inulin, 87, 126, 131, 137, 138, 139, 141, 142, 143, 148

L

livestock, 18, 19, 40, 42, 53, 76, 77, 98, 99, 102, 119, 335, 336, 337, 338, 384, 385, 386, 397, 402

locomotion, 18, 29, 42, 239

M

Mamasani Saanen, 106

meat quality, 18, 19, 38, 40, 41, 42, 45, 52, 53, 111, 117, 118, 124, 148, 149, 201

Milk quality, 56

mortality, 30, 48, 109, 161, 165, 191, 201, 204, 207, 208, 209, 210, 215, 216, 312

myostatin, 17, 18, 19, 43, 44, 45, 47, 48, 49, 50, 51, 52, 53

N

Novel technology, 56, 58, 61, 73

O

organ size, 18, 23, 45

P

parasites, 102, 190, 193, 194, 220, 222, 223, 225, 226, 227, 304, 344, 375

Pneumonia, 152, 158

poultry health, 208

R

rabbit Survey, 344

raw potato starch, 126, 134, 137, 138, 141, 142, 143

refinement, 2, 3, 11, 13, 14

S

skatole, 125, 126, 127, 128, 145, 146, 147, 148, 149

social ethic, 262, 268, 270

Stocker, 151, 152, 155, 158

stress susceptibility, 18

surgery, 1, 2, 4, 6, 7, 8, 12, 15, 272, 341, 342

Synbiotic, 75, 76, 80, 81, 82, 83, 85, 86, 87, 89, 92, 93, 94, 95

T

Testing lab, 56

V

vent pecking, 208, 223

Z

Zoonoses, 281, 282, 303, 305